Properties and Applications of Biomaterials Related to Gels

Properties and Applications of Biomaterials Related to Gels

Editor

Shige Wang

MDPI

Basel • Beijing • Wuhan • Barcelona • Belgrade • Novi Sad • Cluj • Manchester

Editor
Shige Wang
Department of Chemistry
School of Materials and Chemistry
University of Shanghai for
Science and Technology
Shanghai
China

Editorial Office
MDPI
St. Alban-Anlage 66
4052 Basel, Switzerland

This is a reprint of articles from the Special Issue published online in the open access journal *Gels* (ISSN 2310-2861) (available at: www.mdpi.com/journal/gels/special_issues/010DF24A0L).

For citation purposes, cite each article independently as indicated on the article page online and as indicated below:

Lastname, A.A.; Lastname, B.B. Article Title. *Journal Name* **Year**, *Volume Number*, Page Range.

ISBN 978-3-0365-9057-8 (Hbk)
ISBN 978-3-0365-9056-1 (PDF)
doi.org/10.3390/books978-3-0365-9056-1

Contents

About the Editor

Shige Wang

Dr. Shige Wang received his PhD degree in Materials Science and Engineering from the College of Materials Science and Engineering, Donghua University (Shanghai, China), in 2013. He then worked as a Postdoctoral Fellow at the Shanghai Institute of Ceramics, Chinese Academy of Sciences. At present, he is a professor at the University of Shanghai for Science and Technology. His current research interests include two-dimensional transition-metal sulfides for nanomedicinal applications, and polymer hydrogels and nanofiber-based materials for tissue engineering and cancer therapy applications.

Preface

As an excellent biopolymer material, hydrogel has been widely used in drug delivery, tissue engineering, and disease treatment, among other applications. By combining different polymers through physical or chemical changes, the prepared hydrogels can show better performance and be used in various environments.

At present, the exploration of hydrogel in biomedicine seems to be diversified. To adapt to the complex environment in organisms, higher requirements must put forward for hydrogel. Under the premise that hydrogel is biosafe, it must become tougher or more easily degradable according to the purpose of its application, be able to stay in the body for a longer time to exert its effect, or be easily removed. Moreover, certain hydrogels can respond to changes in pH, temperature, biological indicators, or microorganisms in the body. We are interested in the intelligent properties of hydrogels, and the condition in which these properties can be used to facilitate their applications is undoubtedly a question that researchers need to consider carefully.

This reprint consists of 12 papers, of which 2 are editorials and 10 are research papers. We hope to share the experimental and research results in this field and inspire new ideas for the design and applications of hydrogels.

Shige Wang
Editor

gels

MDPI

Review

Advances in Hemostatic Hydrogels That Can Adhere to Wet Surfaces

Wenli Han and Shige Wang *

School of Materials and Chemistry, The University of Shanghai for Science and Technology, No. 516 Jungong Road, Shanghai 200093, China
* Correspondence: sgwang@usst.edu.cn

Abstract: Currently, uncontrolled bleeding remains a serious problem in emergency, surgical and battlefield environments. Despite the specific properties of available hemostatic agents, sealants, and adhesives, effective hemostasis under wet and dynamic conditions remains a challenge. In recent years, polymeric hydrogels with excellent hemostatic properties have received much attention because of their adjustable mechanical properties, high porosity, and biocompatibility. In this review, to investigate the role of hydrogels in hemostasis, the mechanisms of hydrogel hemostasis and adhesion are firstly elucidated, the adhesion design strategies of hemostatic hydrogels in wet environments are briefly introduced, and then, based on a comprehensive literature review, the studies and in vivo applications of wet-adhesive hemostatic hydrogels in different environments are summarized, and the improvement directions of such hydrogels in future studies are proposed.

Keywords: hydrogel; hemostasis mechanism; wet adhesion mechanism; application

check for updates

Citation: Han, W.; Wang, S. Advances in Hemostatic Hydrogels That Can Adhere to Wet Surfaces. *Gels* **2023**, *9*, 2. https://doi.org/10.3390/gels9010002

Academic Editor: Christian Demitri

Received: 23 November 2022
Revised: 11 December 2022
Accepted: 12 December 2022
Published: 22 December 2022

1. Introduction

More than 5.8 million people die from severe trauma worldwide each year and approximately 40% of which are caused by uncontrolled bleeding or its consequences [1]. Massive blood loss from incompressible injuries has been reported to cause hemorrhagic shock coagulopathy, multiple organ failure, life-threatening sepsis, and acidosis. Moreover, interventional diagnosis and surgical treatment are prone to hemorrhage or intraluminal hemorrhage, especially in sites close to the heart, parenchymal organs, vital blood vessels, etc. [2]. Therefore, research into hemostatic technologies and materials with superior performance is critical to reducing adverse side effects and even saving lives.

Recently, exploring fast and effective methods for controlling bleeding in different application environments has always been an important subject of multidisciplinary research [3]. Sutures and staplers are commonly used to close wounds, however, they have the risk of infection, blood exudation, hyperplasia, and keloid formation [4]. For quick hemostasis in superficial wounds, some conventional hemostatic materials, such as tourniquets [5], hemostatic gauze [6], bandages [7], etc., have been utilized extensively. Nevertheless, common materials struggle to meet the demand for quick and efficient hemostasis in cases of intracavitary hemorrhage or injuries involving essential tissues/organs. Additionally, as gauze and bandages are not biodegradable and can cause secondary injury, delayed healing, and additional discomfort, they need to be removed after hemostasis [8,9]. Therefore, various advanced hemostatic powders, hydrogels, sponges, adhesives, and hemostatic agents [10,11], have been extensively explored (Figure 1).

Adhesives can bind different tissues as well as blood vessels together, whereas hemostatic drugs are known to stop bleeding mechanically or by accelerating the coagulation cascade. Additionally, tissue sealants can stop blood dripping from blood vessels [12]. Nevertheless, some of these products are not universally applicable and have obvious disadvantages. For example, fibrinogen and thrombin-based injection solutions (fibrin sealants) may be washed away by the bloodstream and cannot be used effectively in severe or emergency arterial bleeding [13,14]. As emerging medical adhesives, cyanoacrylates may cause allergic reactions, and their rapid curing process is exothermic [15]. Furthermore, such products have not been used inside the body due to concerns about the potential toxicity of the degradation products. As a consequence, designing an ideal hemostasis technique for effective hemostasis in complicated clinical situations remains challenging.

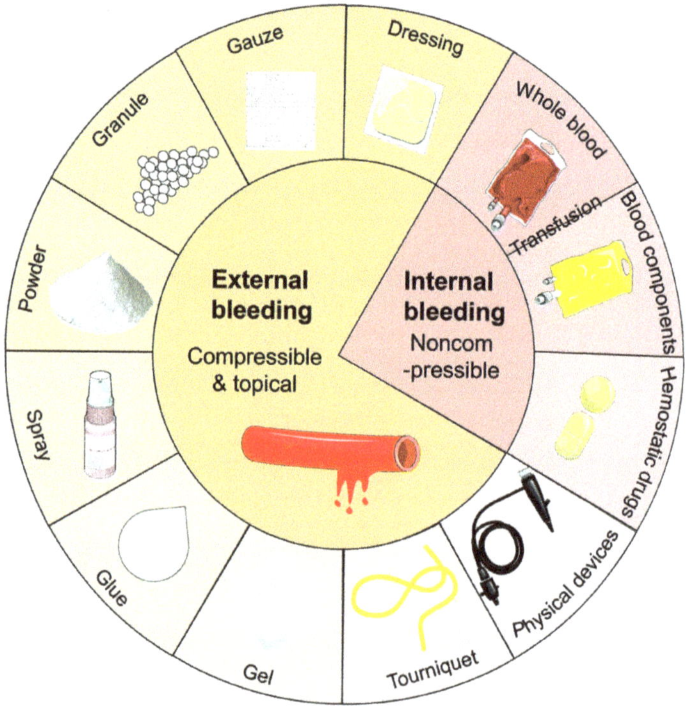

Figure 1. The current methods of hemostasis used for external and internal bleeding management. Figure modified from [11] with permission. Copyright 2020.

Hydrogel-based biomaterials show many advantages over traditional hemostasis methods [16,17]. A hydrogel is a 3D cross-linked hydrophilic polymer with a structure similar to that of the natural extracellular matrix (ECM) [18]. They are widely used in biomedicine, and due to their injectability and fluidity, hydrogels are crucial for achieving quick and lasting hemostasis for a variety of irregular wounds and intraluminal injuries [19]. In addition, the hydrogel-based biomaterial is safe for use in vivo due to its exceptional biocompatibility and biodegradability [20,21]. An ideal polymer hydrogel for hemostatic applications should have the following properties: (i) the hydrogel should be able to effectively promote wound healing and immediately block the bleeding, [22] (ii) the hemostatic hydrogel should have superior mechanical qualities that enable it to quickly adhere and completely seal the damage, especially in humid and dynamic settings, and [23] (iii) the hydrogels must have quick coagulation to prevent the migration of hemostatic hydrogels from the bleeding site [24].

High pressures and blood flow rates may remove or destroy general hydrogels due to their limited underwater adhesion and weak adhesion mechanisms [25]. Therefore, it is crucial to enable hemostatic hydrogels to adhere to tissues and maintain their integrity in humid environments. Moreover, an ideal wet-adhesive hydrogel should completely cover the wound area, minimize surgical scope or complications, carry therapeutic cells and drugs, and release them to the wound site [26]. This review introduces the properties of hemostatic hydrogels that can adhere to wet surfaces, briefly discusses the basic principles of hemostasis, and then focuses on the adhesion mechanism of hydrogels on wet surfaces. This review is anticipated to promote further research progress on wet viscous hemostatic hydrogel and create more possibilities for rapid and effective emergency hemostasis.

2. Properties of Wet Adhesive Hemostatic Hydrogel

The complicated dynamic equilibrium environment of the human body should be adapted when hydrogels are employed to stop bleeding. The ideal adhesive hemostatic hydrogel should possess the properties of strong wetting adhesion, hemostasis, antibacterial, and biocompatibility.

2.1. Hemostatic Properties

Hemostasis is the first stage of wound healing, starting with wound formation and ending with thrombosis. The physiological mechanism of human hemostasis is a complex dynamic process consisting of three processes: vasoconstriction, formation of platelet embolism, and coagulation (Figure 2). Vasoconstriction is a transient reflex contraction that is the first step in preventing blood loss during injury (Figure 2a). Thromboembolism is the second critical stage of hemostasis, involving platelet adhesion, activation, and aggregation into embolization. During the damage, platelets bind to von Willebrand factor (VWF) aggregates via their GP1b receptor and directly adhere to exposed subendothelial collagen via two receptors (glycoprotein VI (GP VI) and integrin $\alpha2\beta1$) [27]. This adhesion triggers an activation process of platelets, which leads to a change in shape and the release of particles rich in substances such as serotonin, thromboxane A2 (TXA2), and adenosine diphosphate (ADP). Because they alter the conformation of GPIIb/IIIa, a receptor on the surface of platelets that enables platelets to bind to fibrinogen and induce platelets to aggregate into the walls of damaged blood arteries (Figure 2b). Primary hemostasis is a brief stage of hemostasis that finally results in the creation of a first embolism. The extrinsic and intrinsic pathways are two separate processes that take place during the second stage of hemostasis [28]. The extrinsic pathway, which starts when a blood artery is damaged, involves the interaction of tissue factor (TF) that is exposed to the blood and coagulation factor VII (FVII). When factor XII comes into contact with a foreign substance having a negatively charged surface, intrinsic pathways are activated. The activation of factor X, an enzyme that assists in the conversion of prothrombin to thrombin (thrombin transforms fibrinogen to fibrin monomers), is the result of these two processes (common pathway). Finally, fibrin monomers self-polymerize into fibrils, which laterally aggregate to create fibers and establish a hydrogel (Figure 2c). The interfiber and interfiber crosslinking caused by factor XIIIa stabilizes the fibrin gel and enhances clot stiffness.

According to cell-based models of hemostasis, the coagulation process has three overlapping stages: initiation, amplification, and propagation. When TF-bearing cells are exposed to flowing blood after vascular damage, the initiation phase begins, which results in the generation of activated factors IX and X as well as thrombin. Such thrombin serves several important roles during the amplification stage. Along with activating platelets, it also causes the cofactors V and VIII to become Va and VIIIa on the surface of the activated platelets. Thrombin activates factor XI on the surface of the platelet during this stage [8]. The proliferation that takes place on the surface of activated platelets during the last stage of this cellular model and eventually results in a burst of thrombin that is sufficient to create the fibrin mesh. Hydrogels can facilitate the hemostasis process on wet surfaces through three typical approaches: i) direct or indirect participation in the coagulation system to activate the physiological hemostatic process; ii) enrichment of the wound site with coagulation components (e.g., polymeric polysaccharides, inorganic zeolite clay, etc.) by physical and chemical means; iii) physical closure of blood vessels using the strong adhesion of the material to the tissue.

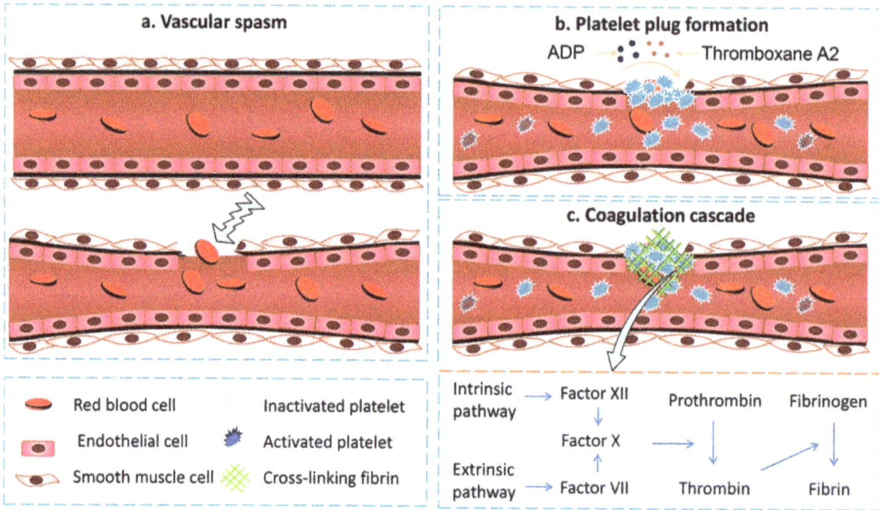

Figure 2. Schematic diagram of the hemostatic process: (**a**) vascular spasm. Immediately after injury, blood vessels are stimulated to constrict to reduce blood loss; (**b**) platelet plug formation. Once platelets are activated, the shape changes, releasing serotonin, ADP, and thromboxane A2, etc., causing more platelets to concentrate at the site of injury to form a thrombus; (**c**) coagulation cascade. Endogenous and exogenous coagulation pathways are activated, involving the interaction of factor (XII) and coagulation factor VII in the blood, with the two pathways eventually converging into a common pathway that activates factor X, which further converts prothrombinogen to thrombin. The generated thrombin activates Factor XIII, which promotes the conversion of fibrinogen into fibrin chains for hemostasis. Figures modified from [11] with permission. Copyright 2020.

2.2. *Adhesion Properties*

2.2.1. Four Theories of Bio-Adhesion

There are four typical bio-adhesion theories, including wetting, mechanical, diffusion, and fracture theories [29] (Figure 3). Wetting is a phenomenon caused by liquid diffusion along a solid surface. Surface tension, capillary force, Van der Waals force, etc., are the basic causes of wet adhesion [30] (Figure 3a). Wet adhesion is an approach in which hydrogels can enhance adherence. According to the wetting hypothesis, capillary actions are brought on by curved water surfaces and tight contact between the interfaces, which results in short-distance interactions [31]. The heterogeneity of the hydrogel is a key element in determining its adhesiveness. Hydrogel adhesions to hydrophobic surfaces are poor in the air, yet they may cling to hydrophilic surfaces quite effectively. This is because water will moisten the surface and create a water meniscus at the border of the contact region. As a consequence, in addition to molecular interactions, capillary adhesions also exist at the interface of the two adhesives [32]. The groups on the polymer can interact with the interface, however, when a hydrogel is submerged in water, pressure is usually needed to remove free water. The interface adhesion is related to the surface energy, therefore, appropriate surface energy is necessary to afford firm adhesion [33]. Inspired by the structures of tree frog toes, Nguyen et al. evaluated the wet stickiness of the contact interface under various circumstances [34]. The findings demonstrated that, in comparison to the surface of the non-patterned plate, the micro pattern cushion's surface boosted the interface adhesion in a wet environment. The effect of contact shape and substrate morphology on adhesion was investigated by Liu et al. [30]. The findings demonstrated that the interstitial liquid's contact geometry, and capillary density are strongly connected to the wet adhesion, and the ratio of capillary and adhesive forces is key to elucidating the mechanism of wet adhesion.

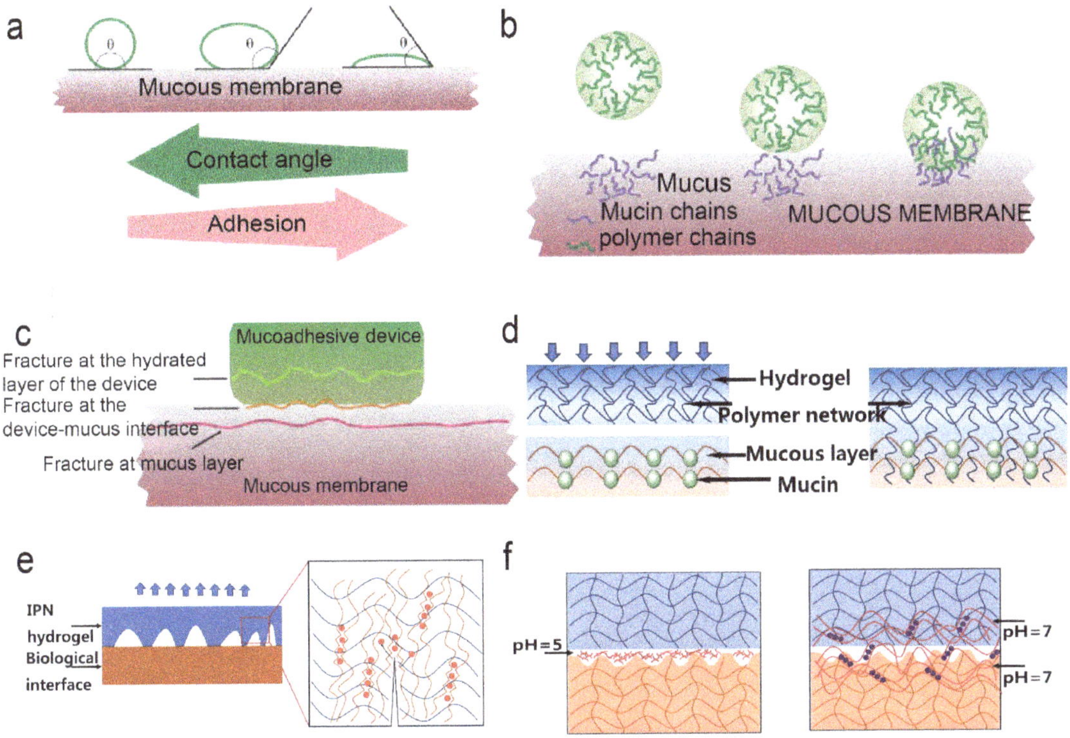

Figure 3. (**a**) Wetting theory. (**b**) Diffusion theory. (**c**) Fracture theory. Figures modified from [35] with permission. (**d**) Mechanical theory. (**e**) The principle of double network hydrogel enhances adhesion. The blue network represents polyacrylamide, the orange network represents alginate, and the red dots represent calcium ions. (**f**) Principle of topological binding. The chitosan chains are dissolved in a solution at pH = 5. The chitosan chains diffuse into the gel at pH = 7, using hydrogen bonds to form a new network. Figures modified from [36] with permission.

On the other hand, a rough interface, in accordance with mechanical theory, increases the surface area that can make contact as well as the viscoelastic and plastic energy dissipation during joint breakage (Figure 3d) [35]. The first bonding hypothesis, known as mechanical interlocking (or mechanical interlocking), was proposed by MAC Bain and Hopkins in 1925 [37] and referred to the interlocking of the adhesive with the microscopic rough surface of the bonded object. Traditional alveolar bone filling methods achieve adhesion between the alveolar bone and the pretreated tooth surface, which is facilitated by mechanical interlocking [38]. Inspired by this phenomenon, bonding and adhesive can be achieved by filling the hydrogel in the pores of the substance. By utilizing the catechol-mediated synergized adhesion and interlocking in the pores of diatom silica, Lee et al. prepared a diatomaceous earth/polysaccharide elastic hydrogel, which showed good biocompatibility and can be strongly adhered to the skin [39]. In a separate study, sodium alginate and acrylamide were employed as raw ingredients by Yuan et al. to create a mechanically interlocked double-net hydrogel with enhanced adhesive capabilities and mechanical properties [40]. Such a hydrogel was competent in blocking the perforation and encouraging the healing of tissues.

In addition, the adhesion of the hydrogel to the mucosal layer can be explained by diffusion theory. For instance, certain bioadhesive polymers may dissolve in mucus mucin, and flexible polymer chains can physically entangle with mucin chains to provide an advantageous adhesion effect (Figure 3b) [35]. According to the theory of diffusion, adhesion is produced by the reciprocal diffusion of molecules between the substrate and the adhesive, therefore, macromolecular chains require sufficient solubility and flexibility for mutual diffusion. Due to the presence of an interpenetrating network structure, the hydrogel can create interdiffusion among polymers to increase its adhesion qualities Feng et al. prepared silk fibroin/konjac glucomannan sponges with an interpenetrating network for wound dressing, which showed high biocompatibility for cell adhesion and proliferation [41]. In another study, Vorwald et al. created fibrin-alginate interpenetrating network hydrogels, which combined the excellent adhesion and stimulatory characteristics of fibrin with the adaptable mechanical properties of alginate for cellular orientation and distribution by combining [42]. The fracture theory was typically used to determine the maximal tensile stress. The toughness of the stiff polymer without flexible chains may also be calculated using this method (Figure 3c).

2.2.2. Influence Factors of Wet Surface Adhesion

In practice, the adhesion of hydrogels involves a complex interaction of chemical, physical and structural factors (Figure 4). Therefore, understanding these factors will help us to design the wet adhesion strategy for a given application. In this section, the chemical, physical, and structural factors that influent the wet surface adhesion of hydrogels are briefly summarized (Table 1).

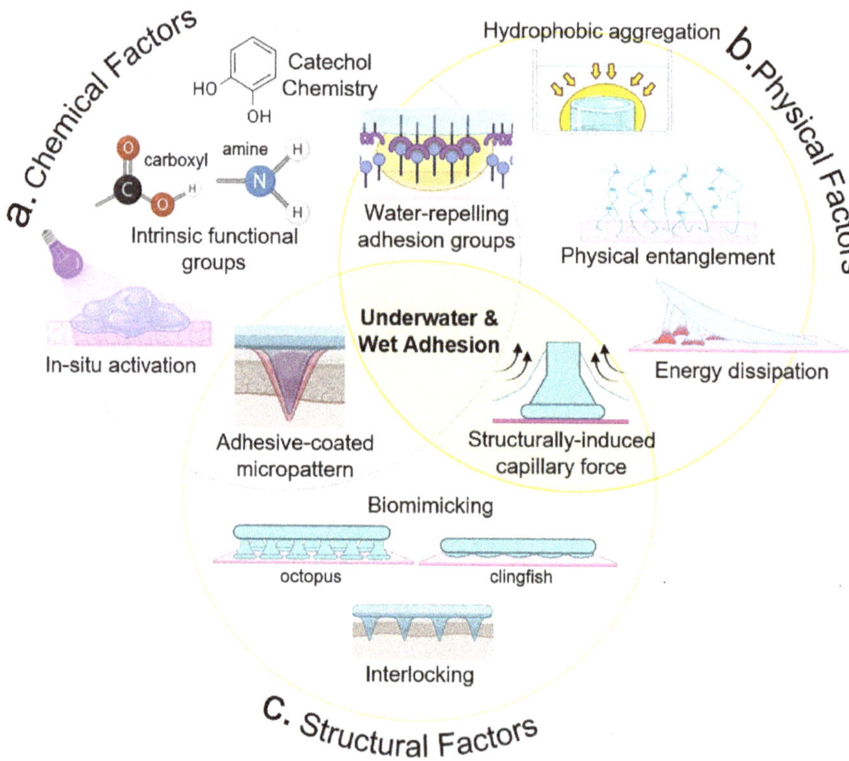

Figure 4. The practical application of hydrogels in wet adhesion involves complex interactions that can be divided into three main categories: chemical, physical, and bionic structural factors. (**a**) Chemical factors, such as catechol chemistry, in situ activation, and reactions of intrinsic groups of the material. (**b**) Physical factors, such as hydrophobic aggregation, physical entanglement, and energy dissipation. (**c**) Biomimetic structural factors, such as micro-patterning and capillary forces. Figure modified from [43] with permission.

(1) Typical Chemical Bond

The adhesion capacity of hydrogels is usually related to the formation of ionic, covalent, and metal coordination bonds between the hydrogels and tissue [44]. Covalent bonding, in which functional groups chemically react with groups on the target surface, is the primary way to enhance adhesion. Covalent bonds have higher underwater adhesion strength than intermolecular forces because they require sharing of electrons. By creating covalent bonds, hydrogels exhibit underwater adhesion and cohesion, making them attractive to substrates containing reactive functional groups [45]. When the hydrogel is used underwater, the presence of chemical or physical crosslinks prevents it from over-swelling and impairs adhesion [46]. However, covalent-based underwater bonding usually involves surface modification or is dependent on the specific adhered surface [47]. Catechol chemistry is one of the most frequently used interactions among the many functional groups. Studies have uncovered facile methods to synthesize catechol polymers and molecules that can be grafted and coated multifunctionally into other materials [48]. Due to their inherent good biocompatibility, catechols are widely used for tissue adhesion in humid environments [49] (Figure 4a). Moreover, the groups in catechol hydrogels (containing ortho-quinone groups) can establish covalent bonds with the nucleophiles on the surface of the tissue and thus adhere to the membranes of biological tissues [50]. Zhou et al. spliced the catechol moiety of dopamine (DA) in the polymer chain of hyaluronic acid hydrogels, which established covalent bonds with nucleophiles (amines, thiols, and hydroxyl groups) and thus adhered to the wet biological tissue [51]. In another study, Ma et al. prepared in situ photo-responsive chitosan (CS) hydrogels based on the imine cross-linking method. After exposing the hydrogels to UV light, *o*-nitrophenyl was converted to the *o*-nitrobenzaldehyde group. This group was further cross-linked with the amino group on the tissue surface, thus forming a covalent bond that allows the wet tissue adhesion [52].

Covalent connections can be further classified into permanent and dynamic covalent bonds in physiological contexts, which have significant advantages in the application of hydrogel bio-adhesion. Permanent covalent bonds can be carbon–sulfur, carbon–carbon junctions, carbon–nitrogen, and silicon–oxygen bonds. Permanent covalent bonds are stable, hard, and irreversible. By irreversibly rupturing covalent or neighboring bonds, these bioadhesive hydrogels can be separated. Adhesives with permanent covalent typically adhered to the surface by reacting with other functional groups. Dynamic covalent bonds (DCBs) can reversibly break and form covalent bonds under specific conditions. DCBs can be formed by the linkage of disulfides, imides, hydrazides, phenylboronic acids, and Diels–Alder reactions. The use of dynamic covalent bonding, combined with the favorable properties of chemical and physical adhesion, allows the design of robust and reversible bioadhesive hydrogel dressings. During the adhesion process, DCBs can assemble a hydrogel network with self-healing capabilities between two adherent gel networks or interfaces. Such gels are widely used for wound closure and wound hemostasis. For example, Yang et al. developed an injectable mussel mucoadhesive self-healing hydrogel with good bioadhesive and hemostatic properties based on C-N single bonds and C-N double bonds [53]. Compared to conventional covalently linked hydrogels, hydrogels containing DCBs exhibited significant self-healing ability and good cytocompatibility. Using aminoglycoside, aldehyde-based hyaluronic acid, and adipic dihydrazide grafted hyaluronic acid as raw materials, Li et al. created the dynamic covalently cross-linked hydrogels. Such hydrogels were formed based on the imine and hydrazone cross-linking and showed strong and long-lasting antimicrobial properties, good biocompatibility, and self-healing ability [54].

(2) Hydrogen Bond

A hydrogen bond is created by the dipole–dipole attraction of two electronegative atoms. The hydrogen bond is an important driving force for hydrogel formation. For example, a novel pH-sensitive hydrogel was prepared by compositing konjac glucomannan, dopamine hydrochloride, L-cysteine hydrochloride, and epigallocatechin acid esters.

Hydrogen bonds and catechol-mediated coordination were responsible for the hydrogel formation. This hydrogel possessed injectability and adhesion properties for potential use in drug delivery and release controllability [55]. In another study, Zhang et al. designed the catechol-modified polylysine/polyacrylamide hydrogel. The numerous hydrogen bonds rendered the hydrogel's strong adhesion, high compressive strength, and effective hemostasis, and can be used to seal bleeding sites [56]. Typically, there existed a counterpart between the hydrogel-wet surface hydrogen bonds and water molecules–induced hydrogen bonds. To enhance the wet surface adhesion of hydrogel, the water molecules-induced hydrogen bonds should be suppressed. A typical way to inhibit hydrogen bonding caused by water molecules is to repel or absorb water molecules from the surface (Figure 5a,b). The water film formed by these water molecules is considered to be the hydration layer. At the molecular level, the hydration layer at the hydrogel-attachment interface can be resisted by hydrophobic interactions. Therefore, hydrophobic solvents and hydrophobic monomers are usually used in hydrogels in order to improve wet adhesion [45,57]. Spraying the hydrophobic solvent on the surface of the hydrogel can also form a thin hydrophobic layer with a small contact angle with water [58]. When external pressure is applied to the hydrogel, the hydration layer can be broken and discharged from the adhered surface. Similarly, some hydrogels contain hydrophobic monomers (e.g., with long carbon chains or aromatic ring structures) that can repel water molecules and stick firmly thereto [59,60].

In contrast, the method of absorbing the hydration layer is simple than repelling the hydration layer. The absorption of the hydration layer is based on the fact that hydrogels contain highly hygroscopic materials [61]. For example, Wang et al. prepared an in situ photocuring hygroscopic hydrogel adhesive by using polyvinylpyrrolidone (PVP, as a hygroscopic polymer), acrylic acid, crosslinker, and photoinitiator [62]. Cong et al. designed an anthracene-based polyethyleneimine underwater adhesion hydrogel, and the high water absorption of polyethyleneimine can promote the absorption of interfacial water molecules and enable the hydrogel to adhere firmly to other substances [63].

(3) Van der Waals Force

Environmental stimuli can easily disrupt some of the adhesions generated based on chemical reactions. By contrast, less reactive physical factors can be used to durably strengthen wet adhesion. Van der Waals forces, electrostatic forces, hydrogen bonding, and hydrophobic interactions are representative examples of non-covalent interactions between interfaces, which affords physical adhesion. Van der Waals force is a weak force, especially in water, therefore, it is not the main driver of the adhesion of hydrogels and other substances. However, van der Waals forces can be aggregated into large volumes of material to provide strong adhesion. For example, geckos can stay and crawl normally on vertical walls due to the aggregated van der Waals forces generated by the contact of their feet with the walls [64]. Yi et al. created a bio-inspired wet hydrogel adhesive using polyethylene glycol hydrogel as a starting material by exploiting the water absorption properties, capillary forces, and van der Waals forces of hydrogels [65]. The mixture has excellent biological application potential and exhibited extraordinary reversible adherence to dry, wet, and submerged substrates. In another study, Sato et al. found that the Van der Waals forces are responsible for the adhesion of submicron silica particles to the surfaces of poly(acrylamide) and poly(dimethyl-acrylamide) hydrogels [66].

(4) Electrostatic Interactions

Adding electrostatic interactions to hydrogels can improve biological wet adherence. Through electrostatic interactions, hydrogels can firmly adhere to the surface of different substances [67]. To create hydrogels with strong adhesion, Huang et al. prepared hybrid hydrogels with ionic and hydrophobic cross-linked networks, with strong adhesion and high toughness properties attributed to the synergistic effect of electrostatic interactions and hydrophobic junctions [68]. Tian et al. designed a strong and stable adhesive hydrogel with synergistic hydrophobic interactions and dynamic electrostatic forces using 2-acrylamido-2-methylpropanesulfonic acid, gelatin, CS, ethyl 2-methoxyacrylate, and acrylic acid as raw materials. The hydrogel could adhere stably and firmly to various tissue surfaces [69]. Song et al. prepared a multifunctional physical hydrogel adhesive using catechol-modified CS and polyvinyl alcohol as raw materials. Dynamic hydrogen bonding and electrostatic interactions improved the persistence and reproducibility of this hydrogel adhesive [70].

(5) Metal Coordination Bond and Ionic Bond

The metal coordination bond and ionic bond also affect the wet surface adhesion. A specific type of covalent link known as a metal coordination bond typically has a stronger binding than a hydrogen bond [71]. The coordination bond is reversible, which gives the hydrogel reversible underwater adhesion characteristics in addition to maintaining the stability of the covalent bond [72]. An ionic bond contains two opposite-charged ions, which are also stronger than hydrogen bonds [73]. In order to create strong and durable ampholyte hydrogels, Huang et al. created a synthetic polyamphiphilic electrolyte hydrogel using the complementary interaction of ionic and metallic coordination bonds [74]. Using acrylamide, sodium alginate, and acrylic acid as raw materials, Liang et al. created an ultra-strong and tough hydrogel with excellent mechanical properties, adhesion, high hardness, toughness, fatigue resistance, and salt resistance. This hydrogel formation was based on the establishment of strong ionic and weak hydrogen bonds [75].

Table 1. Influencing factors of wet surface adhesion.

Influence Factors	Hydrogels	Adhesion Mechanism
Permanent and dynamic covalent bonds	Hyaluronic acid and dopamine polymerized hydrogel [51] In situ photo-responsive chitosan (CS) hydrogel [52]	Dopamine catechol chemical adhesion After exposure to UV light, o-nitrobenzene is converted to o-nitrobenzaldehyde, which is further cross-linked with amino groups on the tissue surface.
Hydrogen bond	Konjac glucomannan, dopamine hydrochloride, L-cysteine hydrochloride, and epigallocatechuic acid ester pH-responsive hydrogels [55] Catechol-modified poly(lysine)/poly(acrylamide) hydrogels [56] Anthracene-based polyethyleneimine underwater adhesion hydrogel [63]	Hydrogen bonding and catechol chemical adhesion Strong adhesion of hydrogel due to large amount of hydrogen bonding The high water absorption of polyethyleneimine promotes the absorption of interfacial water molecules, allowing the hydrogel to adhere firmly to other substances
Van der Waals force	Polyethylene glycol hydrogel [65]	Water absorption properties, capillary and van der Waals forces
Electrostatic interactions	2-Acrylamido-2-methylpropanesulfonic acid, gelatin, CS, ethyl 2-methoxyacrylate, and acrylic acid bonded hydrogels [69] Catechol-modified CS and polyvinyl alcohol were used as raw materials to prepare physical hydrogels [70]	Synergistic hydrophobic and electrostatic interactions Dynamic hydrogen bonding and electrostatic interactions
Metal coordination bond and ionic bond	Acrylamide, sodium alginate, and acrylic acid super tough hydrogel [75]	Strong ionic and weak hydrogen bonding interactions

(6) Biomimemtic Strategies

Research interest in the adhesion abilities of mussels, tree frogs, geckos, snails, teleosts, clingfish, and octopuses has led to the development of biomimetic hydrogel adhesives (Figure 5c–j). Biomimetic adhesions mirror the surface interactions caused by the topological structure of naturally adhesive creatures and are less sensitive to environmental cues. The adhesion of geckos and octopuses to wet surfaces primarily originated from the concurrence of physical forces like negative pressure, capillary force, and mechanical interlocking [76]. The sucker patterns, which exhibited a strong capillary force, were mimicked to elevate the wet adhesion content of polyethylene glycol (PEG) hydrogel [65,77]

(Figure 4c). The hydrogel micropillars expanded when placed on a moist surface, which created a capillary force around the suction cups and moved the micropatterns closer to the substrate for adherence. In another study, a PDA hydrogel was tailored to present the micro-channels and mimic the dynamics of the suction cups of geckos and octopuses. Such a PDA hydrogel was able to contract upon heating and expand when cooled, during which a negative pressure was created to suck the surface water into chambers to enable wet surface adherence [78]. Recently, an endoparasites-inspired hydrogel-forming double-layered adhesive microneedle patch, which is composed of a non-swellable silk fibroin-based core and swellable mussel adhesive protein-based shell, was proposed. The double-layered adhesive microneedle patch showed enhanced tissue insertion capability and superior wound-sealing capacity for wet and/or dynamic external and internal tissues [79]. The cohesion (mechanical properties) of the hydrogel is also closely related to the wet surface adhesion strength, as good mechanical properties support the deformation of the hydrogel without breaking. The duration of adhesion is equally important for practical purposes, as biomedical applications require that the hydrogel remains adherent to the implant site until effective hemostasis, tissue regeneration, or therapeutic delivery is complete. For example, the catechol-mediated wet surface adhesion was prone to disrupt by natural oxidation. Accordingly, protecting it from oxidation is the main strategy to enhance wet adhesion [46].

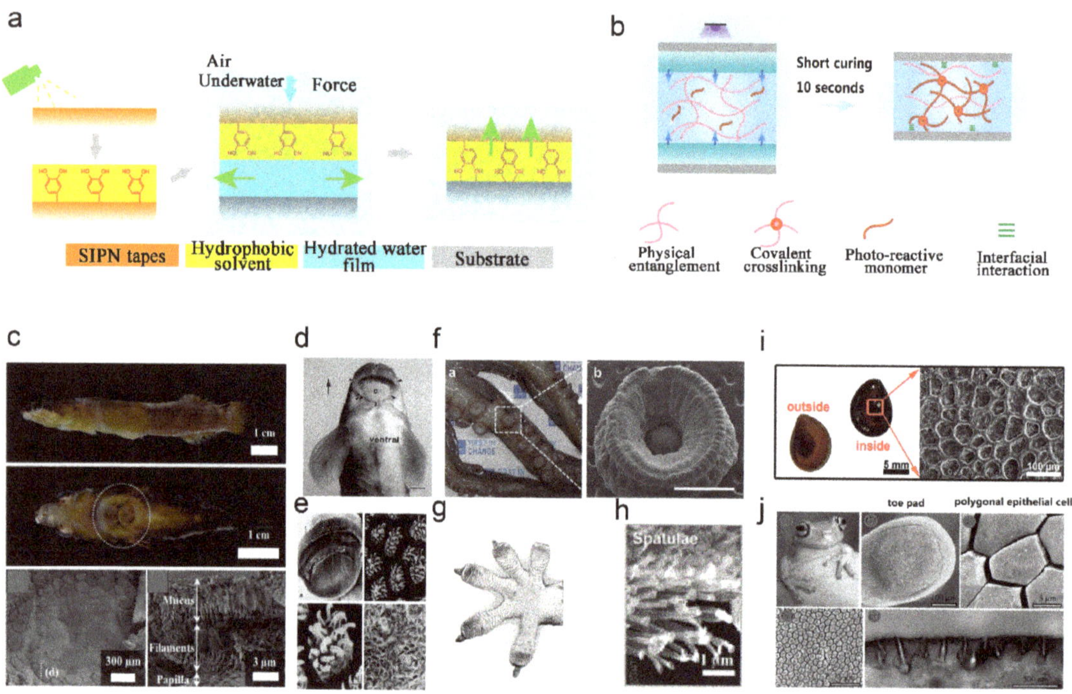

Figure 5. (**a**) Exclusion of interfacial water to enhance adhesion strategy. Figure modified from [58] with permission. (**b**) Absorption of interfacial water and rapid light curing strategies. Figure modified from [62] with permission. (**c**) Clingfish photos and SEM images of adherent disc surface morphology. Reprinted (adapted) with permission from [80]. Copyright {2022} American Chemical Society. (**d**) Teleost photographs and (**e**) SEM images of mouth surface morphology. Figure modified from [81] with permission. (**f**) Photographs(a) of octopus tentacles and SEM images(b) of its suckers (scale bar: 1 mm). Figure modified from [82] with permission. (**g**) Ventral diagram of the gecko foot. (**h**) SEM diagram of foot setae morphology. Figure modified from [83] with permission. (**i**) Photographs of the

gill cover of the snail and SEM images of the surface morphology. Reprinted (adapted) with permission from [84]. Copyright {2020} American Chemical Society. (j) Photographs of tree frogs and SEM and TEM images of the surface morphology. Figure modified from [85] with permission.

3. Application of Wet Adhesion Hemostatic Hydrogels

In this section, the applications of wet adhesion hydrogels for the hemostasis of skin, heart, liver, and other kinds of bleeding were summarized.

3.1. Skin

The skin is the largest multi-layered organ in the human body, consisting of the epidermis and dermis [86]. When the entire epidermis is severely injured, the skin loses its basic defense layer, and microbial infection at the wound site can slow the healing process [87]. Unlike other phospholipid-containing biofilms, the skin surface contains amino, carboxyl, and hydroxyl groups that contribute to the wet adhesion of the hydrogel. The Schiff base reaction can be used to cross-link the aldehyde group in the hydrogel with the amino group on the skin tissue to achieve stronger tissue adhesion. For example, Ma et al. prepared a liquid bandage (LBA), which is an in situ imine cross-linked photoactive CS hydrogel (NB-CMC/CMC hydrogel) [52]. NB-CMC was synthesized by modifying o-nitrobenzyl alcohol (NB) with water-soluble carboxymethyl chitosan (CMC). Under UV irradiation, o-nitrobenzene was converted to an o-nitrosobenzaldehyde moiety and cross-linked with amino groups on the tissue surface, resulting in excellent tissue adhesion. The hemostatic and antimicrobial properties of LBA were correlated with the mass ratio of NB-CMC/CMC. LBA exhibited acceptable biocompatibility and biodegradability, can effectively control bleeding, produce strong tissue adhesion, avoid bacterial infection, and accelerate wound healing (Figure 6a). Other molecules, such as sulfhydryl groups, long-chain alkyl groups, or DA-like compounds, can also be added to the hydrogel to enhance wet surface adhesion. In a study inspired by mussel mucin, a DA modified ε-poly-lysine-polyethylene glycol-based hydrogel (PPD hydrogel) wound dressing was developed in situ using horseradish peroxidase (H_2O_2/HRP) cross-linking method. It was shown that the PPD hydrogel had good wet tissue adhesion properties and exhibited excellent hemostatic effects to accelerate skin wound repair [88] (Figure 6b). In other studies, Lu and co-workers prepared several skin adhesive hydrogels based on the adhesion ability of DA [89,90]. These hydrogels can increase shear viscosity by an average of 10–30 kPa.

In addition, metal ions (e.g. zinc ions, silver ions, calcium ions, etc.) can introduce additional functionality to the hydrogel [91]. For example, Wang et al. developed an injectable and in situ photo-crosslinked hybrid hemostatic hydrogel by combining pectin methacrylate (PECMA) and methacrylate-based gelatin (GelMA). It was shown that the PECMA/GelMA hydrogel has good cytocompatibility and synergizes the hemostatic properties of calcium ions on PECMA, amine residues on GelMA, and a highly porous network to achieve rapid blood absorption and coagulation. An in vitro porcine skin bleeding model confirmed that the hydrogel could be injected directly into the wound and rapidly photo-crosslinked, and reduce the coagulation time by 39%. Importantly, the hydrogel can be easily removed to prevent secondary injury to the wound [92,93]. In another study, Yang et al. developed a photo-crosslinked multifunctional antibacterial and antioxidant hemostatic hydrogel dressing. It contained polyethylene glycol monomethyl ether modified glycidyl methacrylate functionalized CS (CSG-PEG), methacrylamide dopamine (DMA), and zinc ions. In a mouse model with intact skin defects infected with methicillin-resistant Staphylococcus aureus, CSG-PEG/DMA/Zn hydrogel not only adhered well to the wound surface but also showed better hemostasis and promoted wound healing of infected skin tissue defects than the commercially available Tegaderm^{TM} film [94] (Figure 6c). For faster clotting to prevent chronic inflammatory episodes, Chen et al. prepared a novel hemostatic hydrogel by cross-linking inorganic polyphosphate (PolyP) conjugated with polyaspartic hydrazide (PAHP) and PEO_{90} dialdehyde (PEO_{90} DA) [95]. The dynamic nature of the acyl-hydrazone bond allowed the hydrogel to self-repair when damaged by external forces.

The hydrogel simultaneously exhibited excellent tissue adhesion, biocompatibility, antimicrobial activity, and hemostatic efficacy. In a mouse total skin defect model, this hydrogel was loaded with mouse epidermal growth factor (mEGF) to accelerate wound repair and promote the regeneration of fresh tissue.

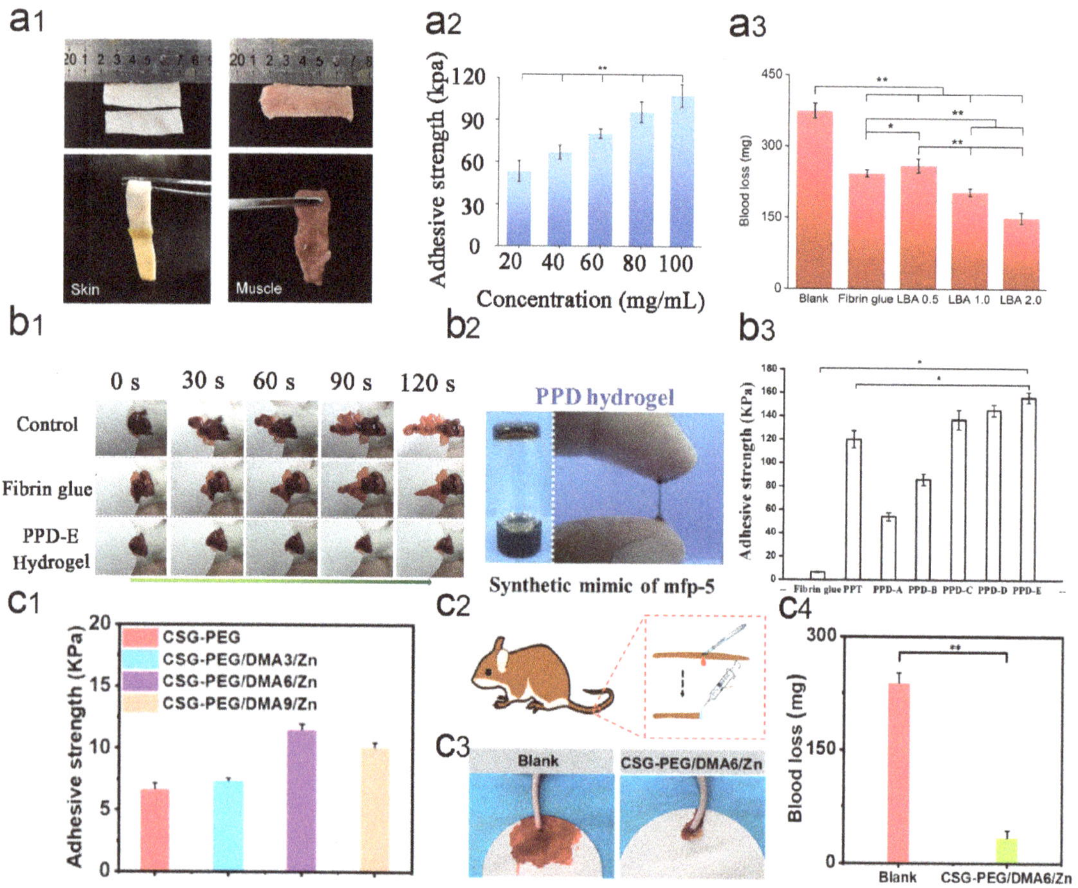

Figure 6. Application of wet-adhesive hemostatic hydrogel in skin wounds. (**a**) Adhesive properties and hemostatic effect of LBA hydrogel. Figure modified from [52] with permission. (**a1**) Photographs of tissue adhesion using LBA in various tissues (pig skin and muscle). (**a2**) Tissue adhesion strength of LBA 1.0 with different precursor solution concentrations. (**a3**) Blood loss during hemostasis with LBA in a rat hepatic hemorrhage model. (**b**) Wet adhesion properties and hemostatic effect of PDD hydrogel. Figure modified from [88] with permission. (**b1**) Total view of livers of bleeding mice treated and untreated with PPD hydrogel, and fibrin glue every 30 seconds for 2 minutes. (**b2**) PPD hydrogels are prepared through an HRP cross-linking reaction. (**b3**) Adhesion strength of PPD hydrogel to porcine tissues. Adhesion of p-nitrophenylchloroformate/PEG/TA hydrogel (PPT) and fibrin glue was used as a control (n = 5, * $p < 0.05$). (**c**) Adhesion and hemostatic properties of CSG-PEG/DMA/Zn hydrogels. Figure modified from [89] with permission. (**c1**) Adhesion strength of hydrogel. (**c2**) Schematic diagram of the mouse tail amputation model. (**c3**) Photographs of blood stains in a mouse tail amputation model. (**c4**) Quantitative results of blood loss in a mouse tail amputation model (n = 4, * $p < 0.05$, ** $p < 0.01$.).

3.2. Heart

Cardiac bleeding can occur as a result of trauma, such as injuries from accidents and wounds to the heart. Certain diseases can also lead to the rupture of blood vessels in the heart. Rapid and strong adhesion to wetted tissue walls and surfaces, high mechanical strength, and good biocompatibility to promote tissue regeneration are central and necessary for rapid hemostasis of cardiac arterial dissection. It has been shown that tannin (TA) contains a large number of benzene rings, which can make hydrogels very sticky in humid environments. When TA interacts with gelatin, CS, filamentous fibrin (SF), and Pluronic F127 (PEO_{99}-PPO_{65}-PEO_{99}), low-swelling hydrogels with good mechanical and wet adhesive capacity were formed [96]. The obtained CS/TA/SF hydrogels showed less bleeding and shorter hemostasis time in various arterial and visceral bleeding models compared to previously reported materials (Figure 7a). Later in another study, Liang et al. created a physicochemical double network cross-linked hydrogel (PCT) using acrylic acid, CS, and TA as the main components [97]. The hydrogels have many active sites on their surfaces, allowing fast, strong and repetitive adhesion to artificial solids and biological tissues (Figure 7b). Due to its amide covalent bond, the hydrogel can act for a longer period in tissue regeneration, and the resulting hydrogel-tissue adhesion interface has strong adhesion even after one month of immersion in a physiological environment. Due to its platelet adhesion and high bursting pressure qualities, this hydrogel can be used for good hemostatic properties at sites of heavy bleeding such as the heart.

Bionic strategies are also effective in wet adhesion hemostasis. It has been shown that the cationic polysaccharide intercellular adhesion contained in staphylococcal biofilms plays a key role in surface adhesion to wet and moving surfaces [99]. Inspired by the strong adhesion mechanisms of biofilms and mussels, Han et al. reported a novel dual bionic adhesion hydrogel (DBAH), which is based on CS-grafted methacrylate (CS-MA), DA, and N-hydroxymethylacrylamide (NMA) [46]. When DBAH was contacted with water, hydrophobic residues (-CH_3) rapidly generated cohesive forces that self-repel water molecules from the substrate surface. At this time, the catechol group of DA and the cationic free amine group NH_3^+ of CS-MA are exposed outward to promote sufficient contact with the adherent matrix for rapid and firm wet tissue adhesion, resulting in excellent hemostasis of DBAH even in the wet and active rabbit heart environment. In another study, Hong et al. designed a photo-responsive biological tissue adhesive that mimics the ECM composition. It consisted of GelMA, N-(2-aminoethyl)-4-(4-(hydroxymethyl)-2-methoxy-5-nitroso)butyramide modified glycosaminoglycan hyaluronic acid (HA-NB), and the polymerization initiator lithium phenyl-2,4,6-trimethylbenzoyl phosphate. Under UV light, this biomolecule-based hydrogel matrix rapidly gelled, adhered, and sealed the bleeding arteries and heart walls. This hydrogel can withstand a blood pressure of up to 290 mm Hg compared to most clinical situations (systolic blood pressure of 60–160 mm Hg). Notably, the hydrogel inhibited high-pressure hemorrhage from a 6-mm diameter cardiac perforation in the porcine heart and a 4- to 5-mm long incisional lesion in the porcine carotid artery [98] (Figure 7c).

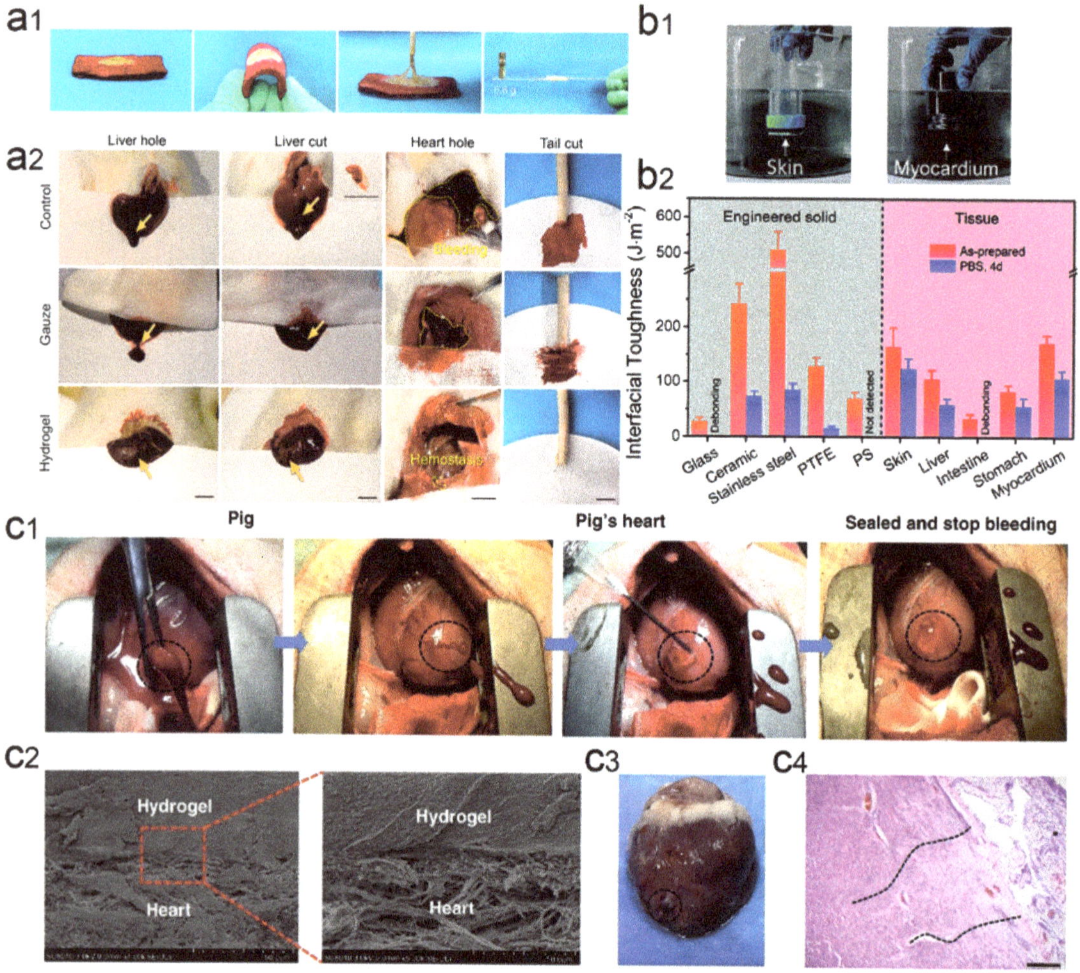

Figure 7. Wet-adhesive hemostatic hydrogel in heart wounds. (**a**) Adhesion and hemostatic properties of CS/TA/SF hydrogels. Figure modified from [96] with permission. (**a1**) The photos show that the hydrogel has strong adhesion to wet pigskin and glass. (**a2**) Hemostatic images of various untreated wounds covered with gauze or hydrogel. (**b**) Strong adhesion properties of PCT hydrogels. Figure modified from [97] with permission. (**b1**) Optical image of direct adhesion of PCT hydrogel between high-density polyethylene substrate and porcine skin or myocardium in PBS environment. (**b2**) Interfacial toughness of PCT hydrogels with different substrate surfaces before and after 4 days of immersion in PBS. (**c**) GelMA/HA-NB hydrogel adhesion and hemostatic properties. Figure modified from [98] with permission. (**c1**) Optical image of rapid hemostatic closure after a heart puncture wound. Blood exudation completely stopped within 10 seconds. (**c2**) Scanning electron micrographs of the interface between a porcine heart puncture wound and hydrogel. (**c3**) Autopsy images of the heart were performed after two weeks of postoperative recovery. (**c4**) Tissue-stained image of the interface between the heart tissue and the matrix gel of a pig heart 2 weeks after postoperative recovery.

3.3. Liver

To achieve rapid and effective hemostasis of liver bleeding, Shou et al. designed a catechol-hydroxybutyl CS (HBCS-C) hydrogel by attaching catechol and hydroxybutyl molecules to a CS backbone. This multifunctional HBCS-C hydrogel showed thermal sensitivity, injectability, tissue adhesion, and biocompatibility [100]. The multiple interactions between catechol hydroxyl/amino groups and tissues allow the biocompatible hydrogel to adhere firmly to the tissue surface. The hydrogel effectively blocked the bleeding in a rat liver hemorrhage model by adhering firmly to the bleeding tissue within 30 seconds (Figure 8a). In another work, Chen et al. created an in situ-generated hemostatic hydrogel (GelMA/oxidized dextran/Borax) for incompressible visceral wound hemostasis and anti-inflammatory applications [101]. The abundant adjacent hydroxyl groups of dextran can be oxidized to aldehyde groups by sodium periodate, which can be further bound to the amino groups of histones by Schiff base reaction, giving dextran good tissue adhesion ability. In addition, sodium tetraborate produces dynamic borate ester linkages when combined with oxidized dextran. Due to the three-layer network structure, the hydrogel exhibits good hemostatic capacity and can withstand high blood pressure exceeding 165 mm Hg, which is higher than the systolic blood pressure threshold for healthy adults (i.e., 120 mm Hg) (Figure 8c).

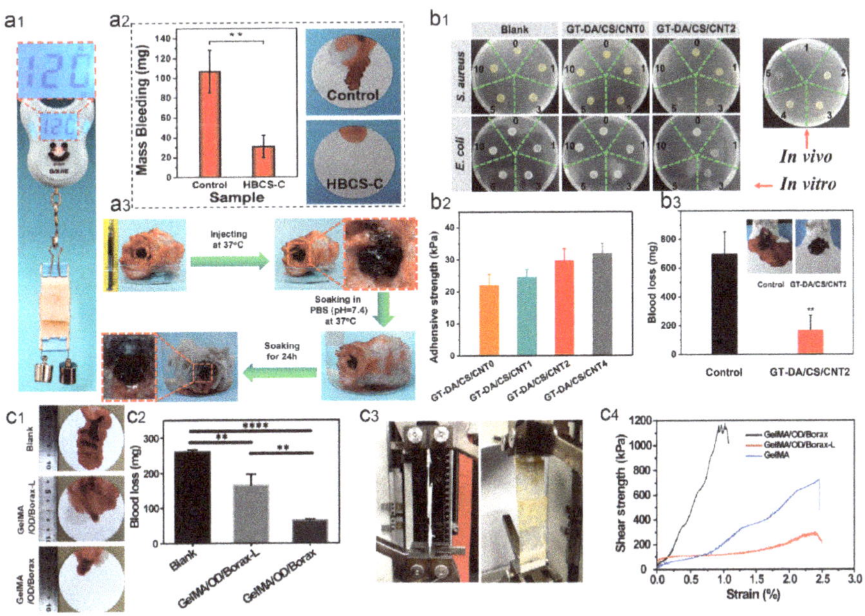

Figure 8. Wet-adhesive hemostatic hydrogel in liver wounds. (**a**) Tissue adhesion and hemostatic effect of HBCS-C hydrogel. Reprinted (adapted) with permission from [100]. Copyright {2020} American Chemical Society. (**a1**) The images show good adhesion of HBCS-C hydrogel on porcine skin (adhesion area of 20 mm × 20 mm and heavy load of 120 g). (**a2**) Total blood loss from liver wounds in hydrogel-treated and untreated rats. (**a3**) Photographs showing the wet bioadhesive behavior and stability of HBCS-C hydrogels in an aqueous environment at 37 °C. (**b**) Antimicrobial, adhesion, and hemostatic properties of GT-DA/CS/CNT hydrogel. Figure modified from [102] with permission. (**b1**) In vitro antibacterial activity of hydrogels induced by NIR illumination, 0, 1, 3, 5, and 10 represent different irradiation times (min). (**b2**) Adhesion strength of the hydrogels after 1 h in the air before testing. (**b3**) Hemostatic ability of GT-DA/CS/CNT hydrogels. (**c**) Adhesion and hemostatic properties of GelMA/oxidized dextran/Borax hydrogels. Figure modified from [101] with permission. (**c1**) Photographs of liver blood loss after different treatments. (**c2**) Blood loss during hemostasis of the liver. (**c3**) Pictures of the modified test method used for the overlapping shear test. (**c4**) Strain–stress curves for the overlapping shear test. (** $p < 0.01$, **** $p < 0.001$).

Various multifunctional injectable hydrogels for hemostasis have been successfully developed, however, these strategies ignore issues such as the ease of removal of these sealants on injured livers and the occurrence of secondary injuries. To address this issue, Bu et al. created a rapidly formed aminolysis tetrapolyethylene glycol (Tetra-PEG) hydrogel sealant which has good mechanical strength and tissue adhesion [103]. The cyclized succinyl ester moiety gave the sealant adjustable solubility and fast decomposition properties. The hydrogel showed high hemostatic activity even in the presence of anticoagulation and exhibited remarkable biocompatibility and utility. The drug loading will allow the hydrogel to accelerate wound healing by enhancing the antibacterial and anti-inflammatory effects and hemostatic effects. Liang et al. used gelatin graft-dopamine (GT-DA) and polydopamine-coated carbon nanotubes (CNT-PDA) to design antibacterial, adhesive, antioxidant, and conductive GT-DA/CS/CNT composite hydrogels by oxidative coupling of catechol moieties using an H_2O_2/HRP catalytic reaction [102]. Then, the antibiotic doxycycline was added to the hydrogel. Together with the photothermal effect of CNT-PDA, the hydrogels gained good antimicrobial action and showed well in vitro and in vivo antimicrobial activity against different microorganisms. In a mouse model of liver hemorrhage, mice in the untreated group drained approximately 700 mg of blood from the liver, while mice in the gelatin-DA/CS/CNT hydrogel group shed only 170 mg of blood (Figure 8b). Furthermore, the GT-DA/CS/CNT hydrogels also exhibited good tissue regeneration capacity, as indicated by the collagen deposition, histomorphometric analysis, and immunofluorescence staining for transforming growth factor (TGF) and a cluster of differentiation 31 (CD31). In a study, He et al. developed a conductive self-healing hydrogel for hemostasis using N-carboxyethyl chitosan (CECS), PF127, and CNT as the main ingredients, and further loaded with moxifloxacin hydrochloride [104]. A mouse liver injury model, a mouse liver incision model, and a mouse tail amputation model were used to evaluate the hemostatic ability of the CECS/PF127/CNT hydrogel. It was found that the hydrogel could adhere to the wound site and form the hydrogel within a short period (75 s), which acted as a stable physical barrier and prevented wound bleeding. In addition, the negative charge of CECS triggered an intrinsic coagulation pathway, which ultimately formed a stable platelet plug during hemostasis. Therefore, the blood loss in the CECS/PF127/CNT hydrogel group was significantly less than that in the control group.

3.4. Other Applications

Many gastric diseases can cause gastric bleeding with gastroscopy as well as with surgical treatment [105]. To solve this problem, He et al. prepared injectable pH-responsive self-healing adhesive hydrogels based on acryloyl-6-aminohexanoic acid (AA) and AA-gN-hydroxysuccinimide (AA-NHS) with efficient self-healing ability, hemostatic properties, and good biocompatibility. With the introduction of AA-NHS as a micro-crosslinker, the hydrogels exhibited enhanced adhesive strength. An in vivo model of porcine gastric bleeding showed that the hydrogel displayed good hemostatic properties by stopping acute arterial bleeding and preventing the bleeding. Gastric wound models showed that hydrogels exhibited excellent therapeutic effects, with type I collagen deposition, α-SMA expression, and angiogenesis significantly promoting wound healing [106]. In another study, Cheng et al. created an oral keratin hydrogel that adheres specifically to ulcers and accelerates ulcer healing [107]. In ethanol-treated rats, approximately 50% of the ulcer-adherent keratin hydrogel was able to remain in the stomach for 12 hours, however, only approximately 18% was left in healthy rats during the same period. In addition, keratin hydrogels prevented epithelial cell damage by gastric acid, reduced inflammation, and accelerated epithelial reconstitution of ethanol-induced gastric ulcers.

Intestinal bleeding is a bleeding disorder that may be caused by inflammation, mechanical injury, vascular lesions, or tumors. Recently, Cui et al. designed a high-strength instant self-adhesive organic-inorganic hybrid (OIH) hydrogel. This OIH hydrogel was composed of N-acryloyl 2-glycine (ACG), a biocompatible glycine derivative vinyl monomer, and the naturally occurring mineral hydroxyapatite (HAp) [108]. The hydrogen bonding of the poly(N-acryloyl 2-glycine) (PACG) side chains, the carboxyl-calcium ion crosslinking, and the PACG chain-HAp physical interactions contribute to the high mechanical properties of the hydrogel. Importantly, this OIH hydrogel exhibited strong adhesion to different substrates, which could be attributed to the synergistic interaction of carboxyl groups and the enhanced contact of PACG chains with the adherent surface.

4. Summary and Outlook

Different substances have been discovered to stop bleeding, however, their effectiveness depends on the type of bleeding and the location of the wound. Although many effective hemostatic bandages, gauze, foams, sprays, adhesives, gels, powders, granules, tourniquets, and tampons have been discovered for externally visible and accessible, and compressible wounds, these uncontrollable and incompressible bleedings remain a challenging problem in treating the internal arterial visceral bleeding and massive surgical blood loss. To stop blood loss, hemostatic hydrogels must firmly adhere to specific tissues under high fluid pressure, moisture, and dynamics while maintaining their good mechanical strength. In recent years, there has been a dramatic increase in the availability of multifunctional polymeric hydrogels to fill the current needs in wound management. In this review, we discussed recent advances in wet-adhesive hydrogel systems as hemostatic agents, sealants, or adhesives, focusing on elucidating the mechanisms of hemostasis, the various factors affecting wet-environment adhesion, and the application of wet-adhesive hemostatic hydrogels in various complex physiological settings such as skin, cardiac and liver wounds, and other mucosal bleeds. Some of the polymer precursors were modified with catechol as well as other groups to provide strong adhesion and good hemostatic properties. These developed hydrogels are not only highly biocompatible but also can effectively exert antiseptic and anti-inflammatory effects when antimicrobial substances are added, thereby promoting wound healing. Moreover, these hydrogels can withstand high blood pressure and have shown advantages over some existing commercial products.

Despite improvements in recent years in research on wet-adhesive hydrogel systems for hemostasis and wound healing, there are still many hurdles that need to be addressed before these hydrogel technologies can be used in the clinic. First, few hydrogels have been able to stop bleeding quickly and effectively in the presence of excess water and heavy bleeding from active tissue. Additional researches are necessary to develop effective injectable hydrogels that can effectively and accurately stop moist, smooth, and dynamic internal tissue and organ bleeding. Second, high-quality raw materials used to prepare wet-adhesive hydrogels should be exploited. Although hemostatic hydrogels prepared by mussel-inspired chemical methods have shown good adhesion and hemostatic properties, substances such as DA are expensive as raw materials for the preparation of these hydrogels, and the potential neurological side effects of DA would limit their mass production and commercialization [109]. There are also low-cost and safe substances considered alternatives to DA, like plant-derived polyphenolic compounds such as TAs containing catechol/pyrogallol molecules. Nevertheless, it should be noted that additional stimuli are required to stimulate the formation of polymer-TA hydrogels [110,111]. Moreover, the incorporation of artificial nanoenzymes with antioxidant properties to fabricate a multifunctional hemostatic hydrogel may also facilitate their translation [112]. Therefore, the development of safe raw materials for viscous hydrogels, as well as the development of new strategies to achieve rapid and effective hemostasis of hydrogels in humid environments, remain to be explored in the future, and it is hoped that this review will inspire future development in this field.

Author Contributions: Conceptualization, S.W. and W.H.; validation, W.H.; investigation, S.W.; writing—original draft preparation, W.H.; writing—review and editing, S.W.; supervision, S.W.; project administration, S.W.; funding acquisition, S.W. All authors have read and agreed to the published version of the manuscript.

Funding: This research and the APC were funded by Shanghai Rising-Star Program, grant number 20QA1407200.

Institutional Review Board Statement: Not applicable.

Informed Consent Statement: Not applicable.

Data Availability Statement: Not applicable.

Conflicts of Interest: The authors declare no conflict of interest.

References

1. Chen, Y.; Wu, L.; Li, P.; Hao, X.; Yang, X.; Xi, G.; Liu, W.; Feng, Y.; He, H.; Shi, C. Polysaccharide based hemostatic strategy for ultrarapid hemostasis. *Macromol. Biosci.* **2020**, *20*, 1900370. [CrossRef] [PubMed]
2. Kushimoto, S.; Kudo, D.; Kawazoe, Y. Acute traumatic coagulopathy and trauma-induced coagulopathy: An overview. *J. Intensive Care* **2017**, *5*, 6. [CrossRef] [PubMed]
3. Simpson, A.; Shukla, A.; Brown, A.C. Biomaterials for hemostasis. *Annu. Rev. Biomed. Eng.* **2022**, *24*, 111–135. [CrossRef] [PubMed]
4. Tummalapalli, M.; Anjum, S.; Kumari, S.; Gupta, B. Antimicrobial surgical sutures: Recent developments and strategies. *Polym Rev.* **2016**, *56*, 607–630. [CrossRef]
5. Robles-Campos, R.; Navarro-Barrios, Á.; Martínez-Caceres, C.; Revilla-Nuin, B.; Brusadin, R.; López-López, V.; López-Conesa, A.; Caballero-Planes, A.; de la Peña-Moral, J.; Parrilla-Paricio, P. The contribution of the deportalized lobe to liver regeneration in tourniquet-ALPPS. *Ann. Surg.* **2020**, *271*, E94–E96. [CrossRef] [PubMed]
6. Schwartz, R.B.; Reynolds, B.Z.; Shiver, S.A.; Lerner, E.B.; Greenfield, E.M.; Solis, R.A.; Kimpel, N.A.; Coule, P.L.; Singletary, J.G.M.; Woods, W. Comparison of two packable hemostatic gauze dressings in a porcine hemorrhage model. *Prehosp. Emerg. Care* **2011**, *15*, 477–482. [CrossRef] [PubMed]
7. Mostafalu, P.; Tamayol, A.; Rahimi, R.; Ochoa, M.; Khalilpour, A.; Kiaee, G.; Yazdi, I.K.; Bagherifard, S.; Dokmeci, M.R.; Ziaie, B.; et al. Smart bandages: Smart bandage for monitoring and treatment of chronic wounds. *Small* **2018**, *14*, 1870150. [CrossRef]
8. Liu, L.; Hu, E.; Yu, K.; Xie, R.; Lu, F.; Lu, B.; Bao, R.; Li, Q.; Dai, F.; Lan, G. Recent advances in materials for hemostatic management. *Biomater. Sci.* **2021**, *9*, 7343–7378. [CrossRef]
9. Cheng, Y.; Zheng, X.; Zhang, L.; Zhao, J.; Hu, L.; Wang, S. Enhanced photothermal and chemotherapy of pancreatic tumors by degrading the extracellular matrix. *Colloids Surf. B* **2023**, *221*, 113010. [CrossRef]
10. Yang, X.; Wang, S.; Zhang, X.; Ye, C.; Wang, S.; An, X. Development of PVA-based microsphere as a potential embolization agent. *Mater. Sci. Eng. C* **2022**, *135*, 112677. [CrossRef]
11. Wang, L.; You, X.; Dai, C.; Tong, T.; Wu, J. Hemostatic nanotechnologies for external and internal hemorrhage management. *Biomater. Sci.* **2020**, *8*, 4396–4412. [CrossRef] [PubMed]
12. Li, D.; Chen, J.; Wang, X.; Zhang, M.; Li, C.; Zhou, J. Recent advances on synthetic and polysaccharide adhesives for biological hemostatic applications. *Front. Bioeng. Biotechnol.* **2020**, *8*, 926. [CrossRef] [PubMed]
13. Seyednejad, H.; Imani, M.; Jamieson, T.; Seifalian, A.M. Topical haemostatic agents. *Br. J. Surg.* **2008**, *95*, 1197–1225. [CrossRef] [PubMed]
14. Chiara, O.; Cimbanassi, S.; Bellanova, G.; Chiarugi, M.; Mingoli, A.; Olivero, G.; Ribaldi, S.; Tugnoli, G.; Basilicò, S.; Bindi, F.; et al. A systematic review on the use of topical hemostats in trauma and emergency surgery. *BMC Surg.* **2018**, *18*, 68. [CrossRef] [PubMed]
15. Spotnitz, W.D. Hemostats, sealants, and adhesives: A practical guide for the surgeon. *Am. Surg.* **2012**, *78*, 1305–1321. [CrossRef] [PubMed]
16. Chen, Z.; Yao, J.; Zhao, J.; Wang, S. Injectable wound dressing based on carboxymethyl chitosan triple-network hydrogel for effective wound antibacterial and hemostasis. *Int. J. Biol. Macromol.* **2023**, *225*, 1235–1245. [CrossRef]
17. Xie, M.; Zeng, Y.; Wu, H.; Wang, S.; Zhao, J. Multifunctional carboxymethyl chitosan/oxidized dextran/sodium alginate hydrogels as dressing for hemostasis and closure of infected wounds. *Int. J. Biol. Macromol.* **2022**, *219*, 1337–1350. [CrossRef]
18. Ouyang, Y.; Zhao, J.; Wang, S. Multifunctional hydrogels based on chitosan, hyaluronic acid and other biological macromoleculesfor the treatment of inflammatory bowel disease: A review. *Int. J. Biol. Macromol.* **2023**, *227*, 505–523. [CrossRef]
19. Wang, T.; Yi, W.; Zhang, Y.; Wu, H.; Fan, H.; Zhao, J.; Wang, S. Sodium alginate hydrogel containing platelet-rich plasma for wound healing. *Colloids Surf. B* **2023**, *221*, 113010. [CrossRef]
20. Yu, L.; Ding, J. Injectable hydrogels as unique biomedical materials. *Chem. Soc. Rev.* **2008**, *37*, 1473–1481. [CrossRef]
21. Zhu, W.; Chuah, Y.J.; Wang, D.A. Bioadhesives for internal medical applications: A review. *Acta Biomater.* **2018**, *74*, 1–16. [CrossRef] [PubMed]
22. Lokhande, G.; Carrow, J.K.; Thakur, T.; Xavier, J.R.; Parani, M.; Bayless, K.J.; Gaharwar, A.K. Nanoengineered injectable hydrogels for wound healing application. *Acta Biomater.* **2018**, *70*, 35–47. [CrossRef] [PubMed]
23. Li, J.; Celiz, A.D.; Yang, J.; Yang, Q.; Wamala, I.; Whyte, W.; Seo, B.R.; Vasilyev, N.V.; Vlassak, J.J.; Suo, Z.; et al. Tough adhesives for diverse wet surfaces. *Science* **2017**, *357*, 378–381. [CrossRef] [PubMed]

24. Xia, X.; Xu, X.; Wang, B.; Zhou, D.; Zhang, W.; Xie, X.; Lai, H.; Xue, J.; Rai, A.; Li, Z.; et al. Adhesive Hemostatic Hydrogel with Ultrafast Gelation Arrests Acute Upper Gastrointestinal Hemorrhage in Pigs. *Adv. Funct. Mater.* **2021**, *32*, 2109332. [CrossRef]
25. Li, Q.; Hu, E.; Yu, K.; Xie, R.; Lu, F.; Lu, B.; Bao, R.; Zhao, T.; Dai, F.; Lan, G. Self-propelling Janus particles for hemostasis in perforating and irregular wounds with massive hemorrhage. *Adv. Funct. Mater.* **2020**, *30*, 2004153. [CrossRef]
26. Xiong, Y.; Zhang, X.; Ma, X.; Wang, W.; Yan, F.; Zhao, X.; Chu, X.; Xu, W.; Sun, C. A review of the properties and applications of bioadhesive hydrogels. *Polym. Chem.* **2021**, *12*, 3721–3739. [CrossRef]
27. Nuyttens, B.P.; Thijs, T.; Deckmyn, H.; Broos, K. Platelet adhesion to collagen. *Thromb. Res.* **2011**, *127*, S26–S29. [CrossRef]
28. Broos, K.; Feys, H.B.; de Meyer, S.F.; Vanhoorelbeke, K.; Deckmyn, H. Platelets at work in primary hemostasis. *Blood Rev.* **2011**, *25*, 155–167. [CrossRef]
29. Singh, I.; Rana, V. Techniques for the assessment of mucoadhesion in drug delivery systems: An overview. *J. Adhes. Sci. Technol.* **2012**, *26*, 2251–2267. [CrossRef]
30. Li, J.; Liu, J.; Ma, C.; Ji, J.; Liu, J. Mechanisms underlying the biological wet adhesion: Coupled effects of interstitial liquid and contact geometry. *J. Bionic Eng.* **2020**, *17*, 448–456. [CrossRef]
31. Li, M.; Xie, J.; Dai, Q.; Huang, W.; Wang, X. Effect of wetting case and softness on adhesion of bioinspired micropatterned surfaces. *J. Mech. Behav. Biomed. Mater.* **2018**, *78*, 266–272. [CrossRef] [PubMed]
32. Gandyra, D.; Walheim, S.; Gorb, S.; Barthlott, W.; Schimmel, T. The capillary adhesion technique: A versatile method for determining the liquid adhesion force and sample stiffness. *Beilstein J. Nanotechnol.* **2015**, *6*, 11–18. [CrossRef] [PubMed]
33. Xie, J.; Li, M.; Dai, Q.; Huang, W.; Wang, X. Key parameters of biomimetic patterned surface for wet adhesion. *Int. J. Adhes. Adhes.* **2018**, *82*, 72–78. [CrossRef]
34. Nguyen, V.P.; Ho, V.A. Wet adhesion of soft curved interfaces with micro pattern. *IEEE Robot. Autom. Lett.* **2021**, *6*, 4273–4280. [CrossRef]
35. Carvalho, F.C.; Bruschi, M.L.; Evangelista, R.C.; Gremião, M.P.D. Mucoadhesive drug delivery systems. *Braz. J. Pharm. Sci.* **2010**, *46*, 1–17. [CrossRef]
36. Wang, H.; Li, X.; Li, M.; Wang, S.; Zuo, A.; Guo, J. Bioadhesion design of hydrogels: Adhesion strategies and evaluation methods for biological interfaces. *J. Adhes. Sci. Technol.* **2022**, *36*, 1–35. [CrossRef]
37. van der Leeden, M.C.; Frens, G. Surface properties of plastic materials in relation to their adhering performance. *Adv. Eng. Mater.* **2002**, *4*, 280–289. [CrossRef]
38. Mehdizadeh, M.; Yang, J. Design strategies and applications of tissue bioadhesives. *Macromol. Biosci.* **2013**, *13*, 271–288. [CrossRef]
39. Lee, J.; Park, E.; Fujisawa, A.; Lee, H. Diatom silica/polysaccharide elastomeric hydrogels: Adhesion and interlocking synergy. *ACS Appl. Mater. Interfaces* **2021**, *13*, 21703–21713. [CrossRef]
40. Yuan, Y.; Wu, H.; Ren, X.; Wang, J.; Liu, R.; Hu, B.; Gu, N. Dual-network hydrogel based on ionic nano-reservoir for gastric perforation sealing. *Sci. China Mater.* **2021**, *65*, 827–835. [CrossRef]
41. Feng, Y.; Li, X.; Zhang, Q.; Yan, S.; Guo, Y.; Li, M.; You, R. Mechanically robust and flexible silk protein/polysaccharide composite sponges for wound dressing. *Carbohydr. Polym.* **2019**, *216*, 17–24. [CrossRef] [PubMed]
42. Vorwald, C.E.; Gonzalez-Fernandez, T.; Joshee, S.; Sikorski, P.; Leach, J.K. Tunable fibrin-alginate interpenetrating network hydrogels to support cell spreading and network formation. *Acta Biomater.* **2020**, *108*, 142–152. [CrossRef] [PubMed]
43. Yin, Y.H.; Pan, B.B.H.; Xin, Z. Underwater and wet adhesion strategies for hydrogels in biomedical applications. *Chem. Eng. J.* **2021**, *431*, 133372. [CrossRef]
44. Varvarenko, S.; Voronov, A.; Samaryk, V.; Tarnavchyk, I.; Nosova, N.; Kohut, A.; Voronov, S. Covalent grafting of polyacrylamide-based hydrogels to a polypropylene surface activated with functional polyperoxide. *React. Funct. Polym.* **2010**, *70*, 647–655. [CrossRef]
45. Fan, X.; Fang, Y.; Zhou, W.; Yan, L.; Xu, Y.; Zhu, H.; Liu, H. Mussel foot protein inspired tough tissue-selective underwater adhesive hydrogel. *Mater. Horiz.* **2020**, *8*, 997–1007. [CrossRef]
46. Li, Z.; Meng, X.; Xu, W.; Zhang, S.; Ouyang, J.; Zhang, Z.; Liu, Y.; Niu, Y.; Ma, S.; Xue, Z.; et al. Single network double cross-linker (SNDCL) hydrogels with excellent stretchability, self-recovery, adhesion strength, and conductivity for human motion monitoring. *Soft Matter* **2020**, *16*, 7323–7331. [CrossRef]
47. Cai, C.; Chen, Z.; Chen, Y.; Li, H.; Yang, Z.; Liu, H. Mechanisms and applications of bioinspired underwater/wet adhesives. *J. Polym. Sci.* **2021**, *59*, 2911–2945. [CrossRef]
48. Zhang, Y.; Zhu, C.; Zhang, Z.; Zhao, J.; Yuan, Y.; Wang, S. Oxidation triggered formation of polydopamine-modified carboxymethyl cellulose hydrogel for anti-recurrence of tumor. *Colloids Surf. B* **2021**, *207*, 112025. [CrossRef]
49. Saiz-Poseu, J.; Mancebo-Aracil, J.; Nador, F.; Busqué, F.; Ruiz-Molina, D. The chemistry behind catechol-based adhesion. *Angew. Chem. Int. Ed. Engl.* **2019**, *58*, 696–714. [CrossRef]
50. Guo, Q.; Chen, J.; Wang, J.; Zeng, H.; Yu, J. Recent progress in synthesis and application of mussel-inspired adhesives. *Nanoscale* **2020**, *12*, 1307–1324. [CrossRef]
51. Zhou, D.; Li, S.; Pei, M.; Yang, H.; Gu, S.; Tao, Y.; Ye, D.; Zhou, Y.; Xu, W.; Xiao, P. Dopamine-modified hyaluronic acid hydrogel adhesives with fast-forming and high tissue adhesion. *ACS Appl. Mater. Interfaces* **2020**, *12*, 18225–18234. [CrossRef] [PubMed]
52. Ma, Y.; Yao, J.; Liu, Q.; Han, T.; Zhao, J.; Ma, X.; Tong, Y.; Jin, G.; Qu, K.; Li, B.; et al. Liquid bandage harvests robust adhesive, hemostatic, and antibacterial performances as a first-aid tissue adhesive. *Adv. Funct. Mater.* **2020**, *30*, 2001820. [CrossRef]
53. Yang, J.; Bai, R.; Suo, Z. Topological adhesion of wet materials. *Adv. Mater.* **2018**, *30*, e1800671. [CrossRef] [PubMed]

54. Li, Z.; Zhou, F.; Li, Z.; Lin, S.; Chen, L.; Liu, L.; Chen, Y. Hydrogel cross-linked with dynamic covalent bonding and micellization for promoting burn wound healing. *ACS Appl. Mater. Interfaces* **2018**, *10*, 25194–25202. [CrossRef]
55. Wang, L.; Li, Y.; Lin, L.; Mu, R.; Pang, J. Novel synthesis of mussel inspired and Fe^{3+} induced pH-sensitive hydrogels: Adhesion, injectable, shapeable, temperature properties, release behavior and rheological characterization. *Carbohydr. Polym.* **2020**, *236*, 116045. [CrossRef]
56. Man, Z.; Sidi, L.; Xubo, Y.; Jin, Z.; Xin, H. An in situ catechol functionalized ε-polylysine/polyacrylamide hydrogel formed by hydrogen bonding recombination with high mechanical property for hemostasis. *Int. J. Biol. Macromol.* **2021**, *191*, 714–726. [CrossRef]
57. Liang, M.; Ge, X.; Dai, J.; Ren, P.; Wei, D.; Xu, L.; Zhang, Q.; He, C.; Lu, Z.; Zhang, T. High-strength hydrogel adhesive formed via multiple interactions for persistent adhesion under saline. *ACS Appl. Bio Mater.* **2021**, *4*, 5016–5025. [CrossRef]
58. Liu, J.; Wang, S.; Shen, Q.; Kong, L.; Huang, G.; Wu, J. Tough underwater super-tape composed of semi-interpenetrating polymer networks with a water-repelling liquid surface. *ACS Appl. Mater. Interfaces* **2021**, *13*, 1535–1544. [CrossRef]
59. Peng, Q.; Wu, Q.; Chen, J.; Wang, T.; Wu, M.; Yang, D.; Peng, X.; Liu, J.; Zhang, H.; Zeng, H. Coacervate-based instant and repeatable underwater adhesive with anticancer and antibacterial properties. *ACS Appl. Mater. Interfaces* **2021**, *13*, 48239–48251. [CrossRef]
60. Cui, C.; Fan, C.; Wu, Y.; Xiao, M.; Wu, T.; Zhang, D.; Chen, X.; Liu, B.; Xu, Z.; Qu, B.; et al. Water-triggered hyperbranched polymer universal adhesives: From strong underwater adhesion to rapid sealing hemostasis. *Adv. Mater.* **2019**, *31*, 1905761. [CrossRef]
61. Pan, F.; Ye, S.; Wang, R.; She, W.; Liu, J.; Sun, Z.; Zhang, W. Hydrogel networks as underwater contact adhesives for different surfaces. *Mater. Horiz.* **2020**, *7*, 2063–2070. [CrossRef]
62. Zhou, Y.; Zhang, C.; Gao, S.; Zhang, B.; Sun, J.; Kai, J.-j.; Wang, B.; Wang, Z. Instant and strong underwater adhesion by coupling hygroscopicity and in situ photocuring. *Chem. Mater.* **2021**, *33*, 8822–8830. [CrossRef]
63. Wang, Z.; Guo, L.; Xiao, H.; Cong, H.; Wang, S. A reversible underwater glue based on photo- and thermo-responsive dynamic covalent bonds. *Mater. Horiz.* **2019**, *7*, 282–288. [CrossRef]
64. Autumn, K.; Sitti, M.; Liang, Y.A.; Peattie, A.M.; Hansen, W.R.; Sponberg, S.; Kenny, T.W.; Fearing, R.; Israelachvili, J.N.; Full, R.J. Evidence for van der Waals adhesion in gecko setae. *Proc. Natl. Acad. Sci. USA* **2002**, *99*, 12252–12256. [CrossRef]
65. Yi, H.; Lee, S.H.; Seong, M.; Kwak, M.K.; Jeong, H.E. Bioinspired reversible hydrogel adhesives for wet and underwater surfaces. *J. Mater. Chem. B* **2018**, *6*, 8064–8070. [CrossRef]
66. Sato, N.; Aoyama, Y.; Yamanaka, J.; Toyotama, A.; Okuzono, T. Particle adsorption on hydrogel surfaces in aqueous media due to van der Waals attraction. *Sci. Rep.* **2017**, *7*, 6099. [CrossRef]
67. Li, J.; Xu, Z.; Xiao, Y.; Gao, G.; Chen, J.; Yin, J.; Fu, J. Macroscopic assembly of oppositely charged polyelectrolyte hydrogels. *J. Mater. Chem. B* **2018**, *6*, 257–264. [CrossRef]
68. Huang, G.; Tang, Z.; Peng, S.; Zhang, P.; Sun, T.; Wei, W.; Zeng, L.; Guo, H.; Guo, H.; Meng, G. Modification of hydrophobic hydrogels into a strongly adhesive and tough hydrogel by electrostatic interaction. *Macromolecules* **2021**, *55*, 156–165. [CrossRef]
69. Tian, G.; Liu, Y.; Yu, M.; Liang, C.; Yang, D.; Huang, J.; Zhao, Q.; Zhang, W.; Chen, J.; Wang, Y.; et al. Electrostatic Interaction-based high tissue adhesive, stretchable microelectrode arrays for the electrophysiological interface. *ACS Appl. Mater. Interfaces* **2022**, *14*, 4852–4861. [CrossRef]
70. Song, F.; Zhang, J.; Lu, J.; Cheng, Y.; Tao, Y.; Shao, C.; Wang, H. A mussel-inspired flexible chitosan-based bio-hydrogel as a tailored medical adhesive. *Int. J. Biol. Macromol.* **2021**, *189*, 183–193. [CrossRef]
71. Xiao, Z.; Li, Q.; Liu, H.; Zhao, Q.; Niu, Y.; Zhao, D. Adhesion mechanism and application progress of hydrogels. *Eur. Polym. J.* **2022**, *173*, 111277. [CrossRef]
72. Ma, X.; Zhou, X.; Ding, J.; Huang, B.; Wang, P.; Zhao, Y.; Mu, Q.; Zhang, S.; Ren, C.; Xu, W. Hydrogels for underwater adhesion: Adhesion mechanism, design strategies and applications. *J. Mater. Chem. A* **2022**, *10*, 11823–11853. [CrossRef]
73. Lee, H.-H.; Kim, Y.-W.; Woo, J.; Park, H.; Hur, K.; Suo, Z.; Sun, J.-Y. Fast healing of ionic bonds in tough hydrogels under an acoustic excitation. *Extreme Mech. Lett.* **2019**, *33*, 100572. [CrossRef]
74. Huang, Y.; Xiao, L.; Zhou, J.; Liu, T.; Yan, Y.; Long, S.; Li, X. Strong tough polyampholyte hydrogels via the synergistic effect of ionic and metal–ligand bonds. *Adv. Funct. Mater.* **2021**, *31*, 2103917. [CrossRef]
75. Liang, Y.; Xue, J.; Du, B.; Nie, J. Ultrastiff, tough, and healable ionic–hydrogen bond cross-linked hydrogels and their uses as building blocks to construct complex hydrogel structures. *ACS Appl. Mater. Interfaces* **2019**, *11*, 5441–5454. [CrossRef]
76. Chen, Y.; Meng, J.; Gu, Z.; Wan, X.; Jiang, L.; Wang, S. Bioinspired multiscale wet adhesive surfaces: Structures and controlled adhesion. *Adv. Funct. Mater.* **2019**, *30*, 1905287. [CrossRef]
77. Sareh, S.; Althoefer, K.; Li, M.; Noh, Y.; Tramacere, F.; Sareh, P.; Mazzolai, B.; Kovac, M. Anchoring like octopus: Biologically inspired soft artificial sucker. *J. R. Soc. Interface* **2017**, *14*, 20170395. [CrossRef]
78. Jung, H.; Kim, M.K.; Lee, J.Y.; Choi, S.W.; Kim, J. Adhesive hydrogel patch with enhanced strength and adhesiveness to skin for transdermal drug delivery. *Adv. Funct. Mater.* **2020**, *30*, 2004407. [CrossRef]
79. Jeon, E.Y.; Lee, J.; Kim, B.J.; Joo, K.I.; Kim, K.H.; Lim, G.; Cha, H.J. Bio-inspired swellable hydrogel-forming double-layered adhesive microneedle protein patch for regenerative internal/external surgical closure. *Biomaterials* **2019**, *222*, 119439. [CrossRef]
80. Tsujioka, K.; Matsuo, Y.; Shimomura, M.; Hirai, Y. A new concept for an adhesive material inspired by clingfish sucker nanofilaments. *Langmuir* **2022**, *38*, 1215–1222. [CrossRef]

81. Das, D.; Nag, T.C. Fine structure of the organ of attachment of the teleost, Garra gotyla gotyla (Ham). *Zoology* **2006**, *109*, 300–309. [CrossRef] [PubMed]
82. Lee, H.; Um, D.-S.; Lee, Y.; Lim, S.; Kim, H.-j.; Ko, H. Octopus-Inspired Smart Adhesive Pads for Transfer Printing of Semiconducting Nanomembranes. *Adv. Mater.* **2016**, *28*, 7457–7465. [CrossRef] [PubMed]
83. Autumn, K.; Peattie, A.M. Mechanisms of adhesion in geckos. *Integr. Comp. Biol.* **2002**, *42*, 1081–1090. [CrossRef] [PubMed]
84. Geng, Z.; Li, Z.; Cui, Z.; Wang, J.; Yang, X.; Liu, C. A novel bionic topography with miR-21 coating for improving bone-implant integration through regulating cell adhesion and angiogenesis. *Nano Lett.* **2020**, *20*, 7716–7721. [CrossRef]
85. Meng, F.; Liu, Q.; Wang, X.; Tan, D.; Xue, L.; Barnes, W.J.P. Tree frog adhesion biomimetics: Opportunities for the development of new, smart adhesives that adhere under wet conditions. *Philos. Trans. R. Soc. A* **2019**, *377*, 31177956. [CrossRef]
86. Dąbrowska, A.K.; Spano, F.; Derler, S.; Adlhart, C.; Spencer, N.D.; Rossi, R.M. The relationship between skin function, barrier properties, and body-dependent factors. *Skin Res. Technol.* **2018**, *24*, 165–174. [CrossRef]
87. Li, S.; Pei, M.; Wan, T.; Yang, H.; Gu, S.; Tao, Y.; Liu, X.; Zhou, Y.; Xu, W.; Xiao, P. Self-healing hyaluronic acid hydrogels based on dynamic Schiff base linkages as biomaterials. *Carbohydr. Polym.* **2020**, *250*, 116922. [CrossRef]
88. Wang, R.; Li, J.; Chen, W.; Xu, T.; Yun, S.; Xu, Z.; Xu, Z.; Sato, T.; Chi, B.; Xu, H. A biomimetic mussel-inspired ε-poly-l-lysine hydrogel with robust tissue-anchor and anti-infection capacity. *Adv. Funct. Mater.* **2017**, *27*, 1604894. [CrossRef]
89. Han, L.; Lu, X.; Liu, K.; Wang, K.; Fang, L.; Weng, L.-T.; Zhang, H.; Tang, Y.; Ren, F.; Zhao, C.; et al. Mussel-inspired adhesive and tough hydrogel based on nanoclay confined dopamine polymerization. *ACS Nano* **2017**, *11*, 2561–2574. [CrossRef]
90. Han, L.; Yan, L.; Wang, K.; Fang, L.; Zhang, H.; Tang, Y.; Ding, Y.; Weng, L.-T.; Xu, J.; Weng, J.; et al. Tough, self-healable and tissue-adhesive hydrogel with tunable multifunctionality. *NPG Asia Mater.* **2017**, *9*, e372. [CrossRef]
91. Zhang, X.; Yao, D.; Zhao, W.; Zhang, R.; Yu, B.; Ma, G.; Li, Y.; Hao, D.; Xu, F.-J. Engineering platelet-rich plasma based dual-network hydrogel as a bioactive wound dressing with potential clinical translational value. *Adv. Funct. Mater.* **2021**, *31*, 2009258. [CrossRef]
92. Wang, J.H.; Tsai, C.W.; Tsai, N.Y.; Chiang, C.Y.; Lin, R.S.; Pereira, R.F.; Li, Y.E. An injectable, dual crosslinkable hybrid pectin methacrylate (PECMA)/gelatin methacryloyl (GelMA) hydrogel for skin hemostasis applications. *Int. J. Biol. Macromol.* **2021**, *185*, 441–450. [CrossRef] [PubMed]
93. Panchami, P.; Prachi, T. Hydrogels differentiated by length scales: A review of biopolymer-based hydrogel preparation methods, characterization techniques, and targeted applications. *Eur. Polym. J.* **2021**, *163*, 110935.
94. Yang, Y.; Liang, Y.; Chen, J.; Duan, X.; Guo, B. Mussel-inspired adhesive antioxidant antibacterial hemostatic composite hydrogel wound dressing via photo-polymerization for infected skin wound healing. *Bioact. Mater.* **2022**, *8*, 341–354. [CrossRef]
95. Chen, D.; Liu, X.; Qi, Y.; Ma, X.; Wang, Y.; Song, H.; Zhao, Y.; Li, W.; Qin, J. Poly(aspartic acid) based self-healing hydrogel with blood coagulation characteristic for rapid hemostasis and wound healing applications. *Colloids Surf. B* **2022**, *214*, 112430. [CrossRef] [PubMed]
96. Qiao, Z.; Lv, X.; He, S.; Bai, S.; Liu, X.; Hou, L.; He, J.; Tong, D.; Ruan, R.; Zhang, J.; et al. A mussel-inspired supramolecular hydrogel with robust tissue anchor for rapid hemostasis of arterial and visceral bleedings. *Bioact. Mater.* **2021**, *6*, 2829–2840. [CrossRef]
97. Liang, M.; Wei, D.; Yao, Z.; Ren, P.; Dai, J.; Xu, L.; Zhang, T.; Zhang, Q. Hydrogel adhesive formed via multiple chemical interactions: From persistent wet adhesion to rapid hemostasis. *Biomater. Sci.* **2022**, *10*, 1486–1497. [CrossRef]
98. Hong, Y.; Zhou, F.; Hua, Y.; Zhang, X.; Ni, C.; Pan, D.; Zhang, Y.; Jiang, D.; Yang, L.; Lin, Q.; et al. A strongly adhesive hemostatic hydrogel for the repair of arterial and heart bleeds. *Nat. Commun.* **2019**, *10*, 2060. [CrossRef]
99. Formosa-Dague, C.; Feuillie, C.; Beaussart, A.; Derclaye, S.; Kucharíková, S.; Lasa, I.; van Dijck, P.; Dufrêne, Y.F. Sticky matrix: Adhesion mechanism of the staphylococcal polysaccharide intercellular adhesin. *ACS Nano* **2016**, *10*, 3443–3452. [CrossRef]
100. Shou, Y.; Zhang, J.; Yan, S.; Xia, P.; Xu, P.; Li, G.; Zhang, K.; Yin, J. Thermoresponsive chitosan/DOPA-based hydrogel as an injectable therapy approach for tissue-adhesion and hemostasis. *ACS Biomater. Sci. Eng.* **2020**, *6*, 3619–3629. [CrossRef]
101. Chen, Z.; Wu, H.; Wang, H.; Zaldivar-Silva, D.; Agüero, L.; Liu, Y.; Zhang, Z.; Yin, Y.; Qiu, B.; Zhao, J.; et al. An injectable anti-microbial and adhesive hydrogel for the effective noncompressible visceral hemostasis and wound repair. *Mater. Sci. Eng. C* **2021**, *129*, 112422. [CrossRef] [PubMed]
102. Liang, Y.; Zhao, X.; Hu, T.; Han, Y.; Guo, B. Mussel-inspired, antibacterial, conductive, antioxidant, injectable composite hydrogel wound dressing to promote the regeneration of infected skin. *J. Colloid Interface Sci.* **2019**, *556*, 514–528. [CrossRef] [PubMed]
103. Bu, Y.; Zhang, L.; Sun, G.; Sun, F.; Liu, J.; Yang, F.; Tang, P.; Wu, D. Tetra-PEG based hydrogel sealants for in vivo visceral hemostasis. *Adv. Mater.* **2019**, *31*, e1901580. [CrossRef]
104. He, J.; Shi, M.; Liang, Y.; Guo, B. Conductive adhesive self-healing nanocomposite hydrogel wound dressing for photothermal therapy of infected full-thickness skin wounds. *Chem. Eng. J.* **2020**, *394*, 124888. [CrossRef]
105. Zhu, Y.; Xu, J.X.; Cheng, J.; Zhang, Z.; Zhu, B.Q.; Chen, T.Y.; Xu, X.Y.; Wang, Y.; Cai, M.Y.; Zhou, P.H. A novel injectable thermo-sensitive binary hydrogels system for facilitating endoscopic submucosal dissection procedure. *United Eur. Gastroenterol. J.* **2019**, *7*, 782–789. [CrossRef]
106. He, J.; Zhang, Z.; Yang, Y.; Ren, F.; Li, J.; Zhu, S.; Ma, F.; Wu, R.; Lv, Y.; He, G.; et al. Injectable self-healing adhesive pH-responsive hydrogels accelerate gastric hemostasis and wound healing. *Nanomicro Lett.* **2021**, *13*, 80. [CrossRef]
107. Cheng, Z.; Qing, R.; Hao, S.; Ding, Y.; Yin, H.; Zha, G.; Chen, X.; Ji, J.; Wang, B. Fabrication of ulcer-adhesive oral keratin hydrogel for gastric ulcer healing in a rat. *Regener. Biomater.* **2021**, *8*, rbab008. [CrossRef]

108. Cui, C.; Wu, T.; Gao, F.; Fan, C.; Xu, Z.; Wang, H.; Liu, B.; Liu, W. An autolytic high strength instant adhesive hydrogel for emergency self-rescue. *Adv. Funct. Mater.* **2018**, *28*, 1804925. [CrossRef]
109. Costa, K.M.; Schoenbaum, G. Dopamine. *Curr. Biol.* **2022**, *32*, R817–R824. [CrossRef]
110. Yan, W.; Shi, M.; Dong, C.; Liu, L.; Gao, C. Applications of tannic acid in membrane technologies: A review. *Adv. Colloid Interface Sci.* **2020**, *284*, 102267. [CrossRef]
111. Baldwin, A.; Booth, B.W. Biomedical applications of tannic acid. *J. Biomater. Appl.* **2022**, *36*, 1503–1523. [CrossRef] [PubMed]
112. Xie, P.; Zhang, L.; Shen, H.; Wu, H.; Zhao, J.; Wang, S.; Hu, L. Biodegradable $MoSe_2$-polyvinylpyrrolidone nanoparticles with multi-enzyme activity for ameliorating acute pancreatitis. *J. Nanobiotechnol.* **2022**, *20*, 113. [CrossRef] [PubMed]

gels

MDPI

Article

Levofloxacin HCl-Loaded Eudragit L-Based Solvent Exchange-Induced In Situ Forming Gel Using Monopropylene Glycol as a Solvent for Periodontitis Treatment

Setthapong Senarat [1], Sarun Tuntarawongsa [2], Nutdanai Lertsuphotvanit [3], Catleya Rojviriya [4], Thawatchai Phaechamud [1,3,5,6,*] and Takron Chantadee [6,7,8,*]

1 Programme of Pharmaceutical Engineering, Faculty of Pharmacy, Silpakorn University, Nakhon Pathom 73000, Thailand; senarat_s@silpakorn.edu
2 Pharmaceutical Intellectual Center "Prachote Plengwittaya", Faculty of Pharmacy, Silpakorn University, Nakhon Pathom 73000, Thailand; tuntarawongsa_s@su.ac.th
3 Program of Pharmaceutical Technology, Department of Pharmaceutical Technology, Faculty of Pharmacy, Silpakorn University, Nakhon Pathom 73000, Thailand; lertsuphotvanit_n@silpakorn.edu
4 Synchrotron Light Research Institute, Mueang District, Nakhon Ratchasima 30000, Thailand; catleya@slri.or.th
5 Department of Industrial Pharmacy, Faculty of Pharmacy, Silpakorn University, Nakhon Pathom 73000, Thailand
6 Natural Bioactive and Material for Health Promotion and Drug Delivery System Group (NBM), Faculty of Pharmacy, Silpakorn University, Nakhon Pathom 73000, Thailand
7 Department of Pharmaceutical Sciences, Faculty of Pharmacy, Chiang Mai University, Chiang Mai 50200, Thailand
8 Center of Excellent in Pharmaceutical Nanotechnology, Chiang Mai University, Chiang Mai 50200, Thailand
* Correspondence: phaechamud_t@su.ac.th (T.P.); takron.chantadee@cmu.ac.th (T.C.); Tel.: +66-53-944310 (T.C.)

Abstract: Solvent exchange-induced in situ forming gel (ISG) is currently an appealing dosage form for periodontitis treatment via localized injection into the periodontal pocket. This study aims to apply Eudragit L and Eudragit S as matrix components of ISG by using monopropylene glycol as a solvent for loading levofloxacin HCl for periodontitis treatment. The influence of Eudragit concentration was investigated in terms of apparent viscosity, rheological behavior, injectability, gel-forming behavior, and mechanical properties. Eudragit L-based formulation presented less viscosity, was easier to inject, and could form more gel than Eudragit S-based ISG. Levofloxacin HCl-loading diminished the viscosity of Eudragit L-based formulation but did not significantly change the gel formation ability. Higher polymer loading increased viscosity, force-work of injectability, and hardness. SEM photographs and μCT images revealed their scaffold formation, which had a denser topographic structure and less porosity attained owing to higher polymer loading and less in vitro degradation. By tracking with fluorescence dyes, the interface interaction study revealed crucial information such as solvent movement ability and matrix formation of ISG. They prolonged the drug release for 14 days with fickian drug diffusion kinetics and increased the release amount above the MIC against test microbes. The 1% levofloxacin HCl and 15% Eudragit L dissolved in monopropylene glycol (LLM15) was a promising ISG because of its appropriate viscosity (3674.54 ± 188.03 cP) with Newtonian flow, acceptable gel formation and injectability (21.08 ± 1.38 N), hardness (33.81 ± 2.3 N) and prolonged drug release with efficient antimicrobial activities against *S. aureus* (ATCC 6538, 6532, and 25923), methicillin-resistant *S. aureus* (*MRSA*) (*S. aureus* ATCC 4430), *E. coli* ATCC 8739, *C. albicans* ATCC 10231, *P. gingivalis* ATCC 33277, and *A. actinomycetemcomitans* ATCC 29522; thus, it is the potential ISG formulation for periodontitis treatment by localized periodontal pocket injection.

Keywords: Eudragit L; in situ forming gel; levofloxacin; monopropylene glycol; solvent exchange; periodontitis

Citation: Senarat, S.; Tuntarawongsa, S.; Lertsuphotvanit, N.; Rojviriya, C.; Phaechamud, T.; Chantadee, T. Levofloxacin HCl-Loaded Eudragit L-Based Solvent Exchange-Induced In Situ Forming Gel Using Monopropylene Glycol as a Solvent for Periodontitis Treatment. *Gels* **2023**, 9, 583. https://doi.org/10.3390/gels9070583

Academic Editor: Damien Dupin

Received: 20 June 2023
Revised: 11 July 2023
Accepted: 13 July 2023
Published: 18 July 2023

1. Introduction

In the present era, approximately 20–50% of the world's population is affected by periodontal diseases [1,2]. Among these individuals, more than 19.8% of adults and 12.2% of the elderly suffer from severe periodontitis, indicated by a periodontal pocket depth exceeding 6 mm [3]. This severe condition increases the risk of infection, inflammation, and tooth loss and negatively impacts the individual's overall quality of life.

Periodontitis, a serious gum infection that destroys soft tissues and bone that supports the teeth, causes a periodontal pocket [4]. When periodontitis develops, pockets form as the inner gum pulls away from the teeth. These pockets collect debris and bacteria, which accelerates infection [5]. The periodontal pocket accumulates various bacteria, mainly pathogens like *Porphyromonas gingivalis* and *Aggregatibacter actinomycetemcomitans*, which have been recognized as the primary cause of advanced disease of periodontitis [6,7]. These pathogens are *Gram*-negative, facultative anaerobes, a nonmotile bacterium often detected in association with localized aggressive periodontitis, a severe infection of the periodontium [8]. Effective periodontal treatment involves the removal or inhibition of bacterial growth on tooth surfaces and within the crevicular pocket through methods like mechanical scaling, root planning, and medications. Antimicrobial agents, including chlorhexidine, tetracyclines (such as tetracycline, doxycycline, and minocycline), metronidazole, ciprofloxacin, vancomycin, and other antimicrobial drugs, are commonly used in periodontal treatment to eliminate infection [9,10]. However, the use of systemic antibiotics for periodontitis treatment necessitates large doses to achieve appropriate concentrations in the gingival crevicular fluid of the periodontal pocket, potentially leading to side effects and concerns about antibiotic resistance in other parts of the body and disrupting normal flora [11–14]. These limitations have prompted the development of local drug delivery systems for periodontal disease treatment. Implementing a controlled release system that delivers antimicrobial agents directly into the crevicular pocket is a more desirable strategy. Local drug delivery restricts the drug to the target site with minimal systemic uptake, allowing for lower doses to be effective and reducing or eliminating harmful side effects [15]. In this study, Eudragit L and Eudragit S were utilized as matrix components in an in situ forming gel (ISG) formulation by using monopropylene glycol as a solvent for loading levofloxacin HCl for periodontitis treatment. The development of worthy local drug delivery systems or devices attracted astonishing attention to overcome the restricted entry for eradication of the periodontopathic microbes in the deep periodontal pocket [16]. The in situ forming gel (ISG) system, induced by solvent exchange, has gained significant attention as a promising injectable drug delivery system for periodontal pocket administration. This is due to its simple administration process and uncomplicated fabrication, which involves the incorporation of a matrix-forming agent to regulate and localize drug release, while also providing effective antimicrobial activity [17]. The ISG system initially consists of a drug dissolved or dispersed in a polymer solution that thereafter solidifies via *a* water-solvent exchange mechanism from a fluid state into a drug-entrapped polymeric gel and matrix over time after injection and exposure to the aqueous fluid in periodontal pocket [17,18]. Atridox® is a commercially marketed ISG containing D,L-lactic acid (PLA) and *N*-methyl-2-pyrrolidone (NMP) loaded with doxycycline hyclate for periodontitis treatment [19]. Doxycycline hyclate-incorporated ISGs for periodontitis treatment using ethyl cellulose and bleached shellac as the polymers have been reported recently [20]. Borneol, lauric acid, and some natural resins (such as propolis, benzoin, and rosin) are currently being applied as matrix-forming agents of ISG for modulating vancomycin HCL and doxycycline hyclate release for periodontitis treatment [21–23]. Eudragit RS has been employed as a polymer for the controlled release of doxycycline hyclate from ISG, in which the addition of peppermint oil and polyethylene glycol could modify drug release [24–28]. All the above-mentioned ISGs use NMP or dimethyl sulfoxide (DMSO) as solvents to dissolve drugs and polymers. The fluid properties of ISG solution state are dependent on Eudragit type and solvent, and therefore, the utilization of Eudragit and solvent in pharmaceutical dosage forms should be considered [29,30]. Increasing Eudragit L concentration in dimethyl sulfoxide (DMSO) has been shown to enhance the viscosity of its solution, and these polymeric solutions exhibited the potential to be developed into ISG for periodontal pocket drug delivery [31].

Eudragit L and S (Figure 1a) are well-recognized as anionic polymers with enteric properties and are fabricated into various dosage forms including microspheres, micro-sponges, nanoparticles, matrix tablets, and coated tablets [32]. Eudragit L and S are known for their pH-dependent solubility. Eudragit L is insoluble in acidic environments but readily dissolving at pH levels above 6 [32], while Eudragit S dissolves at pH 7 or higher [33]. This characteristic is a key feature of the Eudragit line, as it allows for controlled drug release based on varying pH conditions [34]. Eudragit L has been used in the treatment of chronic enteritis as micro-sponges [35]. The strips that are prepared from Eudragit L/Eudragit S are used as intra-pocket delivery systems, and tetracycline HCl and clindamycin are the active agents that are contained within them [36]. These polymers have been utilized to produce nanoparticles that have been sanctioned for clinical application in Europe, Japan, and America. These nanoparticles are employed for the purpose of loading DNA plasmid and low molecular weight heparin [37]. Eudragit S is a negatively charged polymer employed for colonic drug delivery that dissolves at pH 7 or higher. Eudragit® S transdermal film containing metoprolol tartrate may reduce first-pass drug metabolism [33]. Eudragits are commonly acknowledged as nontoxic and nonirritant materials. According to current research, the ingestion of polymethacrylates at a daily dosage of 2 mg per kilogram of body weight is considered to be safe for human consumption. The FDA Inactive Ingredients Guide lists Eudragit as one of its constituents [32]. ISG is employed as a polymer primarily from Eudragit RS and uses NMP as its solvent [24–28]. Nevertheless, the use of Eudragit L and Eudragit S in monopropylene glycol as the ISG system for periodontal treatment have not been investigated. These Eudragits dissolve at pH levels above 6, which might be easy to handle by providing a degradable matrix for drug release without the need for clinical removal.

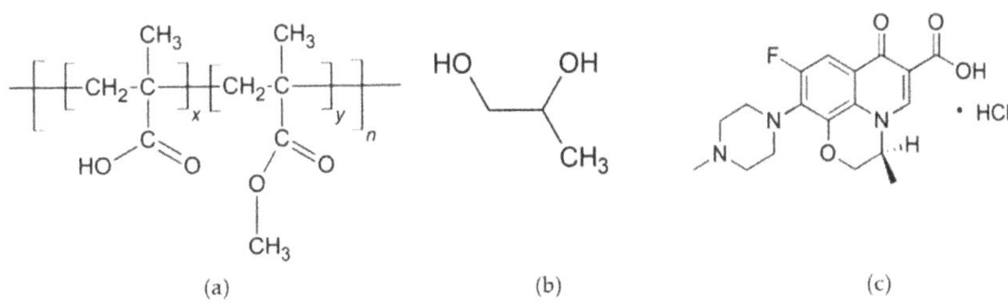

Figure 1. Chemical structure of Eudragit® L and S 100 (**a**), monopropylene glycol (**b**), and levofloxacin HCl (**c**). The chemical structure of Eudragit® L100 and Eudragit® S100, where the ratio of x:y is 1:1 and 1:2, respectively.

Monopropylene glycol is also called propylene glycol, and its chemical formula $C_3H_8O_2$ and structure are given in Figure 1b. Monopropylene glycol is a colorless, odorless, and water-soluble liquid. As an inert ingredient, the Food and Drug Administration (FDA) has given its approval for its use in a variety of products, including foods, tobacco products, and pharmaceuticals [38]. It was selected as a cosolvent in the preparation of in situ gel formulations of mitiglinide calcium for simultaneous extended delivery and enhanced drug bioavailability [39]. It has also been reported as a co-surfactant of an in situ microemulsion-gel in bio-adhesive hydroxypropyl methylcellulose films for zidovudine transdermal administration [40]. In this study, monopropylene glycol is used to dissolve Eudragit L- and Eudragit S-based ISGs since it is safe and has been used in the pharmaceutical field. Therefore, the present research's conceptual framework is interested in applying Eudragit L and Eudragit S as polymers in monopropylene glycol of the ISG system to load antimicrobial agents for periodontitis treatment. Levofloxacin HCl was used as the antimicrobial drug in this study because of its effectiveness at low doses and broad-spectrum antibiotic activities [41]. To determine the appropriateness of using monopropylene glycol, compared to DMSO, we considered experimental results as well. Biocompatibility, on the other hand, is another matter that is based on the premise that their biocompatibility levels are acceptable, as the following reports: DMSO was used as a vehicle in various types of in situ formation systems, such as the in situ forming of implants containing osthole [42], risperidone, and paliperidone [43], and the in situ forming of microparticle loading montelukast sodium [44]. Intravesical instillation of 50% *w*/*w* DMSO (RIMSO-50®) is used as a therapeutic approach for interstitial cystitis. Because it did not cause any toxicity in the chondrocytes, the 7.813% *w*/*v* concentration of DMSO did not have a negative impact on the integrity of the articular cartilage [45]. Additionally, it was discovered that the utilization of a cryopreservation medium made up of fetal bovine serum and DMSO at a concentration of 5% was able to effectively preserve the colony formation and chondrogenic capabilities of synovial mesenchymal stem cells. Moreover, it was discovered that the utilization of a cryopreservation medium made up of fetal bovine serum and DMSO at a concentration of 5% was able to effectively preserve the colony formation and chondrogenic capabilities of synovial mesenchymal stem cells [46].

This study examined and analyzed the impact of varying concentrations of Eudragit on physicochemical properties, matrix-forming behavior, and antimicrobial activities. Furthermore, an investigation was conducted on the physicochemical properties of Eudragit ISG formulations that have been loaded with levofloxacin HCl. This investigation encompasses the apparent viscosity, rheology behavior, injectability, gel forming behavior, mechanical properties, matrix morphology, drug release, and antimicrobial activities of the prepared formulations.

2. Results and Discussion

2.1. Drug-Free Eudragit-Based ISG

2.1.1. Physical Appearance, Viscosity and Rheology, Injectability, Contact Angle, and Mechanical Properties

Figure 2 depicts the effect of Eudragit L and S concentrations on the physical appearance of drug-free ISG formulations. All formulations were clear solutions, as shown in Figure 2B, except SM20 and SM25. This turbidity indicated a limitation in Eudragit S solubility in monopropylene glycol at higher concentrations than 15%. A slight yellowish color was observed in highly concentrated Eudragit L solutions. The viscosity of LM5, LM10, LM15, LM20, SM5, and SM10 formulations were 183.54 ± 2.00 cP, 820.02 ± 31.36 cP, 5292.68 ± 222.78 cP, 26214.29 ± 874.33 cP, 171.24 ± 3.16 cP, and 1884.15 ± 134.47 cP, respectively. The increasing Eudragit concentration resulted in high apparent viscosity of the ISG formulations (Figure 3A). Eudragit L ISG presented significantly less viscosity than Eudragit S ISG ($p < 0.05$). The Eudragit S formulations with polymer concentrations higher than 10% were excluded from viscosity measurement since they were too viscous for measuring. Because of the dominant effect of substance–solvent interaction over substance–substance interaction, a good vehicle for solubilizing the solute should reduce the viscosity of the mixture [47].

Figure 3B depicts the rheological characteristics of developed ISG formulations. The relationship between shear stress and shear rate was linear in the majority of formulations. Thus, the developed formulations displayed Newtonian flow characteristics. Consistent with prior study on the Newtonian behavior of Eudragit RS-based ISG, these results demonstrated the ISG's Newtonian behavior [21]. The increased 3D network formation within Eudragit molecules can result in a significant increase in the viscosity of these formulations [48,49]. In spite of this, the SM10 curve went to a lower shear stress value, showing pseudoplastic flow behavior. Changing the number of molecular entanglements could affect the rheology and structure of entangled polymer systems [50]. Their Newtonian and pseudoplastic flow characteristics have been deemed suitable for injectable dosage forms because an injection through a needle is permissible after providing a compression force to the syringe plunger to eject the fluid through the stainless needle by a physician or dentist [49]. Eudragit L solution dissolved in monopropylene glycol exhibited less viscosity with Newtonian flow, making them more suitable for administration with injection than Eudragit S solution.

Figure 2. Physical appearance of Eudragit L (**A**) and Eudragit S (**B**)-based ISGs comprising different concentrations of polymers.

The administration of ISG should be simple and painless when using the appropriate needle and syringe. Localized ISGs for periodontitis treatment are designed for periodontal pocket delivery with an injection; thus, the developed fluid formulation should be easily expelled from the needle for less pain for the patient [51]. Therefore, it is vital to evaluate injectability, which refers to the amount of force required to expel the product through the needle. The maximum force for injectability of M, LM5, LM10, LM15, LM20, LM25 SM5, SM10, and SM15 formulations were 1.02 ± 0.04 N, 1.24 ± 0.07 N, 1.79 ± 0.05 N, 2.33 ± 0.02 N, 10.66 ± 0.30 N, 30.87 ± 1.68 N, 0.94 ± 0.02 N, 1.71 ± 0.01 N, and 27.74 ± 1.04 N, respectively. The low values of force and work of expulsion indicate ease of injectability. Figure 3C displays the force and work from the injectability determination of the test solutions. Their values increased with Eudragit concentration and corresponded to their apparent viscosity, as previously described. Work of expulsion for SM10 and SM15 were 20.22 ± 0.55 mJ and 305.05 ± 24.98 mJ, respectively. The notable shooting values were seen in a formulation comprising more than 10% of these polymers, especially for Eudragit S solution indicating that they had to be applied with a higher force and energy to expel from the needle. These results revealed that the kind and concentration of Eudragit had a significant impact on the needed injection force and energy. Nonetheless, the results demonstrated that the majority of ISG formulations required a low injection force (50 N). In practice, the total expulsive work required to inject each solution was less than 50 N.mm, suggesting that they met the injection criterion [51,52]. The addition of peppermint oil and polyethylene glycol (PEG) 1500 enhanced the injection performance of drug-free and doxycycline hyclate-loaded Eudragit RS ISG [27,28]. In comparison, the hydrophobic nature of peppermint oil provoked more ease of injection than PEG 1500 for the role of additives, such as peppermint oil and PEG 1500 on ISG [27,28]. The lubricating effect of oil incorporation on the ease of injection of in situ forming systems has been aforementioned [18,53]. For instance, the force and work of injectability of the rosin-based ISG formulations tended to reduce significantly after the addition of lime peel oil due to the lubricating nature of this oil [54]. Nonetheless, the ease of injection of Eudragit L correlated mainly to its lower viscosity. The molecule of polymeric material in solutions expresses the different chain configuration from its different affinity with the used solvents [55].

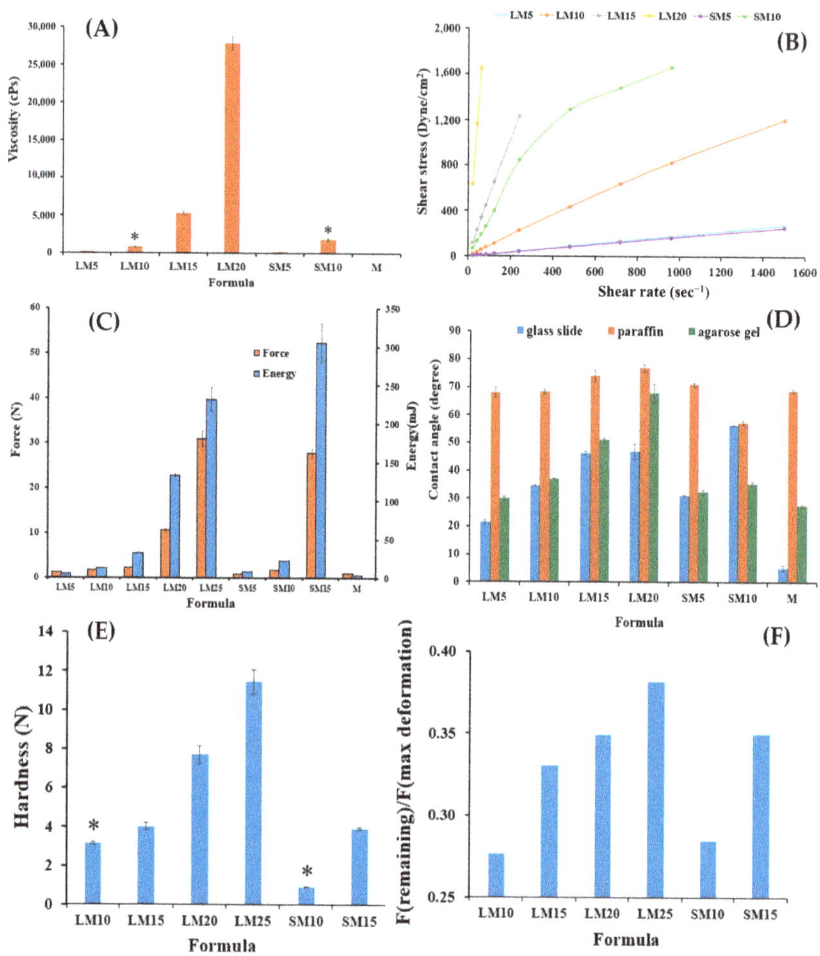

Figure 3. Viscosity (**A**); relationship between shear stress and shear rates (**B**); injection force and energy from injectability test (**C**); contact angle on different surfaces (**D**); hardness (**E**) and adhesiveness (**F**) properties from the mechanical test of Eudragit L and S-based ISG formulations at 25 °C. The data represent triplicates. The asterisk symbol indicates significant difference ($p < 0.05$).

The contact angle could indicate the wettability or spreadability of that liquid on the test surface. The contact angle of monopropylene glycol was lowest on the glass slide (5.08 ± 1.05 degree) (as seen in Figure 3D). The contact angles on glass slide of LM5,LM10, and LM15 were 21.46 ± 0.75 degree, 34.70 ± 0.10 degree, and 46.27 ± 0.72 degree, respectively. Increasing contact angle on three surfaces was evident with polymer concentration dependence of Eudragit solutions due to the higher viscosity retardation of the spreadability of sample droplets on these three surfaces [21]. The agarose model (periodontal mimic) was employed to simulate the gum surface which provided small amount of simulated saliva fluid (PBS) continuously. Since it was found that the choice of release medium, such as water, phosphate buffer, or horse serum, did not have a big effect on phase separation, water uptake rates, or the morphology of the gel [56], PBS was used in this model instead of saliva fluid to reduce the number of variables.

The surface of agarose gel fabricated comprising PBS pH 6.8 was employed to mimic a surface inside the periodontal pocket. The contact angle of Eudragit-based ISGs was found to be notably higher than that of solvent onto this agarose surface (LM10; 36.97 ± 0.22 degree, LM15; 50.98 ± 0.58 degree, LM20; 67.98 ± 3.18 degree) especially higher than on the glass slide, indicating in situ transformation from solution to gel or solid-like Eudragit matrix provoking via a solvent exchange mechanism between monopropylene glycol in the formulation and aqueous solution in prepared agarose gels [18,53,54]. These Eudragit matrices resulting from solvent exchange diminished the spreadability of the formulations [47]. SM10's high viscosity may hinder solvent exchange, and its contact angle on agarose gel was smaller than on the glass slide. Even though they had a high contact angle, their results were less than 90 degrees, indicating adequate wettability on test surfaces [57,58].

Typically, the obtained ISG matrix after solvent exchange should remain for fitting the periodontal pocket and resist the jaw's motion while retaining its shape [59,60]. Figure 3E,F display the mechanical properties presented as hardness and F remaining/F max deformation. The ratio "F remaining/F max deformation" is used to assess the specimen's elasticity/plasticity. A high value indicates high elasticity, while a low value indicates high plasticity [61]. These mechanical properties, especially the adhesive quality of transformed Eudragit ISG matrices, were designed to check the possibility of protecting undesirable slipping out of this dosage form from the periodontal pocket [58,62]. The introduction of Eudragits raised the hardness of the obtained matrix (LM10; 3.15 ± 0.08 N, LM15; 4.03 ± 0.19 N, LM20; 7.70 ± 0.45 N) owing to a greater polymer mass in the matrix. In comparison, the hardness and ratio of F remaining/F max deformation of Eudragit L-based ISGs were apparently and significantly higher than those of Eudragit S-based ISGs at the same concentration ($p < 0.05$). The ratio value of the matrix was significantly less than one, which indicated that the system's behavior was close to plastic deformation. This meant that the system was unable to change its inner structure during the holding time permanently and was likely adaptable in terms of its geometry to dynamic changes in periodontal pocket size and shape with time [61]. This finding corresponds with the previously reported matrix from bleached shellac-based ISG, which was more likely plastic deformation or able to adapt its geometry to dynamic changes [60]. All of the Eudragit-based ISGs displayed adhesive strength properties against the texture analyzer probe, and all of the obtained gels remained attached to the agarose basement throughout the texture analyzer probe pullback.

2.1.2. Gel Formation of Eudragit-Based ISGs

The phase transformation of all selected clear Eudragit-based solutions into a gel and subsequently an opaque matrix-like state over time are depicted in Figure 4A,C. The higher loading of these polymers promoted greater cloudy matrix formation due to the phase separation of Eudragits. From the results mentioned above, LM25 and SM15 exhibited more difficulty injecting through the needle and were more likely to stick in needles than other formulations due to their higher viscosity; therefore, they rapidly transformed into an opaque matrix and settled at the bottom of the PBS of the glass tube. During contact with PBS at pH 6.8, the formulation gradually incurred a solvent exchange, and the outer part initially changed into an opaque skin, while the inner gel phase continually transformed into an opaque solid polymeric matrix over time [49]. Therefore, more incorporation of matrix-forming agents accelerated the phase transformation of ISG. By comparison, there was no complete matrix formation of ISGs comprising ethyl cellulose, bleach shellac, and Eudragit RS of less than 10%, 20%, and 30% *w*/*w*, respectively, since all of them had no sufficient polymeric mass for continuing gel formation [20,24,26,63]. Figure 4B,D show a cross-section view of gel formation under the stereoscope using an agarose well to simulate the periodontal pocket. The phase separation of Eudragit resulted in an opaque layer at the rim of the agarose gel that expanded inward into the formulation. LM15 and SM10 showed slow matrix formation since their dense matrix possibly performs as a barrier and retard solvent exchange. There was a dense network of Eudragit matrixes formed at the interface It slowed down over time since the high tortuosity and low porosity of the former matrix impeded the diffusion of solvent and water [64]. Nevertheless, the high inner polymer loading of LM20, LM25, and SM15 was found to promote Eudragit matrix growth. The mechanistic of Eudragit matrix formation could be explained through solvent exchange, in which the phase separation of dissolved Eudragit was evident into an insoluble solid matrix [26]. Owing to less viscosity, ease of injection, and interesting gel formation, 10, 15, and 20% Eudragit L-based formulations were selected for loading levofloxacin HCl and further investigation as ISGs for a periodontal pocket delivery system.

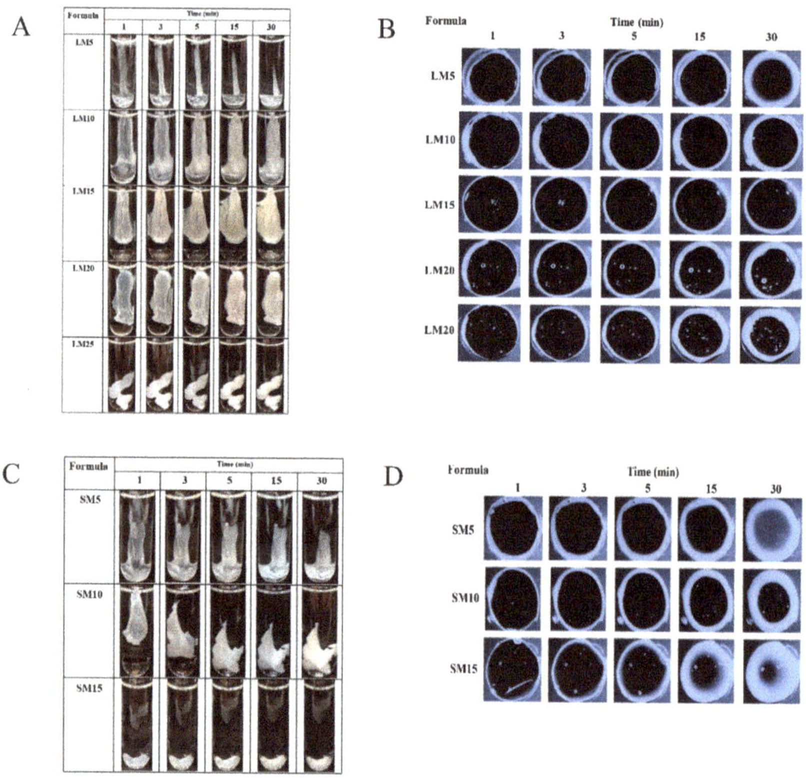

Figure 4. Gel formation after injection into PBS (**A**) and agarose well (**B**) of Eudragit L-based ISG formulations and gel formation after injection into PBS (**C**) and agarose well (**D**) of Eudragit S-based ISG formulations.

2.2. *Levofloxacin HCl-Loaded Eudragit-Based ISG*

2.2.1. Physical Appearance, Viscosity and Rheology, Injectability, Contact Angle, and Mechanical Properties

The 10, 15, and 20% Eudragit L-based formulations were successfully prepared. Slightly yellowish solutions were obtained as presented in Figure 5. The viscosity was found to be lower after drug addition; nevertheless, increasing the Eudragit L concentration enhanced the viscosity (Figure 6A). The hydrophilic drug compound might impede the uncoiled Eudragit molecules and decrease the environmental viscosity. As presented in Figure 6B, all formulations retained Newtonian flow. The viscosity of LLM15 was 3674.54 ± 188.03 cPs. with the Newtonian flow. This flow behavior is considered to be suitable for injection dosage forms and is acceptable after injection [49]. The force and work of injection were enhanced with an increasing Eudragit L concentration (Figure 6C). The force of injection of LLM15 was 21.08 ± 1.38 N. The trend of contact angle was similar to that of drug-free Eudragit-based ISG formulations, denoting that the higher viscosity of systems due to Eudragit L addition enhanced the contact angle value by retarding the spreadability of the droplet. The higher contact angle of solutions, such as lauric acid and borneol dissolved in DMSO, has previously been reported and is related to their higher concentration and viscosity [21,65]. This increase in contact angle on the agarose gel surface was caused by the transformation of Eudragit L into a semisolid state, followed by the solvent exchange process and decreased droplet spreading [66]. Their contact angle values were less than 90°, indicating good wettability of these ISGs on test surfaces [58]. Practically, spreadability is needed for sufficient adhesiveness to protect the undesirable slipping out of the transformed ISG from the periodontal pocket [58,62].

Figure 5. Physical appearance of levofloxacin HCl-loaded Eudragit L-based ISGs comprising different concentrations of polymer.

Figure 6E,F are the mechanical properties of drug-loaded Eudragit L ISGs. The addition of more Eudragit L enhanced the hardness of the obtained drug-loaded matrix, whereas the ratio value of F remaining/F max deformation was less than one, indicating that the system's behavior was similar to plastic deformation. These mechanical behaviors could inform their ability to adapt their geometry in the periodontal pocket, as aforementioned [61]. The hardness of the obtained LLM15 matrix was 33.81 ± 2.3 N. Some investigations reported that the addition of volatile oils improved the adhesion of the Rosin-ISG [67]. The rosin ISG with 10% cinnamon oil added had a higher adhesion property than that of 10% lime peel oil added, although formulations without the oil exhibited similar adhesion values [54]. And hence, the type of additive had a distinct impact on the adhesion properties of the ISG system.

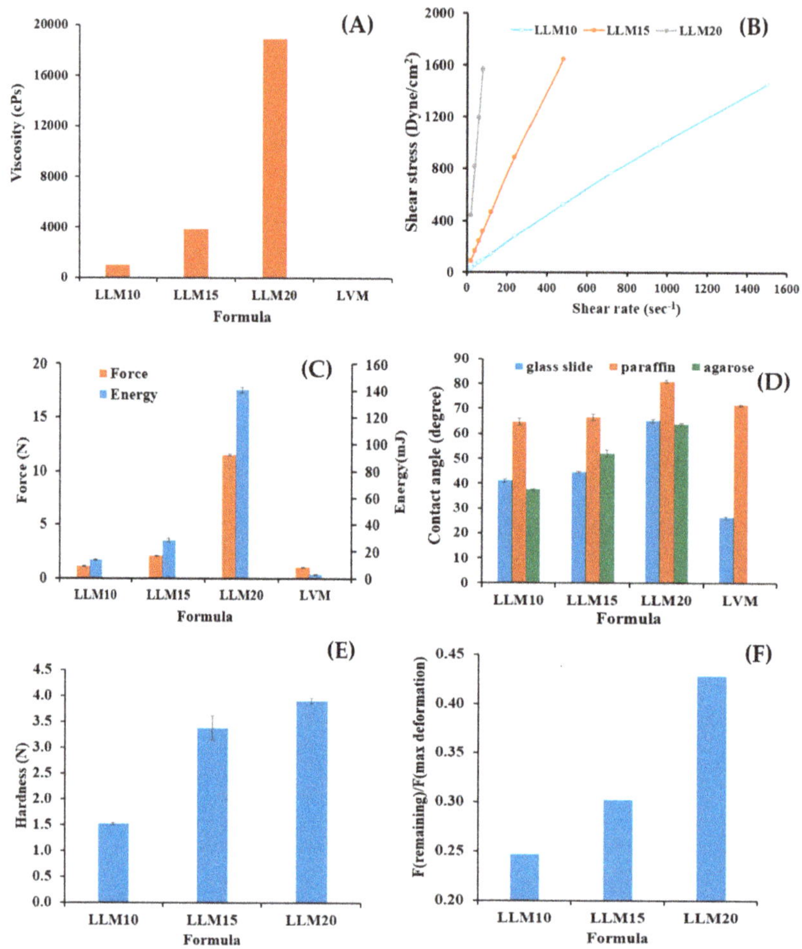

Figure 6. Viscosity (**A**); relationship between shear stress and shear rates (**B**); injection force and energy from injectability test (**C**); contact angle on different surfaces (**D**); hardness (**E**) and adhesiveness (**F**) properties from the mechanical test of levofloxacin HCl-loaded Eudragit L-based ISG formulations at 25 °C. The data are represented in triplicate.

2.2.2. Gel Formation of Levofloxacin HCl-Loaded Eudragit-Based ISG

Figure 7A reveals that the greater cloudy matrix was formed after the injection drug-loaded with higher concentrate Eudragit L solutions or longer PBS exposure time. The phase separation of Eudragit L was due to its lower solubility in PBS than monopropylene glycol. Typically, high miscibility between the solvent and aqueous environment results in a rapid aqueous inward together with solvent outward, and thereafter a self forming process of aqueous insoluble matrix material was obtained [23,60].

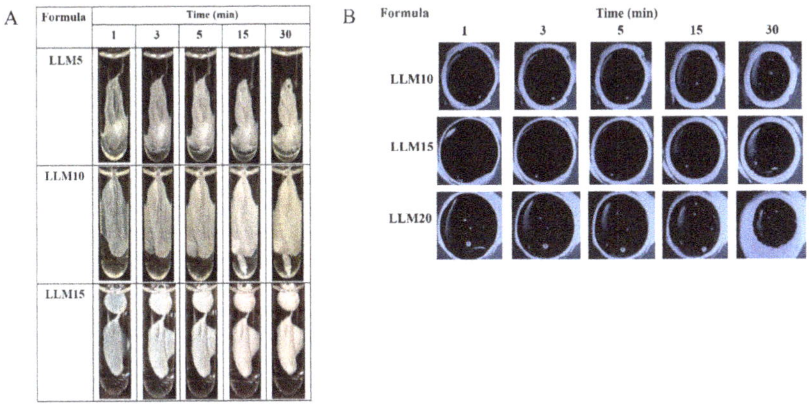

Figure 7. Gel formation after injection into PBS (**A**) and agarose well (**B**) of levofloxacin HCl-loaded Eudragit L-based ISG.

A cross-section view of gel formation in Figure 7B indicated that a slower matrix formation of LLM15 corresponded with that of LM15 (Figure 4B) due to their dense matrix acting as a barrier for matrix growth by declining solvent exchange [64], and LLM20 could achieve a cloudier matrix. The addition of the drug did not affect the gel or matrix formation. In a similar vein, the addition of 1% vancomycin HCl had virtually no impact on the rate of matrix formation of borneol-ISG [21]. These test results confirmed that the injected drug-loaded Eudragit L-based ISG was capable of changing from a solution into a cloudy matrix instantly. The in situ process of Eudragit L-ISG presented after its contact with simulated periodontal fluid; thereafter, monopropylene glycol diffused outward simultaneously from the formulation with the entry of water, resulting in the hardening of cloudy Eudragit L gel into a solid-like matrix over time. The matrix formation of rosin-based ISGs induced by solvent exchange has been reported previously [18,54].

2.2.3. Microscopic Interface Interaction

Figures 8 and 9 display the movement of the fluorescent colors at the interface between agarose gel and formulation or solvent under a fluorescent microscope. LM10 was used as a model formulation in this study since its rather lower amount of polymer loading did not interfere with the emission of fluorescent dyes during its transformation into a gel or matrix-like structure. Figure 8 presents the color movement from Eudragit L-based ISG formulations or monopropylene glycol containing sodium fluorescein (SF) or nile red (NR) (right side) to noncolored agarose gel (left side). The white line indicates the interface between the right component and left agarose gel phase. SF is an orange-red, odorless powder that is freely soluble in water but only sparingly in alcohol [68]. Practically, SF absorbs blue light, with peak excitation occurring at wavelengths between 465 and 490 nm and resulting fluorescence occurring at yellow-green wavelengths between 520 and 530 nm [69]. When in neutral to an alkaline solution, it exhibits an intense yellow-green fluorescence under ultraviolet light [69]. The SF in the SF-loaded LM10 formulation did not express its green color in monopropylene glycol, but a dark band green was found at the interface, indicating that some SF diffused into the interface (Figure 8). However, the Eudragit L matrix formation impeded SFs further diffusion into agarose gel sufficiently to express its green color. Moreover, the water movement from agarose gel was insufficient to induce the emission of SF as bright green color generation in the formulation. Nile red is a hydrophobic probe that is intensely fluorescent [70]. Its fluorescence appears in organic solvents and preferentially dissolves in hydrophobic compounds, such as lipids, but is quenched in water [71]. NR-loaded LM10 exhibited red color from the NR emission initially dispersed into the agarose gel that might be caused by its diffusion along with monopropylene glycol, then it was absorbed onto the Eudragit L matrix; nonetheless, the mixing and diluting of water in agarose gel decreased NR solubility and diffused backward into the formulation, resulting in red color fading in agarose side while a darker red was seen in the formulation as more time passed. The occurred Eudragit L matrix was found as a white band at the interface at 1 min similar to the test with SF-loaded LM10. The higher intensity of fluorescence colors at equilibrium made it difficult to see this matrix band and its expansion. However, it seemed to prohibit SF and NR diffusion into agarose gel effectively. The fast diffusion of green color SF from monopropylene glycol into agarose was observed. Although SF did not express its color in monopropylene glycol, it abruptly presented a green color after being dissolved in a water phase. It was sudden green since the water from agarose (left side) faster migrate into the SF-loaded monopropylene glycol than that of migrate into SF-loaded LM10. As well as, NR-loaded systems, the rapid movement of water into the NR-loaded monopropylene glycol promptly reduced NR solubility and yielded in the absence of its color. While NR-loaded LM10 showed the red since the water influx was retarded and NR could present its emission color in the remained-monopropylene glycol.

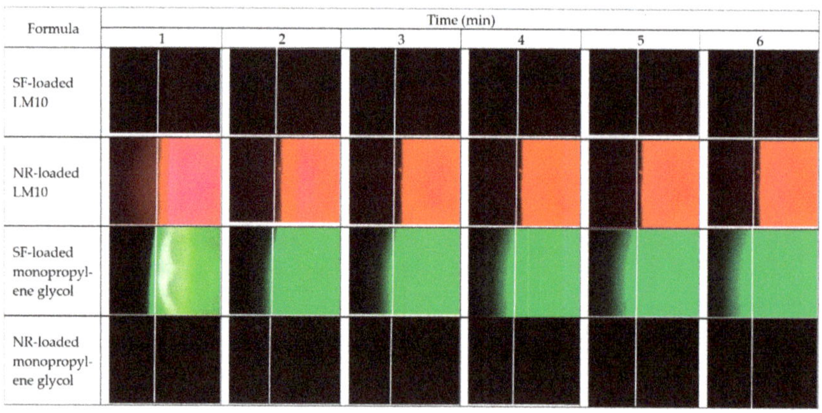

Figure 8. Interface interaction between noncolored agarose gel (the left side) against Eudragit L-based ISG formulation or mono propylene glycol containing SF or NR (the right side) at different time intervals under an inverted fluorescent microscope at a magnification of 400×.

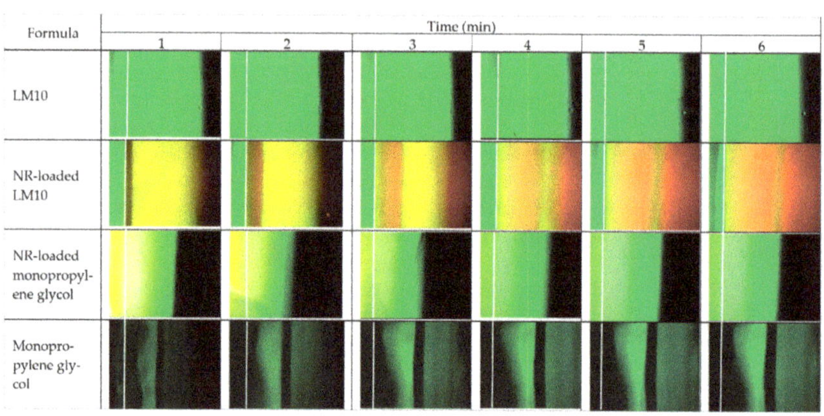

Figure 9. Interface interaction between sodium fluorescence-loaded agarose gel (the left side) against Eudragit L-based ISG formulation and monopropylene glycol without or containing NR (the right side) at different time intervals under the inverted fluorescent microscope at a magnification of 400×.

The microscopic interface interaction was also evaluated using SF-loaded agarose gel, as seen in Figure 9. The white line indicates the initial interface between two phases. The opaque expanding band occurred after SF-loaded agarose gel (left side) contacted with LM10 (right side), confirmed the matrix formation at the interface, and moved inward into the formulation as time passed, as also previously mentioned in the cross-section view of gel formation. These images also indicated that SF in formulation or solvent did not show its bright green color. For NR-loaded LM10, the initial Eudragit L matrix as a yellow band encompassed some NR at a narrow orange band near agarose gel and it diffused backward with solvent to accumulate in the formulation side and presented its red color more clearly. As aforementioned, the NR absorbed onto Eudragit L matrix could diffuse backward into the formulation, leading to the red color fading on the agarose side and a darker red in the formulation. The apparent green color was observed after SF-loaded agarose gel was contacted with NR-loaded monopropylene glycol, with the absence of clear red color. The initial NR red color appeared on the left side of the agarose due to its initial solvent diffusion. However, this solvent could be diluted after rapid mixing with water from the agarose gel. After adequate dilution, NR in the formulation was insoluble and could not express its color. The Eudragit L matrix of NR-loaded LM10 efficiently retarded the water movement into the formulation; therefore, there was enough solvent-maintained NR solubility to pronounce its red color. The rapid movement of SF with sufficient water from agarose into monopropylene glycol generated the brighter green color of this solvent over time. The back band should be the accumulation of monopropylene glycol at the interface with solvent font during contact time, and SF could not express its bright green color properly. Thus, applying fluorescence probes like SF and NR could provide crucial information such as solvent movement ability and matrix formation of ISG.

2.2.4. Drug Content and Release of Levofloxacin HCl-Loaded Eudragit-Based ISGs

The in vitro drug release was conducted in PBS at pH 6.8 to simulate the periodontal environment. The amount of drug content of LVM, LLM10, LLM15, and LLM20 samples were 102.52 ± 1.74, 100.85 ± 0.94, 98.77 ± 1.02, and 101.04 ± 0.86, respectively. After the preparation of the formulations, their amounts of drug content was determined before the in vitro drug release test using the cup method to mimic the environment of the periodontal pocket [54,67]. Previous researchers applied this method to study the drug release behavior of in situ forming drug delivery systems [20,21,63]. Figure 10 presents levofloxacin HCl release profiles from Eudragit L-based ISGs and LVM [control group]. There was a rapid levofloxacin HCl release from LVM, the control group which have no polymer. It released the drug completely in one day.

Sustainable drug release was attained from three developed ISGs. LLM20 had a more efficient prolongation of drug release than LLM15 and LLM10, respectively. They showed an initial burst drug release during the first day and gradually liberated the remaining drug during 1–14 days. The diffusion outward of monopropylene glycol into the release medium promoted a phase inversion and the formation of the beginning porous rubbery gel structure. These transformations occurred as a result of a burst release of drugs deposited on the surface layer. In the case of ISG using hydrophobic solvents such as triacetin and ethyl benzoate for dissolving PLGA, it exhibited slow gelation and significantly reduced the burst drug release [72]. This initial fast drug release behavior potentially promotes the high drug concentration to the therapeutic level above MIC for periodontitis treatment and, subsequently, the sustained release to maintain the drug level at the target site. Thermosensitive poloxamer 407 combined with chitosan gel showed an initial burst where about 60–70% of levofloxacin HCl was released within 6–7 h [73]. The thermosensitive gel, comprising 0.6% w/v gellan gum and 14% w/v pluronic F127, extended the release of levofloxacin HCl up to 3–4 h [74]. Therefore, these hydrophilic polymeric gels could insufficiently prolong the release of this drug. Doxycycline hyclate released from 15%, 25%, 30%, and 35% w/w Eudragit RS-loaded ISGs at 91%, 79%, 71%, and 68% at 54 h (2.25 days), respectively, since this polymer is a swellable, water-insoluble polymer and forms into different loose, swollen membrane structures in aqueous for drug release [24]. The commercial product for periodontitis treatment, such as Atridox®, uses 5% or 10% doxycycline hyclate as the antibacterial compound loaded into a 33.03% poly (D, L-lactide) solution sustained the drug release for at least 7 days [19,75]. In this study, levofloxacin HCl-loaded Eudragit L-based ISG formulations could control drug release for two weeks by utilizing a lower concentration of polymer than Atridox®. The developed ISGs present a fruitful alternative as a sustainable local drug periodontal pocket delivery system for improved patient compliance due to reduced drug administration frequency [16,61]. Moreover, from the literature review, the minimum inhibitory concentration (MIC) of levofloxacin HCl against *S. aureus*, *E. coli*, and *P. gingivalis* was 0.12 μg/mL [76], ≤0.12 μg/mL [77], and 8 μg/mL [78], respectively. The concentration of whole drug release from this study was 43.75 μg/mL (100% release); therefore, the 18.29% drug release or after around 4 h, all formulations could achieve the MIC of levofloxacin HCl against *P. gingivalis* and other test microbes.

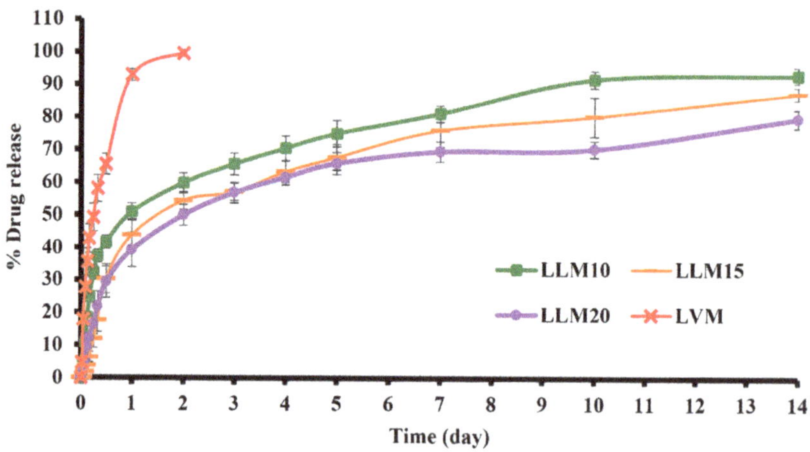

Figure 10. Drug release from Levofloxacin HCl-loaded Eudragit L-based ISG formulations (n = 3) (mean ± S.D).

The estimated parameters from the release profile fitting to the zero order, first order, Higuchi's, and power law equations including regression coefficient (r^2) value and model selection criteria (msc) are presented in Table 1. All of the release profiles were also consistent with the Korsmeyer–Peppas equation, with particularly high r^2 and msc values. k is the constant incorporating structural and geometric characteristics of the device, and n is the release exponent revealing the drug release mechanism. Drug released from LVM fit well with a first-order equation (r^2 and msc was 0.9744 and 3.3211, respectively) indicating drug concentration gradient-dependent behavior or a fraction of remaining drug in the solution. The r^2 and msc and diffusion exponent value (n) obtained from the drug release profile of levofloxacin HCl-loaded Eudragit L-based ISGs close to 0.45 signified that they complied best with Higuchi's equation. Thus, their kinetic of drug release was Fickian diffusion. Although their release profiles were well fitted with Higushi's equation, increasing the polymer concentration significantly decreased the k value in Higushi, because the drug diffusion into the release medium was retarded by the denser Eudragit L matrix. The release of hydrophilic drugs from other polymer or fatty acid-based ISGs also mostly fitted well with Higuchi's equation obeying Fickian diffusion kinetics [23,55].

Table 1. Regression coefficient (r^2) value, msc, and diffusion exponent value (n) obtained from drug release profile of levofloxacin HCl-loaded Eudragit L-based ISG formulations fitting different mathematic equations.

Formula	Zero Order		First Order		Higuchi's		Korsmeyer–PEPPAS				
	r^2	msc	r^2	msc	r^2	msc	r^2	msc	k ± SD	n ± SD	Release Mechanism
LLM10	0.8048	1.3742	0.8683	1.9759	0.9058	2.5144	0.9770	3.5961	3.344 ± 1.837	0.444 ± 0.050	Fickian diffusion
LLM15	0.7118	0.9164	0.8872	2.0854	0.8945	2.0844	0.9503	2.6867	2.603 ± 0.666	0.367 ± 0.028	Fickian diffusion
LLM20	0.7941	1.2387	0.9090	2.3402	0.9188	2.3419	0.9320	2.3732	1.856 ± 0.819	0.444 ± 0.055	Fickian diffusion

2.2.5. Scanning Electron Microscopy (SEM)

Figure 11 presents the SEM photographs that exhibited the surface and cross-section topographies of dried remnants of drug-free Eudragit L-based ISGs (LM10, LM15, and LM20). The apparent scaffold topography was evident for LM10 both on the surface and cross-section. As a scaffold containing polymeric fibrous networks, this topography characteristic confirmed the readily solvent exchange promoting polymer separation. Nevertheless, the larger polymer loading initiated a denser surface while the inner larger fibrous occurred in the cross section-part. It seemed to have many spherical polymer particles connected and formed into a continuous fibrous network. The size of the void inside the structure was decreased with increasing polymer concentration. This obtained topographic characteristic was similar to that of the dried remnant of Eudragit RS-based ISGs as previously reported [24,27,28]. Eudragit L dissolves in the pH range of 5.5 to 6.0 and can be used to deliver medication to the ileum, while Eudragit S can provide release in the pH range of 7.0 and, hence, deliver the medication to the large intestine [32]. SEM photomicrograph of these developed ISGs, taken after the release study, presented that pores had occurred throughout the inner matrix with different densities depending on polymer content. As a result, the formation of both pores and scaffolds structure of ISGs signified the involvement of drug diffusion with solvent from a matrix or in situ forming scaffold into release medium [79]. The porosity and pore-connectivity were demonstrated to indicate how solvent migrated which subsequent construction of Eudragit scaffold and drug release. Nonetheless, the dissolution of Eudragit L in PBS 6.8 could occur continuously, and this process depended on the polymer content. The high-loading polymer content yielded a denser structure and lowered matrix dissolution.

An SEM image of levofloxacin HCl powder revealed the crystalline rod-shaped-like particles (Figure 12A). The SEM photographs of the surface and cross-section topographies of levofloxacin HCl-loaded 10%, 15%, and 20% Eudragit L-based ISG remnants after 7 days of drug release are presented in Figure 12B. Because no drug crystals appeared in the remnants, the tiny amount of drug might disperse molecularly in a polymer matrix, and some dispersed drugs had already dissolved from the matrix. The enhanced concentration of Eudragit L remarkably promoted denser surface and inner topographies of the obtained polymer matrix. LLM20 achieved greater drug release prolongation than LLM15 and LLM10 due to a denser topographic matrix of ISG with higher polymer loading. By comparison, the porosity of drug-loaded ISG remnants was noticeably less than that of drug-free ISG remnants. The introduction of this hydrophilic drug could enhance the polarity of the formulation and might promote solvent exchange; therefore, the increased rate of phase separation of the polymer could induce a smaller fibrous formation. Furthermore, as aforementioned in Section 2.2.1, the inclusion of levofloxacin HCl decreased the viscosity of prepared ISGS. The nucleation and growth mechanism of the polymer in the dilute phase gives rise to the phase separation process, and the spinodal de-mixing that occurs during the membrane formation process is responsible for the interconnectivity that develops between the membranes [80]. Changes in solution viscosity caused by the addition of a hydrophilic drug could alter the phase separation rate, which ultimately causes the differences in matrix morphology as aforementioned for the addition of polyvinyl pyrrolidone (PVP) for producing polyethersulfone hollow fiber membranes [81]. The morphology of this membrane was evaluated as a function of the viscosity of the spinning solution, where the addition of PVP decreased the viscosity of the solution, favoring the high speed of precipitation and the formation of a specific type of pores similar to a finger in the membrane [81,82]. As shown in Figure 12B, the lower viscosity owing to drug loading that is mentioned in Section 2.2.1 may induce a more rapid phase separation of Eudragit L from solution and a less pronounced porosity and finger-like appearance of drug-loaded ISG remnants. It was reported that high polymeric concentration enhanced the gel resistance to aqueous influx and solvent outflux due to the less porous structure [83] Hence, these SEM photographs emphasized the retardation of drug release of Eudragit L matrices, as distinguished previously in Section 2.2.4. The presented morphology of these remnants revealed the distinct porous structure demonstrating the involvement of solvent exchange as responsible mechanistic for modulating the drug release from Eudragit L-based ISG matrices.

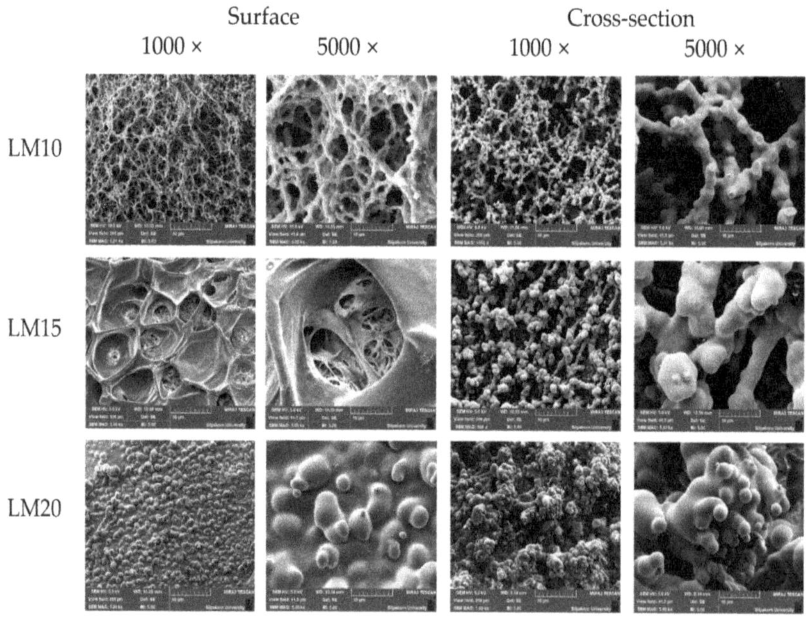

Figure 11. SEM images of surface and cross-section of freeze-dried remnant of Eudragit L-based ISG formulations after release test for 7 days at magnification of 1000× and 5000×.

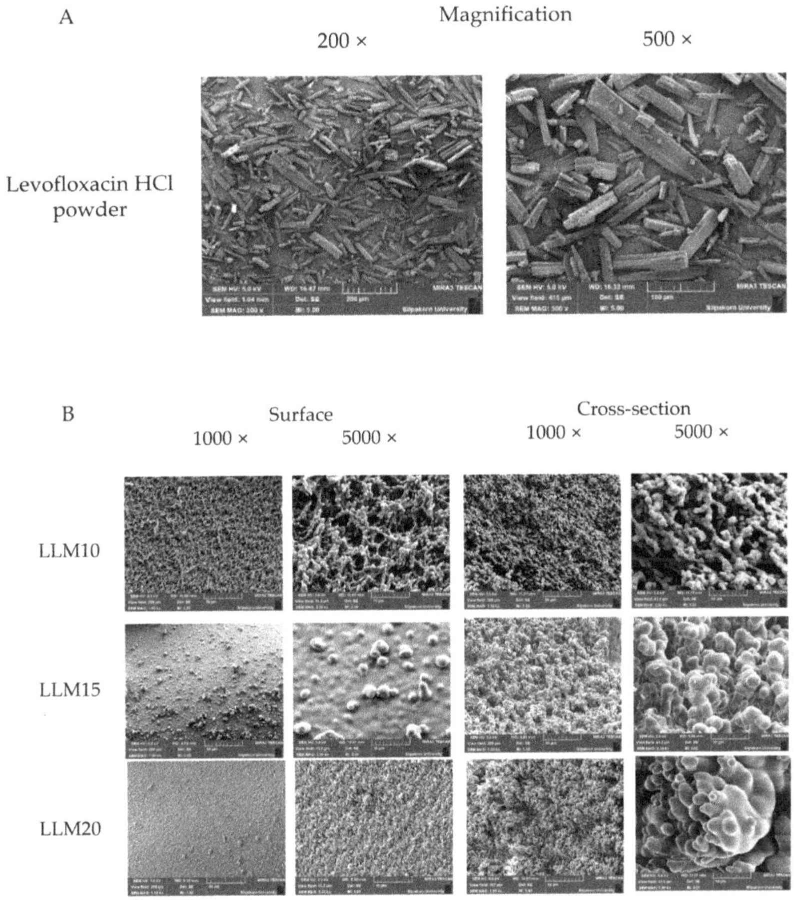

Figure 12. SEM images of levofloxacin HCl powder (**A**); surface and cross-section of a freeze-dried remnant of levofloxacin HCl-loaded Eudragit L-based ISG formulations at a magnification of 1000× and 5000× (**B**).

2.2.6. In Vitro Degradation

The behavior for in vitro degradation representing the % mass loss of three drug-loaded formulations is shown in Table 2. This mass loss is pronounced by the diffusion of the drug and monopropylene glycol as well as the dissolution of Eudragit L following an interchange with the release medium. The degradation of drug delivery systems could be described by their weight loss or molecular weight loss of carriers such as polymers [84]. ISG formulation containing a higher concentration of Eudragit L presented a less % mass loss, as shown in this table. On days 1 and 3, the three ISGs had significantly different values ($p < 0.05$), whereas that of LLM10 was significantly higher than that of LLM20 ($p < 0.05$) on days 5 and 7. The solvent exchange after immersion of these ISGs into the release medium initiated the formation of the porous matrix structure. Thus, the foremost mass loss from ISGs was owing to the diffusion of monopropylene glycol and also the drug release amount, which depended on the porosity and tortuosity of the Eudragit L matrices, as depicted and described in the previous SEM section. Typically, the degradability of polymeric structures such as Eudragit L is not due to enzyme or acidic conditions. Thus, the dominant mass loss that appeared in this investigation was from solvent leakage from the depot matrix and the gradual dissolution of Eudragit L mass. The complete mass loss was achieved at 14 days, as reported in Table 2, confirming that Eudragit L dissolution occurred gradually. The higher amount of matrix-forming agent enhanced the denser matrix structure and higher tortuosity, making it more difficult for monopropylene glycol and drug molecules to diffuse into the external medium. Thereafter, this phenomenon diminished in vitro degradation as well as prolongation of drug release. The steeper slope of the drug release profile during the early hours of this study likely occurred owing to the rapid release of the drug loaded onto or near the surface of Eudragit L matrix, followed by a sustained release of the drug distributed in the inner polymeric matrix. The composite material comprising polyurethane and Eudragit L as nanofiber fabricated by electrospinning technique exhibited adequate mechanical properties and in vitro cell biocompatibility, implying that the material is suitable for use in applications involving drug-eluting stent covers [34]. The incorporation of DNA plasmid for gene delivery and low molecular weight heparin delivery in nanoparticles fabricated by this group of polymers has been given the clinical go-ahead in the United States of America, Japan, and Europe. These nanoparticles have a low level of toxicity [37]. Eudragit L 100 is an anionic copolymer-based methacrylic acid and methyl methacrylic acid [32]. Its targeted drug release area is the jejunum, and it dissolves at a pH above 6 [32]. Eudragit L is a pH-dependent polymer soluble that dissolves continuously and slowly at a pH above 6. The amount of polymer present influenced the rate of degradation. Both the rate of degradation and the amount of degradation were increased due to the low polymer content. Because of this result, it was clear that this system would be useful for developing periodontal pocket drug delivery systems due to the fact that it degrades on its own in vitro over time.

Table 2. Mass loss from in vitro degradation test of levofloxacin HCl-loaded Eudragit L-based ISG formulations. The superscripts (a–d) in the column represent a significant difference within the tested formulations ($p < 0.05$).

Formula	Weight Loss (%)				
	Day 1	Day 3	Day 5	Day 7	Day 14
LLM10	85.39 ± 0.32^{a}	91.96 ± 0.98^{b}	94.53 ± 1.46^{c}	94.96 ± 1.21^{d}	100.00 ± 0.00
LLM15	80.22 ± 2.90^{a}	86.95 ± 1.97^{b}	87.91 ± 0.54	92.57 ± 1.92	100.00 ± 0.00
LLM20	72.81 ± 1.85^{a}	82.02 ± 1.46^{b}	85.95 ± 1.98^{c}	90.21 ± 1.80^{d}	100.00 ± 0.00

2.2.7. X-ray Computed Microtomography (μCT)

μCT is used for interiorly visualizing a scanned physical solid object and obtaining information on its 3D geometries and properties. This method is based on the idea that the internal features of a solid object have variations in their density and/or atomic composition [85]. It has been applied to measure implant porosity and pore size in vitro [86]. The μCT images of LLM15 and LLM20 remnants obtained from a synchrontron light source are illustrated in Figure 13. The rather fragile LLM10 was not examined since it could not retain its feature for long period during assessment with synchrontron light. The 3D volume and the cross-section with the voids inside the LLM15 and LLM20 remnants are presented in 3D. The LLM20 remnant appears to have a less void structure than that of LLM15. Because of the striking contrast that exists primarily between solid phases and air, this method can be used to determine the porosity of the object that is the focus of the investigation [85]. Furthermore, the simultaneous analysis of the 3D structure showed that the %porosity of LLM20 was less than that of LLM15. This outcome was consistent with in vitro release behavior and topography (by SEM). Hence, the illustrated data from μCT was useful as supporting evidence. Previously, the μCT-imaging was applied for checking longitudinal quantification of degradation and intra-articular biocompatibility of hydrogel based on acyl-capped triblock copolymer poly[ε-caprolactone-co-lactide)-b-poly (ethylene glycol]-b-poly[ε-caprolactone-co-lactide] [87].

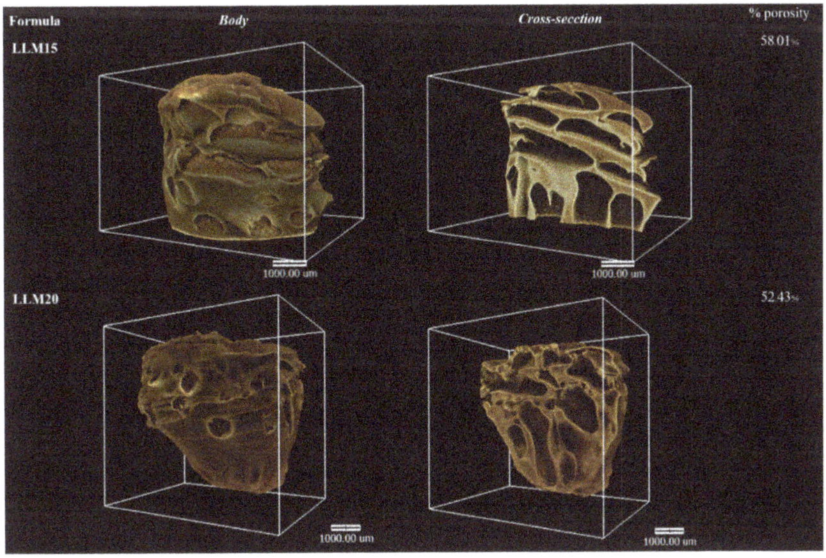

Figure 13. X-ray tomography image and %porosity using X-ray tomography of freeze-dried remnant after drug release test for 7 days of levofloxacin HCl-loaded Eudragit L-based ISG formulations.

2.2.8. Antimicrobial Activities

The inhibition zone diameter of solvents, drug-free, and levofloxacin HCl-loaded Eudragit L ISGs against *S. aureus* (ATCC 6538, 6532, and 25923), *methicillin-resistant S. aureus (MRSA)* (*S. aureus* ATCC 4430), *E. coli* ATCC 8739, *C. albicans* ATCC 10231, *P. gingivalis* ATCC 33277, and *A. actinomycetemcomitans* ATCC 29522 from antimicrobial activities test via cup agar diffusion method is shown in Table 3. These tested bacterial and fungal species are typically associated with periodontitis disease. The main obligate anaerobe pathogen bacteria of adult periodontitis have been identified as *Prophyromonas gingivalis* and *Aggregatibacter actinomycetemcomitans* [6,7]. In spite of this, more research needs to be carried out to determine the specific roles that each of these species plays, either on its own or in conjunction with other species, in the pathogenesis of periodontal breakdown. *Staphylococcus aureus* could be occasionally isolated from the periodontal pockets of patients with aggressive periodontitis, and *Escherichia coli* is sometimes declared as a microorganism in patients with periodontitis [88]. In contrast to the vast majority of other types of infections, all of the organisms that are thought to cause periodontal disease are native to the oral flora, *Candida albicans* courses of refractory periodontitis [89].

From this study, monopropylene glycol could slightly inhibit the growth of all test microbes, as presented in Table 3. It has been previously reported for antimicrobial effectiveness against three organisms, namely *S. mutans*, *E. faecalis*, and *E. coli*, and the bactericidal activity was at a concentration of 50%, 25%, and 50%, respectively [90]. Typically, organic solvents, such as monopropylene glycol, have the potential of disturbing lipids in the cell walls of microbes; thus, they could prohibit microbe growth. The drug-free Eudragit L-based ISG tended to inhibit microbe growth less than its solvent with polymer concentration dependence because the higher viscous and grater matrix formation noticeably impeded the movement of monopropylene glycol from ISG into inoculated agar media. Therefore the high polymeric concentration resulted in less antimicrobial activity. It was clearly confirmed since the LM15 and LM20 have no antimicrobial activity against *S. aureus* 25923, while solvent and LM10 showed inhibition zone. This role of the polymer in the retardation of solvent diffusion and antimicrobial activities has also been previously noticed in drug-free borneol-based ISGs [21].

Table 3. Clear zone diameter of monopropylene glycol, drug-free, and levofloxacin HCl-loaded Eudragit L-based ISG formulations against *S. aureus* (ATCC 6538, 6532, and 25923), methicillin-resistant *S. aureus (MRSA)* (*S. aureus* ATCC 4430), *E. coli* ATCC 8739, *C. albicans* ATCC 10231, *P. gingivalis* ATCC 33277, and *A. actinomycetemcomitans* ATCC 29522 (n = 3).

Formula	Clear Zone Diameter (mm.) Mean ± S.D.							
	S. aureus 6538	*S. aureus* 4430	*S. aureus* 6532	*S. aureus* 25923	*E. coli* 8739	*C. albicans* 10231	*P. gingivalis* ATCC 33277	*A. actinomycetemcomitans* ATCC 29522
MP	12.7 ± 0.5	13.0 ± 0.8	12.3 ± 0.5	10.3 ± 0.5	14.7 ± 0.5	18.7 ± 1.2	17.0 ± 2.2	26.3 ± 0.5
LM10	11.7 ± 0.5	10.7 ± 0.5	11.7 ± 0.5	9.8 ± 0.2	12.0 ± 0.8	16.7 ± 0.5	12.0 ± 0.8	23.7 ± 0.5
LM15	10.5 ± 0.4	10.7 ± 0.5	10.7 ± 0.5	-	12.0 ± 1.4	16.0 ± 0.8	15.0 ± 1.4	23.3 ± 0.5
LM20	10.3 ± 1.2	9.7 ± 0.5	11.3 ± 1.2	-	12.7 ± 1.2	15.0 ± 0.8	13.3 ± 1.2	22.3 ± 0.5
LVM	26.3 ± 0.9 [a]	25.3 ± 0.9 [b]	26.0 ± 0.8 [c]	25.3 ± 0.5 [d]	25.3 ± 0.9 [e]	20.0 ± 1.6 [f]	26.0 ± 0.8 [g]	>40
LLM10	26.7 ± 0.5	23.3 ± 0.5	23.3 ± 1.2	23.3 ± 0.5	23.3 ± 0.9	15.3 ± 2.1	25.3 ± 1.2	>40
LLM15	25.3 ± 0.5	22.7 ± 0.5 [b]	21.7 ± 0.5 [c]	22.7 ± 1.2 [d]	22.3 ± 0.5 [e]	16.0 ± 0.8 [f]	23.3 ± 1.2	>40
LLM20	24.3 ± 0.5 [a]	21.3 ± 0.9 [b]	20.3 ± 0.5 [c]	20.7 ± 0.5 [d]	22.0 ± 0.8 [e]	14.7 ± 0.9 [f]	21.3 ± 0.5 [g]	>40

The superscripts (a–g) in the column represent a significant difference within the tested formulations ($p < 0.05$).

Because of the absence of polymeric matrix and free drug spreading, the LVM solution acted as a control group with the largest diameter of the inhibition zone and was significantly larger than the drug-free formulations ($p < 0.05$) against almost all of the test microbes (Table 3). All drug-loaded ISGs demonstrated a significantly larger inhibition diameter than free-drug-loaded ISGs ($p < 0.05$) indicated the antimicrobial activity against all microbes, except the activity against *C. albicans*. Levofloxacin HCl have no activity on fungus [41]; therefore, LLM10, LLM15, and LLM20 were no different in the clear zone, compared to LM10-20. It was suggested that antifungal activity against C. albicans was mainly caused by the solvent because their clear zones were close to that of the solvent. Moreover, there was also a tendency for smaller inhibition clear zone with Eudragit L concentration dependent indicating retardation of the drug diffusion into inoculated media. These obtained results conformed with the viscosity of ISGs and their drug-release behaviors. In addition, all of the developed drug-loaded ISGs could inhibit *S. aureus* 25923. Hence, the Eudragit L matrices after phase transformation delayed drug movement with sustained drug release and decreased inhibition zone diameter. Some higher polymer-loaded ISGs, such as LLM15 and LLM20, had a significantly smaller inhibition zone diameter than LVM ($p < 0.05$), as presented in Table 3. Similar results were formerly obtained for the matrices that occurred from ISG phase transition fabricated using borneol [21], natural resins [18,22,54], cholesterol [91], polymers [53,92], and saturated fatty acids [23] as the matrix-forming agents of in situ forming systems. The inhibition zone diameter of drug-loaded formulations against *A. actinomycetemcomitans* ATCC 29,522 was larger than 40 mm; therefore, it could not observe and compare the efficacy between formulations. Nonetheless, this result demonstrated that they potently inhibited this pathogen. Therefore, the effective levofloxacin HCl ISG emerged as an attractive localized dosage form for periodontitis treatment in a controlled drug-release manner. Basically, the use of medicated in situ forming gel exhibited better clinical outcomes than scaling and root planning alone, which could be associated with the presence of monopropylene glycol which has antimicrobial activities but also performs as a good vehicle for ISG to deliver the drug into the periodontal pocket. LLM15 exhibited an appropriate viscosity with acceptable injectability and still retained its scaffold structure, as seen in μCT-imaging and SEM. Moreover, this ISG sustained the release of the drug above the MIC against the main pathogen, such as *P. gingivalis*. Thus, LLM15 emerged as the promising ISG formulation for periodontitis treatment. Eudragit L and monopropylene glycol are both safe and non-toxic. Although the safety data and applications of these compounds indicate the probability of both substances for injectable dosage forms, their safety and therapeutic efficacy of developed LLM15 ISG should be further determined in a clinical experiment.

2.2.9. FOURIER Transform Infrared Spectroscopy (FTIR)

The analysis of the intact MP using Fourier transform infrared (FTIR) spectra revealed highly pronounced and distinct bands at 1650 cm^{-1} and 3000 cm^{-1}. On the other hand, Eudragit L demonstrated the presence of a carbonyl band (C=O) between the wavelengths of 1720 and 1710 cm^{-1}. There was no discernible difference in the peaks of either LM20 or LLM20, indicating that their trends were identical to those of the unaltered levofloxacin HCl (Figure 14). As a direct result of this, the conclusion that there was no interaction between Eudragit L and the solvent and LV was confirmed by the findings.

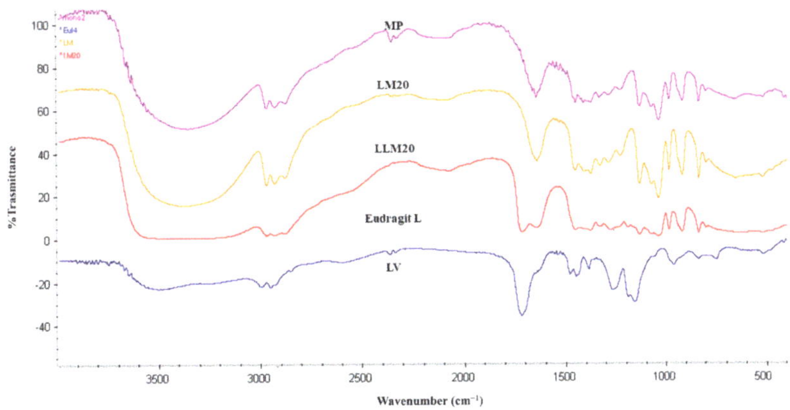

Figure 14. FTIR spectra of MP, LVM, LM20, LLM20, Eudragit L (raw material).

3. Conclusions

We attempted to apply monopropylene glycol as a solvent to dissolve Eudragit L and Eudragit S, as well as levofloxacin HCl of solvent exchange-induced ISG. Eudragit L-based formulation in monopropylene glycol was selected for drug incorporation and further investigation as ISG for a periodontal pocket delivery system due to its good physical characteristics, including less viscosity, and more ease of injection with gel formation ability. The phase transition occurred from the solution of gel and matrix-like structures subsequently some time after ISG exposure to the aqueous environment of PBS and agarose gel. The diffusion of ISGs monopropylene glycol in water led to the hardening of cloudy Eudragit L gel into a solid-like matrix over time. Interface interaction study provided crucial information, such as solvent movement ability and matrix formation of ISG, by tracking fluorescence dyes. All developed levofloxacin HCl-loaded Eudragit L-based ISGs prolonged the drug release for 2 weeks via Fickian drug diffusion kinetics and the release amounts reached the MIC against test microbes. SEM and µCT images revealed their scaffold formation, which had a denser topographic structure and less porosity attained owing to higher polymer loading less in vitro degradation. LLM15 exhibited an appropriate viscosity, Newtonian flow, acceptable gel formation, and injectability, and prolonged drug release for 14 days with efficient antimicrobial activities against *S. aureus* (ATCC 6538, 6532, and 25923), methicillin-resistant *S. aureus (MRSA) (S. aureus* ATCC 4430), *E. coli* ATCC 8739, *C. albicans* ATCC 10231, *P. gingivalis* ATCC 33277, and *A. actinomycetemcomitans* ATCC 29522; Therefore, it has the potential to be an ISG formulation used in the treatment of periodontitis. Even though there are safety data associated with a variety of medical applications of monopropylene glycol and Eudragit L, the newly developed LLM15 ISG is essential for further clinical experiments to evaluate the safety and therapeutic efficacy of the combination.

4. Materials and Methods

4.1. Materials

Both Eudragit® L and Eudragit® S came from EVONIK RohmGsmbH, which is located in Darmstadt, Germany. The antimicrobial agent levofloxacin HCl was generously provided by Siam Pharmaceutical Co., located in Bangkok, Thailand, and was put to use in this study. We visited P.C. Drug in Bangkok, Thailand, in order to purchase monopropylene glycol, which was then put to use as the solvent. During the investigation of the gel formation process, agarose with lot number H7014714 from Vivantis in Selangor Darul Ehsan, Malaysia, was utilized. For antimicrobial research, the following agars were utilized as media: sheep blood agar and chocolate agar (both from the Ministry of Public Health in Nonthaburi, Thailand), tryptic soy agar and tryptic soy broth (both from DifcoTM in Detroit, MI, USA). Sabouraud dextrose agar (SDA) and Sabouraud dextrose broth (SDB) (both manufactured by Difco in Detroit, MI, USA) were used in the antifungal test. Components of the phosphate-buffered saline solution included potassium dihydrogen orthophosphate (lot no. E23W60, Ajax Finechem, New South Wales, Australia) and sodium hydroxide (lot no. AF310204, Ajax Finechem, New South Wales, Australia). Both of these were obtained from Ajax Finechem (PBS). Test microbes, *S. aureus* ATCC 6538, *E. coli* ATCC 8739, *Candida albicans* ATCC 10231 (Department of Medical Sciences, Ministry of Public Health, Nonthaburi, Thailand), *P. gingivalis* ATCC 33277, and *A. actinomycetemcomitans* ATCC 29522 (MicroBiologics Inc., St. Cloud, MN, USA) were purchased from Thai Can Biotech Co. *P. gingivalis* ATCC 33277 and *A. actinomycetemcomitans* ATCC 29522 were two of the anaerobe bacteria that were cultured on sheep blood agar and chocolate agar, respectively, in this study. For the purpose of developing the two-dimensional structure of Eudragit L 100, ACD/ChemSketch was utilized. As a water-soluble fluorescent dye and a lipophilic fluorescence probe, respectively, sodium fluorescein (Carl Roth GmbH, Karlsruhe, Germany) and Nile red powder (SIGMA-Aldrich Chemie GmbH, Steinheim, Germany) were used for the interfacial interaction that occurred during phase transformation.

4.2. Preparation of In Situ Forming Gel

Both Eudragit L and S were separately dissolved in monopropylene glycol at a range of different concentrations over the course of an overnight stir session in an airtight container. In order to investigate the physical properties, systems were selected in which the components had completely dissolved, the liquid was transparent and flowable, and there was no formation of solid matter or gel. Table 4A,B detail the constituents that make up drug-free ISG. The polymers and levofloxacin HCl at a concentration of one percent were prepared using the aforementioned method. The components of the drug-loaded ISGs are outlined in Table 4C.

4.3. Viscosity and Rheology Characterization

The apparent viscosity values were measured using a viscometer with a cone-plate called the RM 100 CP2000 plus, which was manufactured by Lamy Rheology Instruments Company and is located in Champagne-au-Mont-d'Or, France. To carry out this measurement, a series of measurements at varying shear rates were taken at regular 15-s intervals during the equilibration time. This instrument was also used to test the rheological behavior of the substance. At room temperature, the shear stress of the formulations was determined by conducting the experiment at a variety of shear rates. The experiments were carried out three times to ensure accuracy.

4.4. Contact Angle

The goniometer (FTA 1000, First Ten Angstroms, Newark, CA, USA) was used to determine the contact angle of the formula on various surfaces, such as glass, paraffin, and agarose gel, at a time point of 5 s after a pump out rate of 1.9 L/s. The manufacturer of the goniometer is First Ten Angstroms. After that, an estimate of the contact angle was derived from the very first automatic image of a droplet that was taken in triplicate (n = 3).

Table 4. Composition formula of 5–25% various Eudragit L, S in monopropylene glycol (A,B); and levofloxacin HCl-loaded 10–20% Eudragit L-based in monopropylene glycol.

A			
Formula	Amount (% *w*/*w*)		
	Eudragit L	MP	
LM5	5	95	
LM10	10	90	
LM15	15	85	
LM20	20	80	
LM25	25	75	
B			
Formula	Amount (% *w*/*w*)		
	Eudragit S	MP	
SM5	5	95	
SM10	10	90	
SM15	15	85	
SM20	20	80	
SM25	25	75	
C			
Formula	Amount (% *w*/*w*)		
	Levofloxacin HCl	Eudragit L	MP
LLM10	0.5	10	89.5
LLM15	0.5	15	84.5
LLM20	0.5	20	79.5
LVM	0.5	-	95.5

4.5. Injectability

Using the compression mode of a texture analyzer, a measurement of injectability was carried out in order to determine the precision of injection through the stainless needle (TA.XT plus, Stable Micro Systems, Godalming, UK). This apparatus was utilized to measure the pressure exerted during the injection of test liquid from a 1 mL syringe that was coupled to an 18-gauge needle. This device's upper probe pressed the syringe plunger with a constant force of 0.1 N and a speed of 1.0 mm/second until it reached the base of the syringe barrel. After analyzing force-displacement profiles, the maximum injection force, along with its associated energy for injection, was logged. Three separate trials of these experiments were carried out.

4.6. Mechanical Properties

A texture analyzer was utilized in order to investigate the formulation's mechanical properties after it had been prepared (TA.XT Plus, Stable Micro Systems Ltd., Godalming, UK). After setting with a complete phase transformation into Eudragit matrix for 72 h, the prepared 150 microliters formulation was added into 0.6 percent *w*/*w* agarose gel. This process took place for 72 h. After that, an analytical probe from the previously described instrument was lowered into the polymer matrix at a rate of 0.5 mm per second. After maintaining this position for sixty seconds, the probe was then moved upward at a rate of ten millimeters per second. The amount of force that was checked at the point where the probe was able to penetrate the polymer matrix to its maximum depth was indicated as the maximum deformation force or the hardness of the material. Additionally, the amount of force that was checked at the point where the probe was able to move upward between the surface of the sample and the probe was indicated as the adhesion force. The maximum deformation force, also known as F max deformation, is the force that is measured at the point where the probe has penetrated the specimen the farthest, whereas the remaining force, also known as F remaining, is a force that is measured after the specimen has been held for sixty seconds. A measure for the elasticity and plasticity of the specimen was the ratio "F remaining/F max deformation". Values of elasticity that are high indicate a material with high elasticity, while values of elasticity that are low indicate a material with high plasticity [61]. The experiments were carried out in triplicate.

4.7. Gel Formation Study

The apparent gel formation at macroscopic level was performed by injected the prepared ISG through an 18-gauge needle into PBS (pH 6.8) medium. The cloudy gel or matrix-like formation at various times was observed (1, 3, 5, 15, and 30 min). Moreover, another gel formation behavior investigation at macroscopic level was also investigated. The in situ formation in an agarose well was also investigated under a stereoscope. This experiment was set to simulate the change of formulation within a periodontal pocket To begin, an agarose gel with a height of one centimeter was prepared by dissolving 0.6 percent agarose in PBS with a pH of 6.8. The gel was then poured onto Petri dishes for the setting process. The settle agarose was drilled using a stainless cylinder cup (7.6 mm diameter) to create an agarose well, the cylindrical well with a capacity of 300 microliters that simulated a periodontal pocket. For the purpose of testing, a prepared formulation containing 150 microliters was inserted into the well which is a simulated periodontal pocket. After being exposed to PBS from the agarose gel, the phase separation of Eudragit leads to the generation of a thicker, cloudier gel or matrix-like substance over time. Under a stereo microscope (SZX10, Olympus Corp., Tokyo, Japan), photographs of the morphological Eudragit gel or matrix were taken simultaneously at 1, 3, 5, 15, and 30 min. The software used for these photographs was the SZX10 Series software (Olympus Corp., Tokyo, Japan).

4.8. Interfacial Phenomena of Formulation-Aqueous Phase

Both a plain agarose gel with a weight-to-volume ratio of 0.6 percent and an agarose gel loaded with 0.4 μg/mL of SF were prepared. The LL10M, 0.4 μg/mL SF-loaded MP, 0.4 μg/mL SF-loaded LL10M, 3 μg/mL nile red (NR) -loaded MP, and 3 μg/mL nile red (NR) -loaded L10M were prepared. The different levels of color intensity after the emission of these fluorescent dyes were the root cause of the difference in concentration between them. Based on preliminary inspections, it was found that their concentration could be clearly seen at an intensity that was sufficient to observe the change in color or movement. In order to investigate the interfacial interaction between an aqueous phase and a solvent or ISG formulation, the SF-free and SF-loaded agarose gels were prepared, and then 50 microliters of prepared test samples were dropped close to that agarose gel. The interface interaction was investigated using an inverted fluorescent microscope (TE-2000U, Nikon, Kaw, Japan) by capturing the image using a blue (B2A) filter with excitation at 450–490 nm for probing the green color of SF and using a green (G2A) filter with excitation of 510–560 nm for tracking the red color of NR. Both of these filters were used in conjunction with the appropriate wavelengths of illumination.

4.9. Drug Content and In Vitro Drug Release Studies

Using a UV-Visible spectrophotometer (Cary 60 UV-Vis, Model G6860A, Agilent, Selangor, Malaysia), a standard curve was constructed to estimate the amount of drug in the LVM, LLM10, LLM15, and LLM20 samples (n = 6). The in vitro drug release behavior of levofloxacin HCl from a developed formulation comprising 10, 15, and 20 percent Eudragit L was undertaken to employ the cup method to mimic the drug liberation behavior from the periodontal pocket. The drug release profiles that were obtained were compared with those obtained from the control formulation, which consisted of levofloxacin HCl dissolved in monopropylene glycol at a concentration of one percent by weight. In a nutshell, 0.4 g of the refined dosage form was carefully placed into a cylindrical-shaped porcelain cup with 50 mL of PBS with a pH of 6.8 and heated to 37 degrees Celsius (diameter of 1 cm and higher of 1.2 cm). After that, it was placed in a rotation shaker (Model NB-205, N-Biotek, Gyeonggi-do, Republic of Korea) for 14 days at a temperature of 37 degrees Celsius and a rotational speed of 50 revolutions per minute. After that, 5 mL of release fluid was taken, and it was subsequently replaced with 5 mL of fresh PBS. The results of the study on the release of levofloxacin HCl were converted into a percentage of the cumulative drug amount using a UV-Visible spectrophotometer (Cary 60 UV-Vis, Model G6860A, Agilent, Selangor, Malaysia). The experiments were carried out three times to ensure accuracy.

Their drug release data were then fitted with a variety of mathematical models, such as zero order, first order, Higuchi's, and Korsmeyer–Peppas models, using the DD-Solver software application, which is an add-in program for Microsoft Excel. This enabled the researchers to determine the kinetics of drug release (Microsoft Corporation, Redmond, WA, USA). The estimated parameters, in particular the r^2 and model selection criteria (msc) values, were reported. Additionally, the release exponent (n-value) from the Korsmeyer–Peppas equation was used to indicate the mechanism of drug release. The following equations were used for mathematical model [93–95]. The zero-order model: the following equation indicates the release rate of the drug remains constant over time, independent of the drug concentration.

$$Q_t = Q_0 + k_0 t \tag{1}$$

where Q_t is the total amount of drug release at given time (t), Q_0 is initial drug release, k_0 is the release rate constant, and t is the given time.

The first-order model is another commonly employed mathematical model for describing drug release behavior. It assumes that the release rate is proportional to the remaining drug concentration as following equation.

$$\log Q_1 = \log Q_0 + \frac{k_1 t}{2.303} \tag{2}$$

where Q_1 is the amount of active agent released on time (t), Q_0 is the initial amount of drug dissolved, and k_1 is the first-order constant.

The Higuchi model assumes that drug release from polymer matrix systems is solely diffusion-controlled and that the drug is uniformly distributed in non-degradable planar systems. The Higuchi's model is given below.

$$Q_t = k\sqrt{t} \tag{3}$$

where Q_t is the total amount of drug release at given time (t), k is the release rate constant, and t is given time.

The Korsmeyer–Peppas model is widely used to assess the release mechanism. It incorporates a parameter 'n' to characterize the release mechanism, with values indicating different release kinetics. For swelling system, Fickian diffusion is indicated when the value of n is close to 0.45. If the value exceeds 0.45 but is lower than 0.89, this indicates non-Fickian diffusion or anomalous transport, and case II transport is indicated when n is close to 0.89. Fickian diffusion is a drug transport mechanism where structure relaxation is slower than diffusion. When relaxation exceeds diffusion, case-II transport occurs. Anomalous diffusion (non-Fickian transport) occurs when this drug release rate is close to structure relaxation [96–99].

$$\log \frac{M_{(i-1)}}{M_\infty} = \log K + n\log(t-1) \tag{4}$$

where M_∞ is the amount of drug at the equilibrium state, M_i is the amount of drug released over time (t), K is the constant of incorporation of structural modifications and geometrical characteristics of the system (also considered the release velocity constant), n is the exponent of release (related to the drug release mechanism) in function of time t, and l is the latency time.

4.10. Scanning Electron Microscopy (SEM)

After seven days of drug release testing in PBS with a pH of 6.8, the Eudragit-based ISG remnants were washed with 200 mL of distilled water and then freeze-dried with the assistance of a freeze dryer (TriadTM Labconco, Kansas City, MO, USA). The following process took place after the remnants had been stored in a desiccator for one week. The dried Eudragit-based ISG remnants were then coated with gold before being examined using the SEM method and admitted of comparison with intact levofloxacin HCl powder. At an accelerating voltage of 15 kV, a scanning electron microscope (SEM) TESCAN MIRA3 from Brno-Kohoutovice, Czech Republic, was used to observe the surface and cross-sectional morphologies of a dried Eudragit-based ISG remnant from a developed ISG system.

4.11. In Vitro Degradation Test

The in vitro degradability of forming matrices was carried out by determining the mass loss of the system after the drug release test. The initial weight of the sample and that after the release test at 14 days were recorded and calculated (n = 3) as follows:

$$\%\ \text{mass loss} = \left(\frac{W_i - W_t}{W_i}\right) \times 100 \quad (5)$$

where

W_i= initial weight of the sample
W_t = weight of remained sample at a specific time

4.12. X-ray Imaging and X-ray Tomographic Microscopy

The remnants of LL15M and LL20M were collected from drug release study, then washed with distilled water. They were dried together using a freeze dryer. The dried systems were kept in a desiccator for 7 days. The dried systems were scrutinized with X-ray imaging and X-ray tomographic microscopy at the X-ray tomographic microscopy (XTM) beamline, Synchrotron Light Research Institute (SLRI), Thailand. The following conditions were set: Generation of X-ray beam was archived by 2.2-T multipole wiggler at the 1.2-GeV Siam Photon Source facility (150 mA). The synchrotron radiation with X-ray tomographic microscopy (SRXTM) inspections was performed using a filtered polychromatic X-ray beam at a mean energy of 11.5 kV with a source-to-sample distance of 34 m. Detection system was equipped with a 200 μm thick scintillator (YAG: Ce, Crytur, Turnov, Czech Republic), lens-coupled X-ray microscope, and the CMOs camera (PCO edge 5.5, 2560 pixels, 16 bits) (Optique Peter, Lentilly, France). Isotropic voxel size of 3.61 μm was used. X-ray projection was normalized using the Octopus reconstruction (TESCAN). The tomographic volumes were achieved via Drishti software (National Computational Infrastructure's VizLab). The porosity was determined using Octopus Analysis (TESCAN) [100].

4.13. Antimicrobial Activities

The antimicrobial activities of monopropylene glycol, drug-free, and levofloxacin HCl-loaded Eudragit-based ISG formulations were evaluated against standard microbes (*S. aureus* DMST 6532, *MRSA S. aureus* ATCC 4430, *S. aureus* ATCC 6532, and *S. aureus* ATCC 25923), *E. coli* ATCC 8739, *C. albicans* 10231, *P. gingivalis* and *A. actinomycetemcomitans*) using the agar diffusion assay (cylinder plate method). For this bioactivity test, bacteria inocula were incubated for 36 h in tryptic soy broth (TSB), whereas sabouraud dextrose broth (SDB) was employed for *C. albicans*. The turbidity of broth suspensions of organisms was calibrated using the 0.5 McFarland standard. Consequently, the attained broth suspensions of *S. aureus* DMST 6532, *S. aureus* ATCC 4430, *S. aureus* ATCC 6532, *S. aureus* ATCC 25923, and *E. coli* ATCC 8739 were swab-spread on the tryptic soy agar plates, whilst sheep blood agar and chocolate agar were used as media for anaerobic antimicrobial testing of *P. gingivalis* and *A. actinomycetemcomitans*, respectively. The inoculum of calibrated *C. albicans* was swab spread on sabouraud dextrose agar (SDA). The 200 microliters aliquot sample was filled into a sterilized cylindrical cup that was already placed on the surface of the swabbed agar before incubation at 37 °C for 24 h. In the case of antibacterial measurement against *P. gingivalis* and *A. actinomycetemcomitans*, the incubation was conducted in an anaerobic incubator (Forma Anaerobic System, Thermo Scientific, Ohio, USA) at 37 °C for 72 h. To indicate and compare the antimicrobial activities (n = 3), the diameter (mm) of the inhibition zone was measured individually with a ruler.

4.14. Fourier Transform Infrared (FTIR) Spectroscopy

In addition to that, an FTIR spectrophotometer was used in order to record the spectra of the samples. The sample was mixed with potassium bromide, and then it was compressed with a plunger and die in a hydraulic press (pressure of 5 tons). The obtained pellet was placed inside of an FTIR chamber after it had been mounted. The sample's transmittance as a percentage was measured and recorded. In the range of 400–4000 cm^{-1}, the spectra were collected at a resolution of 2 cm^{-1}.

4.15. Statistical Analysis

SPSS for Windows (version 11.5) was employed for the statistical analysis. The difference between the experiments was determined by using Student's *t*-test. The results obtained were statistically significant because of the *p*-value, which was found to be less than 0.05.

Author Contributions: Experiments were conceived of and planned by S.S., S.T., T.C. and N.L., who were also responsible for sample preparation and evaluations. X-ray imaging and X-ray tomographic microscopy were both performed in the study that C.R. and S.S. carried out. S.S., N.L., S.T. and T.P. all made contributions to the interpretation of the results, with S.S. taking the lead in terms of writing the manuscript. Writing, reviewing, and editing were all tasks that T.C., S.T. and T.P. contributed to. T.P. and T.C. were a part of the planning process as well as the supervision of the work. Each author was responsible for providing constructive criticism and made significant contributions to the development of the research, the analysis, and the manuscript. The final, published version of the manuscript has been reviewed and accepted by all of the authors. All authors have read and agreed to the published version of the manuscript.

Funding: This research received no external funding.

Institutional Review Board Statement: Not applicable. This research did not involve humans or animals for investigation.

Informed Consent Statement: Not applicable.

Data Availability Statement: The data presented in this study are available on request from the corresponding author.

Acknowledgments: The authors would like to express their appreciation to the Synchrotron Light Research Institute, Center of Excellent in Pharmaceutical Nanotechnology, and Faculty of Pharmacy at Chiang Mai University for their generous support and facilitation of this study. In addition, we would like to express our gratitude to the Research and Creative Fund, which is part of the Silpakorn University Faculty of Pharmacy, for providing us with the opportunity to use their space and for their careful proofreading of this paper.

Conflicts of Interest: The authors declare no conflict of interest.

References

1. Fiorillo, L. Oral health: The first step to well-being. *Medicina* **2019**, *55*, 676. [CrossRef] [PubMed]
2. Nazir, M.A. Prevalence of periodontal disease, its association with systemic diseases and prevention. *Int. J. Health Sci.* **2017**, *11*, 72–80.

3. Nazir, M.; Al-Ansari, A.; Al-Khalifa, K.; Alhareky, M.; Gaffar, B.; Almas, K. Global prevalence of periodontal disease and lack of its surveillance. *Sci. World J.* **2020**, *2020*, 2146160. [CrossRef] [PubMed]
4. National Institute of Dental and Craniofacial Research. *Periodontal (Gum) Disease Cause, Symptoms and Treatments*; NIH: Bethesda, MD, USA, 2013.
5. Könönen, E.; Gursoy, M.; Gursoy, U.K. Periodontitis: A Multifaceted Disease of Tooth-Supporting Tissues. *J. Clin. Med.* **2019**, *8*, 1135. [CrossRef]
6. Papapanou, P.N.; Susin, C. Periodontitis epidemiology: Is periodontitis under-recognized, over-diagnosed, or both? *Periodontology* **2017**, *75*, 45–51. [CrossRef] [PubMed]
7. Sanz, M.; Ceriello, A.; Buysschaert, M.; Chapple, I.; Demmer, R.T.; Graziani, F.; Herrera, D.; Jepsen, S.; Lione, L.; Madianos, P.; et al. Scientific evidence on the links between periodontal diseases and diabetes: Consensus report and guidelines of the joint workshop on periodontal diseases and diabetes by the International diabetes Federation and the European Federation of Periodontology. *Diabetes Res. Clin. Pract.* **2018**, *137*, 231–241. [CrossRef]
8. Kikuchi, T.; Hayashi, J.I.; Mitani, A. Next-generation examination, diagnosis, and personalized medicine in periodontal disease. *J. Pers. Med.* **2022**, *12*, 1743. [CrossRef]
9. Genco, R.J. Antibiotics in the treatment of human periodontal diseases. *J. Periodontol.* **1981**, *52*, 545–558. [CrossRef]
10. Vyas, S.P.; Sihorkar, V.; Mishra, V. Controlled and targeted drug delivery strategies towards intraperiodontal pocket diseases *J. Clin. Pharm. Ther.* **2000**, *25*, 21–42. [CrossRef]
11. Pitcher, G.R.; Newman, H.N.; Strahan, J.D. Access to subgingival plaque by disclosing agents using mouthrinsing and direct irrigation. *J. Clin. Periodontol.* **1980**, *7*, 300–308. [CrossRef]
12. Slots, J.; Rams, T.E. Antibiotics in periodontal therapy: Advantages and disadvantages. *J. Clin. Periodontol.* **1990**, *17*, 479–493 [CrossRef] [PubMed]
13. Zilberman, M.; Elsner, J.J. Antibiotic-eluting medical devices for various applications. *J. Control. Release* **2008**, *130*, 202–215. [CrossRef] [PubMed]
14. Rams, T.E.; Degener, J.E.; van Winkelhoff, A.J. Antibiotic resistance in human chronic periodontitis microbiota. *J. Periodontol.* **2014**, *85*, 160–169. [CrossRef] [PubMed]
15. Rajeshwari, H.R.; Dhamecha, D.; Jagwani, S.; Rao, M.; Jadhav, K.; Shaikh, S.; Puzhankara, L.; Jalalpure, S. Local drug delivery systems in the management of periodontitis: A scientific review. *J. Control. Release* **2019**, *307*, 393–409. [CrossRef]
16. Nair, S.C.; Anoop, K.R. Intraperiodontal pocket: An ideal route for local antimicrobial drug delivery. *J. Adv. Pharm. Technol. Res.* **2012**, *3*, 9–15.
17. Batool, F.; Agossa, K.; Lizambard, M.; Petit, C.; Bugueno, I.M.; Delcourt-Debruyne, E.; Benkirane-Jessel, N.; Tenenbaum, H.; Siepmann, J.; Siepmann, F.; et al. In-situ forming implants loaded with chlorhexidine and ibuprofen for periodontal treatment: Proof of concept study in vivo. *Int. J. Pharm.* **2019**, *569*, 118564. [CrossRef]
18. Chuenbarn, T.; Sirirak, J.; Tuntarawongsa, S.; Okonogi, S.; Phaechamud, T. Design and comparative evaluation of vancomycin HCl-loaded rosin-based in situ forming gel and microparticles. *Gels.* **2022**, *8*, 231. [CrossRef]
19. Costa, J.V.; Portugal, J.; Neves, C.B.; Bettencourt, A.F. Should local drug delivery systems be used in dentistry? *Drug Deliv. Transl. Res.* **2022**, *12*, 1395–1407. [CrossRef]
20. Phaechamud, T.; Mahadlek, J.; Chuenbarn, T. In situ forming gel comprising bleached shellac loaded with antimicrobial drugs for periodontitis treatment. *Mater. Des.* **2016**, *89*, 294–303. [CrossRef]
21. Lertsuphotvanit, N.; Santimaleeworagun, W.; Narakornwit, W.; Chuenbarn, T.; Mahadlek, J.; Chantadee, T.; Phaechamud, T. Borneol-based antisolvent-induced in situ forming matrix for crevicular pocket delivery of vancomycin hydrochloride. *Int. J. Pharm.* **2022**, *617*, 121603. [CrossRef]
22. Ei Mon Khaing, T.I. Wichai Santimaleeworagun, Yaowalak Phorom, Tiraniti Chuenbarn, Thawatchai Phaechamud. Natural-resin in-situ-forming gels: Physicochemical characteristics and bioactivities. *Pharm. Sci. Asia* **2021**, *48*, 461–470. [CrossRef]
23. Chantadee, T.; Santimaleeworagun, W.; Phorom, Y.; Chuenbarn, T.; Phaechamud, T. Vancomycin HCl-loaded lauric acid *in situ*-forming gel with phase inversion for periodontitis treatment. *J. Drug Deliv. Sci. Technol.* **2020**, *57*, 101615. [CrossRef]
24. Phaechamud, T.; Jantadee, T.; Mahadlek, J.; Charoensuksai, P.; Pichayakorn, W. Characterization of antimicrobial agent loaded eudragit RS solvent exchange-induced in situ forming gels for periodontitis treatment. *AAPS PharmSciTech* **2017**, *18*, 494–508. [CrossRef]
25. Phaechamud, T.; Thurein, S.M.; Chantadee, T. Role of clove oil in solvent exchange-induced doxycycline hyclate-loaded Eudragit RS in situ forming gel. *Asian J. Pharm. Sci.* **2018**, *13*, 131–142. [CrossRef] [PubMed]
26. Senarat, S.; Wai Lwin, W.; Mahadlek, J.; Phaechamud, T. Doxycycline hyclate-loaded in situ forming gels composed from bleached shellac, Ethocel, and Eudragit RS for periodontal pocket delivery. *Saudi Pharm. J.* **2021**, *29*, 252–263. [CrossRef] [PubMed]
27. Phaechamud, T.; Mahadlek, J.; Tuntarawongsa, S. Peppermint oil/doxycycline hyclate-loaded Eudragit RS in situ forming gel for periodontitis treatment. *J. Pharm. Investig.* **2018**, *48*, 451–464. [CrossRef]
28. Lwin, W.; Puyathorn, N.; Senarat, S.; Mahadlek, J.; Phaechamud, T. Emerging role of polyethylene glycol on doxycycline hyclate-incorporated Eudragit RS in situ forming gel for periodontitis treatment. *J. Pharm. Investig.* **2019**, *50*, 81–94. [CrossRef]
29. Senarat, S.; Phaechamud, T.; Mahadlek, J.; Tuntarawongsa, S. Fluid properties of various Eudragit® solutions in different solvent systems for periodontal pocket injection. *Mater. Today Proc.* **2022**, *65*, 2399–2406. [CrossRef]

30. Tuntarawongsa, S.; Mahadlek, J.; Senarat, S.; Phaechamud, T. Eudragit® RL in 2-pyrrolidone as antisolvent-based in-situ forming matrix. *Mater. Today Proc.* **2022**, *52*, 2534–2538. [CrossRef]
31. Mahadlek, J.; Tuntarawongsa, S.; Senarat, S.; Phaechamud, T. In situ solvent removal-based Eudragit®L/dimethyl sulfoxide forming gel for periodontal pocket drug delivery. *Mater. Today Proc.* **2021**, *52*, 2394–2399. [CrossRef]
32. Patra, C.N.; Priya, R.; Swain, S.; Kumar Jena, G.; Panigrahi, K.C.; Ghose, D. Pharmaceutical significance of Eudragit: A review. *Future J. Pharm. Sci.* **2017**, *3*, 33–45. [CrossRef]
33. Malipeddi, V.R.; Awasthi, R.; Ghisleni, D.D.; de Souza Braga, M.; Kikuchi, I.S.; de Jesus Andreoli Pinto, T.; Dua, K. Preparation and characterization of metoprolol tartrate containing matrix type transdermal drug delivery system. *Drug Deliv. Transl. Res.* **2017**, *7*, 66–76. [CrossRef] [PubMed]
34. Aguilar, L.E.; Unnithan, A.R.; Amarjargal, A.; Tiwari, A.P.; Hong, S.T.; Park, C.H.; Kim, C.S. Electrospun polyurethane/Eudragit® L100-55 composite mats for the pH dependent release of paclitaxel on duodenal stent cover application. *Int. J. Pharm.* **2015**, *478*, 1–8. [CrossRef] [PubMed]
35. Sareen, R.; Nath, K.; Jain, N.; Dhar, K.L. Curcumin loaded microsponges for colon targeting in inflammatory bowel disease: Fabrication, optimization, and in vitro and pharmacodynamic evaluation. *BioMed Res. Int.* **2014**, *2014*, 340701. [CrossRef]
36. Amato, M.; Santonocito, S.; Polizzi, A.; Tartaglia, G.M.; Ronsivalle, V.; Viglianisi, G.; Grippaudo, C.; Isola, G. Local delivery and controlled release drugs systems: A new approach for the clinical treatment of periodontitis therapy. *Pharmaceutics.* **2023**, *15*, 1312. [CrossRef]
37. Chen, G.L.; Cai, H.Y.; Chen, J.P.; Li, R.; Zhong, S.Y.; Jia, X.J.; Liu, X.F.; Song, B.B. Chitosan/alginate nanoparticles for the enhanced oral antithrombotic activity of clam heparinoid from the clam coelomactra antiquata. *Mar. Drugs* **2022**, *20*, 136. [CrossRef]
38. FDA. *Generally Recognized as Safe*; 21 Cfr 184.1666; FDA: Silver Spring, MD, USA, 1982.
39. Mahmoud, D.B.; Shukr, M.H.; ElMeshad, A.N. Gastroretentive cosolvent-based in situ gel as a promising approach for simultaneous extended delivery and enhanced bioavailability of mitiglinide calcium. *J. Pharm. Sci.* **2019**, *108*, 897–906. [CrossRef]
40. Souza de Araujo, G.R.; de Oliveira Porfírio, L.; Santos Silva, L.A.; Gomes Santana, D.; Ferreira Barbosa, P.; Pereira dos Santos, C.; Narain, N.; Vitorino Sarmento, V.H.; de Souza Nunes, R.; Ting, E.; et al. In situ microemulsion-gel obtained from bioadhesive hydroxypropyl methylcellulose films for transdermal administration of zidovudine. *Colloids Surf. B* **2020**, *188*, 110739. [CrossRef]
41. Bansal, M.; Mittal, N.; Yadav, S.K.; Khan, G.; Mishra, B.; Nath, G. Clinical evaluation of thermoresponsive and mucoadhesive Chitosan in situ gel containing Levofloxacin and Metronidazole in the treatment of periodontal pockets—A split-mouth, clinical study. *J. Pierre Fauchard Acad.* **2016**, *30*, 6–14. [CrossRef]
42. Zhang, X.; Yang, L.; Zhang, C.; Liu, D.; Meng, S.; Zhang, W.; Meng, S. Effect of Polymer Permeability and Solvent Removal Rate on In Situ Forming Implants: Drug Burst Release and Microstructure. *Pharmaceutics* **2019**, *11*, 520. [CrossRef]
43. Wang, L.; Wang, A.; Zhao, X.; Liu, X.; Wang, D.; Sun, F.; Li, Y. Design of a long-term antipsychotic in situ forming implant and its release control method and mechanism. *Int. J. Pharm.* **2012**, *427*, 284–292. [CrossRef] [PubMed]
44. Ahmed, T.A.; Ibrahim, H.M.; Samy, A.M.; Kaseem, A.; Nutan, M.T.H.; Hussain, M.D. Biodegradable Injectable In Situ Implants and Microparticles for Sustained Release of Montelukast: In Vitro Release, Pharmacokinetics, and Stability. *Am. Assoc. Pharm. Sci. J.* **2014**, *15*, 772–780. [CrossRef] [PubMed]
45. Elmoazzen, H.Y.; Poovadan, A.; Law, G.K.; Elliott, J.A.W.; McGann, L.E.; Jomha, N.M. Dimethyl sulfoxide toxicity kinetics in intact articular cartilage. *Cell Tissue Bank.* **2007**, *8*, 125–133. [CrossRef] [PubMed]
46. Fujisawa, R.; Mizuno, M.; Katano, H.; Otabe, K.; Ozeki, N.; Tsuji, K.; Koga, H.; Sekiya, I. Cryopreservation in 95% serum with 5% DMSO maintains colony formation and chondrogenic abilities in human synovial mesenchymal stem cells. *BMC Musculoskelet. Disord.* **2019**, *20*, 316. [CrossRef]
47. Phaechamud, T.; Senarat, S.; Puyathorn, N.; Praphanwittaya, P. Solvent exchange and drug release characteristics of doxycycline hyclate-loaded bleached shellac in situ-forming gel and -microparticle. *Int. J. Biol. Macromol.* **2019**, *135*, 1261–1272. [CrossRef]
48. dos Santos, J.; da Silva, G.S.; Velho, M.C.; Beck, R.C. Eudragit®: A versatile family of polymers for hot melt extrusion and 3D printing processes in pharmaceutics. *Pharmaceutics* **2021**, *13*, 1424. [CrossRef]
49. Zhang, Q.; Fassihi, M.A.; Fassihi, R. Delivery considerations of highly viscous polymeric fluids mimicking concentrated biopharmaceuticals: Assessment of injectability via measurement of total work done "WT". *AAPS PharmSciTech* **2018**, *19*, 1520–1528. [CrossRef]
50. Seo, Y.P.; Seo, Y. Effect of molecular structure change on the melt rheological properties of a polyamide (Nylon 6). *ACS Omega* **2018**, *3*, 16549–16555. [CrossRef]
51. Rein, S.M.T.; Lwin, W.W.; Tuntarawongsa, S.; Phaechamud, T. Meloxicam-loaded solvent exchange-induced in situ forming beta-cyclodextrin gel and microparticle for periodontal pocket delivery. *Mater. Sci. Eng. C* **2020**, *117*, 111275. [CrossRef]
52. Thurein, S.M.; Lertsuphotvanit, N.; Phaechamud, T. Physicochemical properties of β-cyclodextrin solutions and precipitates prepared from injectable vehicles. *Asian J. Pharm. Sci.* **2018**, *13*, 438–449. [CrossRef]
53. Chuenbarn, T.; Chantadee, T.; Phaechamud, T. Doxycycline hyclate-loaded Eudragit® RS PO in situ-forming microparticles for periodontitis treatment. *J. Drug Deliv. Sci. Technol.* **2022**, *71*, 103294. [CrossRef]
54. Khaing, E.M.; Mahadlek, J.; Okonogi, S.; Phaechamud, T. Lime peel oil-incorporated rosin-based antimicrobial in situ forming gel. *Gels* **2022**, *8*, 169. [CrossRef] [PubMed]
55. Phaechamud, T.; Setthajindalert, O. Antimicrobial in-situ forming gels based on bleached shellac and different solvents. *J. Drug Deliv. Sci. Technol.* **2018**, *46*, 285–293. [CrossRef]

56. Do, M.P.; Neut, C.; Metz, H.; Delcourt, E.; Siepmann, J.; Mäder, K.; Siepmann, F. Mechanistic analysis of PLGA/HPMC-based in-situ forming implants for periodontitis treatment. *Eur. J. Pharm. Biopharm.* **2015**, *94*, 273–283. [CrossRef]
57. Hoang, C.; Nguyen, A.K.; Nguyen, T.Q.; Fang, W.; Han, B.; Hoang, B.X.; Tran, H.D. Application of dimethyl sulfoxide as a therapeutic agent and drug vehicle for eye diseases. *J. Ocul. Pharmacol. Ther.* **2021**, *37*, 441–451. [CrossRef]
58. Zheng, Z.-J.; Ye, H.; Guo, Z.-P. Recent progress in designing stable composite lithium anodes with improved wettability. *Adv. Sci.* **2020**, *7*, 2002212. [CrossRef]
59. Mei, L.; Huang, X.; Xie, Y.; Chen, J.; Huang, Y.; Wang, B.; Wang, H.; Pan, X.; Wu, C. An injectable in situ gel with cubic and hexagonal nanostructures for local treatment of chronic periodontitis. *Drug Deliv.* **2017**, *24*, 1148–1158. [CrossRef]
60. Senarat, S.; Charoenteeraboon, J.; Praphanwittaya, P.; Phaechamud, T. Phase behavior of doxycycline hyclate-incorporated bleached shellac in-situ forming gel/microparticle after solvent movement. *Key Eng. Mater.* **2020**, *859*, 21–26. [CrossRef]
61. Do, M.P.; Neut, C.; Delcourt, E.; Seixas Certo, T.; Siepmann, J.; Siepmann, F. In situ forming implants for periodontitis treatment with improved adhesive properties. *Eur. J. Pharm. Biopharm.* **2014**, *88*, 342–350. [CrossRef]
62. Golmaghani-Ebrahimi, E.; Bagheri, A.; Fazli, M. The influence of temperature on surface concentration and interaction energy between components in binary liquid systems. *J. Chem. Thermodyn.* **2020**, *146*, 106105. [CrossRef]
63. Phaechamud, T.; Mahadlek, J. Solvent exchange-induced in situ forming gel comprising ethyl cellulose-antimicrobial drugs. *Int. J. Pharm.* **2015**, *494*, 381–392. [CrossRef] [PubMed]
64. Kolawole, O.M.; Cook, M.T. In situ gelling drug delivery systems for topical drug delivery. *Eur. J. Pharm. Biopharm.* **2023**, *184*, 36–49. [CrossRef] [PubMed]
65. Puyathorn, N.; Sirirak, J.; Chantadee, T.; Phaechamud, T. Phase separation and intermolecular binding energy of ibuprofen in some organic solvents. *Mater. Today Proc.* **2022**, *65*, 2303–2308. [CrossRef]
66. Ershad, A.L.; Rajabi-Siahboomi, A.; Missaghi, S.; Kirby, D.; Mohammed, A.R. Multi-analytical framework to assess the in vitro swallowability of solid oral dosage forms targeting patient acceptability and adherence. *Pharmaceutics* **2021**, *13*, 411. [CrossRef] [PubMed]
67. Khaing, E.M.; Intaraphairot, T.; Mahadlek, J.; Okonogi, S.; Pichayakorn, W.; Phaechamud, T. Imatinib mesylate-loaded rosin cinnamon oil-based in situ forming gel against colorectal cancer cells. *Gels* **2022**, *8*, 526. [CrossRef] [PubMed]
68. Baddam, D.O.; Ragi, S.D.; Tsang, S.H.; Ngo, W.K. Ophthalmic fluorescein angiography. *Methods Mol. Biol.* **2023**, *2560*, 153–160. [CrossRef]
69. Gusmão, M.R.; Alves, T.C.; Lemes, A.P.; Bettiol, G.M.; Pedroso, A.d.F.; Barioni Junior, W.; Oliveira, P.P.A.; Grego, C.R. Sodium fluorescein as an internal tracer on the location of bovine urine patches in pastures. *Grass Forage Sci.* **2016**, *71*, 305–314. [CrossRef]
70. Alemán-Nava, G.S.; Cuellar-Bermudez, S.P.; Cuaresma, M.; Bosma, R.; Muylaert, K.; Ritmann, B.E.; Parra, R. How to use Nile Red, a selective fluorescent stain for microalgal neutral lipids. *J. Microbiol. Methods* **2016**, *128*, 74–79. [CrossRef]
71. Martinez, V.; Henary, M. Nile red and nile blue: Applications and syntheses of structural analogues. *Chemistry.* **2016**, *22*, 13764–13782. [CrossRef]
72. Ahmed, T. Review: Approaches to develop PLGA based in situ gelling system with low initial burst. *Pak. J. Pharm. Sci.* **2015**, *28*, 657–665.
73. Bansal, M.; Mittal, N.; Yadav, S.K.; Khan, G.; Gupta, P.; Mishra, B.; Nath, G. Periodontal thermoresponsive, mucoadhesive dual antimicrobial loaded in-situ gel for the treatment of periodontal disease: Preparation, *in-vitro* characterization and antimicrobial study. *J. Oral Biol. Craniofacial Res.* **2018**, *8*, 126–133. [CrossRef] [PubMed]
74. Ashish, G.; Rahul, Y.; Mukesh, R.; Prakash, M. Formulation and evaluation of in situ gel containing ciprofloxacin hydrochloride in the treatment of periodontitis. *Asian J. Pharm. Clin. Res.* **2017**, *10*, 154. [CrossRef]
75. Javali, M.A.; Vandana, K.L. A comparative evaluation of atrigel delivery system (10% doxycycline hyclate) Atridox with scaling and root planing and combination therapy in treatment of periodontitis: A clinical study. *J. Indian Soc. Periodontol.* **2012**, *16*, 43–48. [CrossRef] [PubMed]
76. Speck, S.; Wenke, C.; Feßler, A.T.; Kacza, J.; Geber, F.; Scholtzek, A.D.; Hanke, D.; Eichhorn, I.; Schwarz, S.; Rosolowski, M.; et al. Borderline resistance to oxacillin in *Staphylococcus aureus* after treatment with sub-lethal sodium hypochlorite concentrations. *Heliyon* **2020**, *6*, e04070. [CrossRef]
77. Rattanaumpawan, P.; Nachamkin, I.; Bilker, W.B.; Roy, J.A.; Metlay, J.P.; Zaoutis, T.E.; Lautenbach, E. High fluoroquinolone MIC is associated with fluoroquinolone treatment failure in urinary tract infections caused by fluoroquinolone susceptible *Escherichia coli*. *Ann. Clin. Microbiol. Antimicrob.* **2017**, *16*, 25. [CrossRef]
78. Conrads, G.; Klomp, T.; Deng, D.; Wenzler, J.S.; Braun, A.; Abdelbary, M.M.H. The antimicrobial susceptibility of *porphyromonas gingivalis*: Genetic repertoire, global phenotype, and review of the literature. *Antibiotics* **2021**, *10*, 1438. [CrossRef]
79. Poursamar, S.A.; Azami, M.; Mozafari, M. Controllable synthesis and characterization of porous polyvinyl alcohol/hydroxyapatite nanocomposite scaffolds via an in situ colloidal technique. *Colloids Surf. B* **2011**, *84*, 310–316. [CrossRef]
80. Machado, P.S.T.; Habert, A.C.; Borges, C.P. Membrane formation mechanism based on precipitation kinetics and membrane morphology: Flat and hollow fiber polysulfone membranes. *J. Membr. Sci.* **1999**, *155*, 171–183. [CrossRef]
81. Dias, R.; Medeiros, V.; Silva, B.; Araujo, E.; Lira, H. Study of the influence of viscosity on the morphology of polyethersulfone hollow fiber membranes/additives. *Mat. Res.* **2019**, *22*, e20180913. [CrossRef]

82. Pawar, I.A.; Joshi, P.J.; Kadam, A.D.; Pande, N.B.; Kamble, P.H.; Hinge, S.P.; Banerjee, B.S.; Mohod, A.V.; Gogate, P.R. Ultrasound-based treatment approaches for intrinsic viscosity reduction of polyvinyl pyrrolidone (PVP). *Ultrason. Sonochem.* **2014**, *21*, 1108–1116. [CrossRef]
83. Ranch, K.M.; Maulvi, F.A.; Koli, A.R.; Desai, D.T.; Parikh, R.K.; Shah, D.O. Tailored doxycycline hyclate loaded in situ gel for the treatment of periodontitis: Optimization, in vitro characterization, and antimicrobial studies. *AAPS PharmSciTech* **2021**, *22*, 77. [CrossRef] [PubMed]
84. Tahtat, D.; Mahlous, M.; Benamer, S.; Nacer Khodja, A.; Larbi Youcef, S. Effect of molecular weight on radiation chemical degradation yield of chain scission of γ-irradiated chitosan in solid state and in aqueous solution. *Radiat. Phys. Chem.* **2012**, *81*, 659–665. [CrossRef]
85. Zhang, P.; Lee, Y.I.; Zhang, J. A review of high-resolution X-ray computed tomography applied to petroleum geology and a case study. *Micron* **2019**, *124*, 102702. [CrossRef] [PubMed]
86. Liu, Y.; Xie, D.; Zhou, R.; Zhang, Y. 3D X-ray micro-computed tomography imaging for the microarchitecture evaluation of porous metallic implants and scaffolds. *Micron* **2021**, *142*, 102994. [CrossRef]
87. Zhang, Z.; Ni, J.; Chen, L.; Yu, L.; Xu, J.; Ding, J. Biodegradable and thermoreversible PCLA-PEG-PCLA hydrogel as a barrier for prevention of post-operative adhesion. *Biomaterials* **2011**, *32*, 4725–4736. [CrossRef]
88. Yacoubi, D.A.; Bouziane, D.; Leila, M.; Bensoltane, A. Microbiological Study of Periodontitis in the West of Algeria. *World J. Med. Sci.* **2010**, *5*, 7–12.
89. Rajendiran, M.; Trivedi, H.M.; Chen, D.; Gajendrareddy, P.; Chen, L. Recent development of active ingredients in mouthwashes and toothpastes for periodontal diseases. *Molecules* **2021**, *26*, 2001. [CrossRef]
90. Nalawade, T.M.; Bhat, K.; Sogi, S.H. Bactericidal activity of propylene glycol, glycerine, polyethylene glycol 400, and polyethylene glycol 1000 against selected microorganisms. *J. Int. Soc. Prev. Community Dent.* **2015**, *5*, 114–119. [CrossRef]
91. Phaechamud, T.; Setthajindalert, O. Cholesterol in situ forming gel loaded with doxycycline hyclate for intra-periodontal pocket delivery. *Eur. J. Pharm. Sci.* **2017**, *99*, 258–265. [CrossRef]
92. Siepmann, J.; Siepmann, F. Mathematical modeling of drug dissolution. *Int. J. Pharm.* **2013**, *453*, 12–24. [CrossRef]
93. Wang, F.; Saidel, G.M.; Gao, J. A mechanistic model of controlled drug release from polymer millirods: Effects of excipients and complex binding. *J. Control. Release* **2007**, *119*, 111–120. [CrossRef] [PubMed]
94. Siepmann, J.; Siepmann, F. Mathematical modeling of drug delivery. *Int. J. Pharm.* **2008**, *364*, 328–343. [CrossRef] [PubMed]
95. Bruschi, M.L. (Ed.) 5—Mathematical models of drug release. In *Strategies to Modify the Drug Release from Pharmaceutical Systems*; Woodhead Publishing: Delhi, India, 2015; pp. 63–86. [CrossRef]
96. Ritger, P.L.; Peppas, N.A. A simple equation for description of solute release I. Fickian and non-fickian release from non-swellable devices in the form of slabs, spheres, cylinders or discs. *J. Control. Release* **1987**, *5*, 23–36. [CrossRef]
97. Thomas, N.L.; Windle, A.H. A theory of case II diffusion. *Polymer* **1982**, *23*, 529–542. [CrossRef]
98. Paarakh, M.P.; Jose, P.A.; Setty, C.M.; Peterchristoper, G.V. Release Kinetics—Concepts and applications. *IJPRT* **2018**, *8*, 12–20. [CrossRef]
99. Costa, P.; Sousa Lobo, J.M. Modeling and comparison of dissolution profiles. *Eur. J. Pharm. Sci.* **2001**, *13*, 123–133. [CrossRef]
100. Aziz, I.H.; Abdullah, M.M.A.B.; Mohd Salleh, M.A.A.; Yoriya, S.; Chaiprapa, J.; Rojviriya, C.; Li, L.Y. Microstructure and porosity evolution of alkali activated slag at various heating temperatures. *J. Mater. Res. Technol.* **2020**, *9*, 15894–15907. [CrossRef]

gels

Article

Development and Evaluation of Nanoformulations Containing Timur Oil and Rosemary Oil for Treatment of Topical Fungal Infections

Afeefa Noor [1,2], Shahid Jamil [3], Tariq Waece Sadeq [3,4], Muath Sheet Mohammed Ameen [3] and Kanchan Kohli [1,2,*]

1. Lloyd Institute of Management and Technology, Greater Noida 201306, India
2. Department of Pharmaceutics, School of Pharmaceutical Education and Research (SPER), Jamia Hamdard University, New Delhi 110062, India
3. Department of Pharmacy, College of Pharmacy, Knowledge University, Kirkuk Road, Erbil 44001, Iraq
4. Pharmacy Department, Erbil Medical Technical Institute, Erbil Polytechnic University, Ebril 44001, Iraq

* Correspondence: kanchan.kohli@lloydpharmacy.edu.in

Abstract: The pervasiveness of fungal infections is an issue for skin health globally, and there are a reported 40 million cases in developed and developing countries. Novel drug delivery systems provide better therapeutic efficacy over conventional drug therapy due to their lower side effects and toxicity. Furthermore, combinations of essential oils can represent alternative therapies for fungal infections that are resistant to synthetic drugs. This study is aimed at developing Timur oil into a nanoemulgel and evaluating its antifungal effects. The development of the formulation involved the preparation of a nanoemulsion by the titration method, followed by its evaluation for various physicochemical properties. The antifungal activity of the nanoemulgel against *Candida albicans* was evaluated. The zone of inhibition was determined using the disk diffusion method. The results show that the developed nanoemulgel has a particle size of 139 ± 6.11 nm, a PDI of 0.309, and a zeta potential of −19.12 ± 2.73 mV. An in vitro drug release study showed a sustained release of 70 ± 0.289% of the drug over a period of 24 h. The % drug permeation across the skin was found to be 79.11 ± 0.319% over 24 h. However, the amount of drug retained in the skin was 56.45 μg/g. The flux for the nanoemulgel was found to be 94.947 $\mu g/cm^2/h$, indicating a better permeability profile. The nanoemulgel formulation showed a zone of inhibition of 15 ± 2.45 mm, whereas the 1% ketoconazole cream (marketed preparation) exhibited a zone of inhibition of 13 ± 2.13 mm. The results of this study suggest that developed nanoemulgel containing Timur oil and rosemary oil has the potential to be used for treating topical fungal infections caused by *Candida albicans*.

Keywords: Timur oil; rosemary oil; nanoemulgel; fungal infections

check for updates

Citation: Noor, A.; Jamil, S.; Sadeq, T.W.; Mohammed Ameen, M.S.; Kohli, K. Development and Evaluation of Nanoformulations Containing Timur Oil and Rosemary Oil for Treatment of Topical Fungal Infections. *Gels* **2023**, *9*, 516. https://doi.org/10.3390/gels9070516

Academic Editor: Esmaiel Jabbari

Received: 1 June 2023
Revised: 16 June 2023
Accepted: 20 June 2023
Published: 26 June 2023

1. Introduction

Topical fungal infections greatly affect skin health in developing and developed countries. *Candida* species are the most common fungi for superficial skin infections. Fungal infections infect the skin superficially and then enter the deeper layers of the skin by desquamation; hence, agents that can permeate deep into the skin are needed to treat these infections [1]. Topical therapies are the preferred treatment for dermal fungal infections because of their targeted action and reduced side effects [1,2]. Various synthetic drugs such as itraconazole, ketoconazole, and clotrimazole are used as conventional topical drugs for fungal infections. Although they have advantages such as localized effects, increased bioavailability, and good patient compliance, conventional drugs also have disadvantages, such as the potential to trigger allergies and eczema [1]. The repeated use of synthetic drugs also increases toxicity and resistance [3].

The utilization of essential oils to combat microbial infections is gaining importance. Timur oil (*Zanthoxylum armatum*), of the family *Rutaceae*, is an aromatic, annual herb, mostly growing in the hot valleys of the subtropical Himalayas, from Indus areas to Bhutan, up to an altitude of 2400 m. It also grows in the Khasi Hills from 700 m to 1000 m. Essential oils and fatty acids are thought to be the main constituents of this medicinal plant. The essential oils obtained from fruits of *Zanthoxylum armatum* have antifungal, antibacterial, and anthelmintic properties. Numerous scientific investigations have shown the antifungal [4] and mosquito-repelling properties [5] of Timur oil. According to Tiwary et al. [6], the main ingredients of the essential oil of *Zanthoxylum armatum* are linalool (57%) and limonene (19.8%). The linalool content in essential oils has been shown to mainly account for their antifungal properties [7]. However, limonene has modest antibacterial activity while also exhibiting effective antimicrobial action [4]. Prakash et al. [8] reported that the essential oil of a different variety of Zanthoxylum possesses a wide range of antifungal activity against molds such as *Aspergillus flavus, niger, terreus, candidus, sydowi, fumigates, Alternaria alternate, Cladosporium cladosporioides, Curvular ialunata, Fusarium nivale, Penicillium italicum,* and *Trichoderma virdie.* Zanthoxylum alatum showed a zone of inhibition 18 mm in diameter against *Alternaria alternate* [9].

Rosemary is a member of the mint family [10], and it possesses various antioxidants such as carnosic acid, carnosol, and rosmarinic acid [11,12]. Additionally, it has significant concentrations of flavonoids, phenolic acid, diterpene, and triterpene. Furthermore, studies have revealed that rosemary oil exerts antifungal [10], antibacterial, anti-inflammatory [13] antimicrobial [14], and antileishmanial activities [15]. The protective/beneficial effects of rosemary essential oil on candidiasis have been explored in a few studies [16].

Recent findings have revealed that natural herbal products can be used to treat major health issues such as cancer, microbial infections, and cardiovascular disorders. Therefore, novel drug delivery systems are required to tackle the challenges associated with these natural compounds to increase their effectiveness [17]. Nanoemulsion is a novel approach to producing stable preparations with increased solubility. Producing nanoemulsions is a good technique for developing formulations that contain essential oils as active ingredients for antifungal activity [18]. The spontaneous emulsification method (titration method) for nanoemulsion is very simple. It includes gently swirling while combining oil, water, surfactant, and co-surfactant in an ideal ratio [19]. To determine this ideal ratio, a phase diagram can be constructed using the aqueous titration technique. This method produces particles that range in size from 20 to 200 nm [20,21]. The hydrophilic character of the formulation makes it simple to remove the NEG from the application site when the necessary efficacy has been achieved. The thixotropic NEG facilitates easy spreadability at the target site and prolongs retention at the application site due to its mucoadhesive property [22,23]. In order to increase the thickness, lower the interfacial tension, and improve the stability, a nanoemulsion is added to the hydrogel matrix, such as a carbomer (Carbopol), to create a nanoemulgel, which functions as a drug reservoir [24]. The nanoemulgel increases the perforation of the oil into the skin [25]. When oily particles exit the skin's gel matrix intact and enter its layers, they reach their bodily targets [26].

The study aimed to use a rosemary-mediated nanoemulgel containing Timur oil as the main oil to effectively treat topical fungal infections. In this study, Timur oil is nanoemulsified in Smix and rosemary oil is used as carrier oil on the basis of its miscibility with Timur oil. Considering geographical factors as important parameters for herbal products, the extraction of two varieties of Timur seeds from different geographical sources (India and Nepal) was carried out to determine the percentages of linalool in both varieties by GC analysis. As this combination was new, there has not yet been research on combining these two essential oils for antifungal activity. In this study, Tim–Ros–NEG is the developed rosemary-mediated nanoemulgel containing Timur oil and its topical antifungal activity was evaluated against *C. albicans*, compared with that of a 1% ketoconazole cream (marketed preparation).

2. Results and Discussion

2.1. Extraction of Timur Oil and Characterization

In order to develop a nanoformulation containing Timur oil and rosemary oil, Timur oil was extracted from the fresh seeds of *Zanthoxylum Armatum* (Timur) of Indian and Nepali origin, with a 56% percentage of active constituents (linalool) of Nepali origin; the yield was 1.5 percent/200 g. The Timur oil was characterized by FTIR, HPTLC, and GC. The FTIR functional group was identified, and the results are shown in Figure 1 and Table 1. The quantification of linalool was carried out by GC, and the area for linalool (Standard) and the area for linalool present in the Timur oil were found to be 272,960,345 and 51,550,363, respectively. The retention times for linalool (Standard) and linalool in the extracted Timur oil were observed to be 15.583 and 15.531 (Table 2), respectively. HPTLC analysis showed that the linalool content in the Nepali variety (56%) was higher than that in the Indian variety (36%). Therefore, the variety from Nepal was selected for further studies. In order to study the release of the developed formulation, solubility studies for Timur and rosemary oil were carried out in distilled water, methanol, ethanol, toluene, and phosphate buffer at pH values of 5.5 and 7.4. Figure 2 represents the maximum solubility of rosemary oil in buffer at pH 7.4, while Timur oil was soluble in acetone and buffer at pH 5.5 and 7.4. It was observed that the rosemary oil and Timur oil have significantly ($p < 0.05$) higher solubility in pH 7.4 phosphate buffer compared to the other solvents.

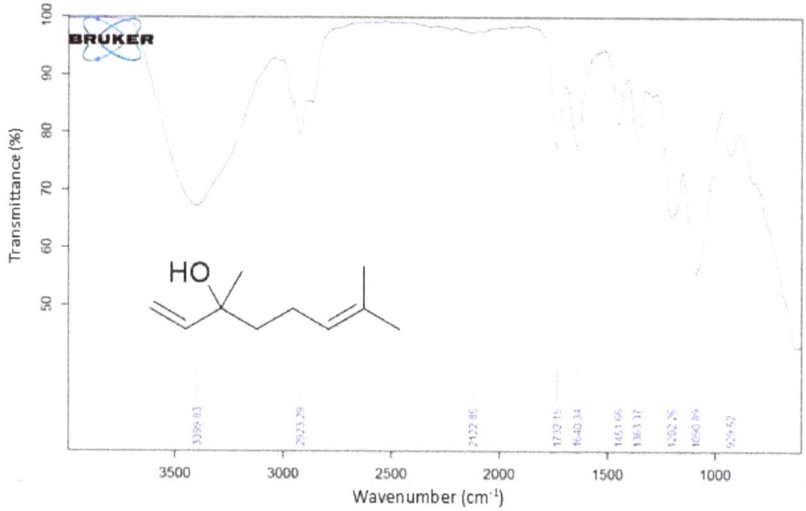

Figure 1. FT-IR Spectra of Timur oil and chemical structure of linalool.

Table 1. Observed wavenumbers of different functional groups in the sample.

Sample No.	Functional Group	Wavenumber (cm^{-1})	Observed (cm^{-1})
1	OH stretching	3550–3200	3399.83
2	C-H Stretching	3000–2840	2923.29
3	C=C Stretching	1680–1630	1640.34
4	C-H bending	1600–1400	1451.66
5	C-O stretching	1124–1087	1090.89

Table 2. GC analysis of Timur oil.

Parameter	Linalool Content in Timur Oil (Nepali Variety)	Standard (Linalool)
Area	51,550,363	272,960,345
Retention time	15.531	15.583

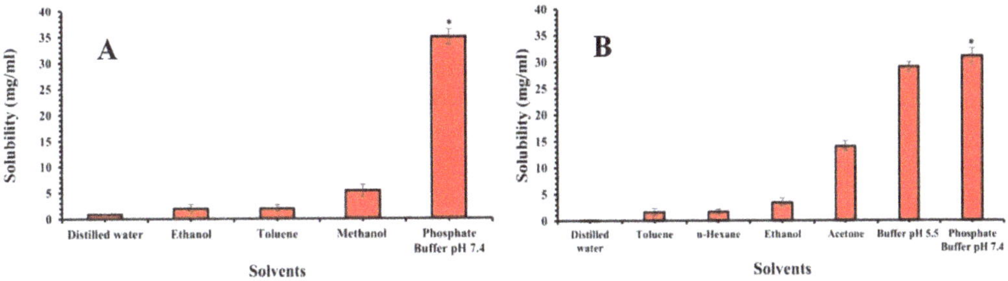

Figure 2. (**A**) Mean solubility (mg/mL) of rosemary oil at 25 °C. (**B**) Mean solubility (mg/mL) of Timur oil at 25 °C. An asterisk (*) indicates that the value is significantly different at $p < 0.05$ compared to the other group.

2.2. *Construction of Pseudoternary Phase Diagram*

Timur oil, rosemary oil, Transcutol P (cosurfactant), and Tween 80 (surfactant) were combined at various concentrations to create ternary phase diagrams to identify the best formulation, which resulted in a nanoemulsion with a PDI of 0.3 and a droplet size of less than 200 nm (Figure 3). Pseudoternary phase diagrams were constructed by preparing various Smix ratios (1:0, 1:1, 1:2, 2:1, 3:1, and 4:1). First, placebo formulations were prepared using the selected oil/surfactant/co-surfactant in the above-mentioned Smix ratio. After the formation of the placebo, drug loading was performed for the placebo. Timur oil was used as the oil phase, Tween 80 as the surfactant, Transcutol P as the co-surfactant, and distilled water as the aqueous phase during phase diagram formation. The Smix ratio of 4:1 occupied the largest area out of all the ratios. Thus, it was determined that the 4:1 Smix ratio produced the largest possible nanoemulsion area.

An O/W nanoemulsion was prepared by using the titration method. It is well established that, the higher the quantity of the surfactant, the greater the potential for toxicity toward the skin and irritation. Therefore, an attempt was made to increase the oil concentration and decrease the concentration of the Smix. Therefore, from the titration chart, the suitable oil/Smix ratios were selected, which were 5:5, 6:4, and 7:3. Out of these, the 5:5 oil/Smix ratio was selected.

Physicochemical Characterization of Nanoemulsion

The prepared nanoemulsion was examined for globule size using a Malvern Zetasizer [27]. The droplet size of the developed nanoemulsion (Smix ratio: 5:5) was 93.12 nm. In contrast, the polydispersity index of the same formulation was 0.243, which shows a better distribution of the globules of the dispersed phase into the dispersion system (Figure 4). The ideal range of the PDI lies between 0.1 and 0.5.

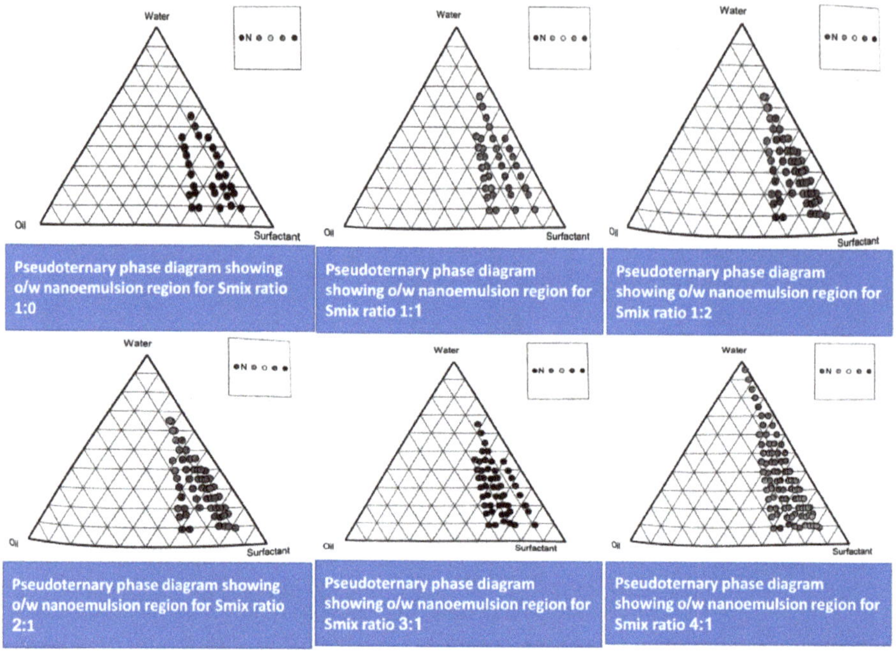

Figure 3. Pseudoternary phase diagram showing o/w nanoemulsion regions for the Smix ratio (1:0, 1:1, 1:2, 2:1, 3:1, and 4:1).

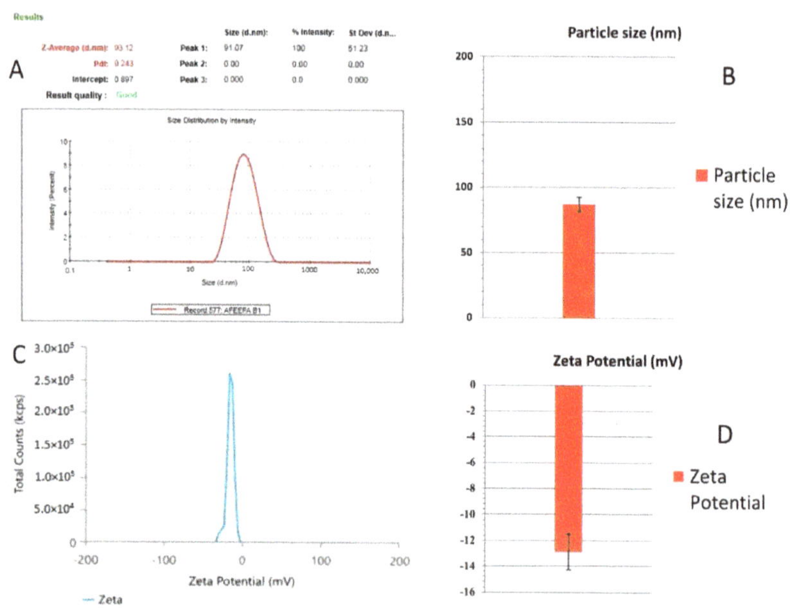

Figure 4. (**A**) Droplet size of developed nanoemulsion. (**B**) Bar diagram for droplet size of developed nanoemulsion. (**C**) Zeta potential of developed nanoemulsion. (**D**) Bar diagram for zeta potential of developed nanoemulsion.

An appropriate zeta potential ensures that the nanoformulation system is intense and stable [27,28]. Nanoemulsions typically have a zeta potential value between +100 mV and −100 mV. Neutral nanoemulsions correspond to those with zeta potential values between −10 and +10 mV [29]. The developed nanoemulsion formulation showed a zeta potential value of −14.32 mv, indicating good stability (Figure 4).

2.3. Fabrication of Timur Oil Nanoemulgel Formulation

Five nanoemulgel formulations were prepared using the same drug concentration but different concentrations of polymer (Carbopol-940), i.e., 0.5, 1.0, 1.5, 1.8, and 2.0% *w*/*v*, and the results were determined. It was found that, upon increasing the polymer concentration from 0.5% *w*/*v*, the particle size, as well as the polydispersity index of the respective formulations, decreased but only up to a polymer concentration of 1.8% *w*/*v*, and beyond this concentration, the particle size and PDI further increased (shown in Figure 5). Therefore, it was discovered that a Carbopol-940 concentration of 1.8% *w*/*v* was appropriate for creating an optimal nanoemulgel formulation. It was observed that the nanoemulgel containing 0.5% *w*/*v* polymer concentration significantly ($p < 0.05$) reduced the droplet size compared to the nanoemulgel containing 1%, 1.5%, 1.8% & 2% *w*/*v* polymer concentrations.

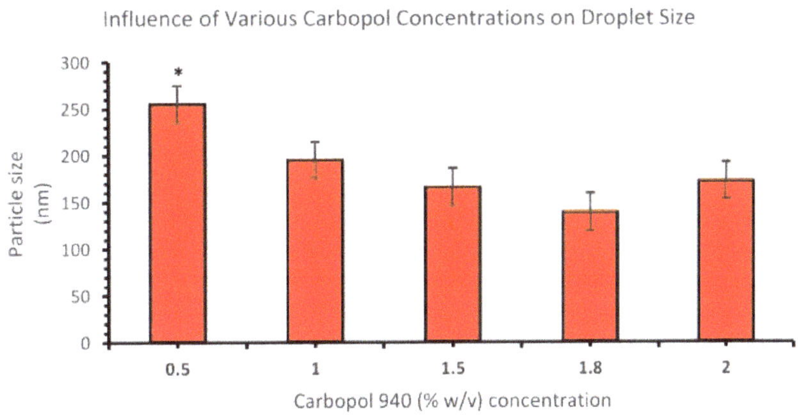

Figure 5. Influence of various Carbopol concentrations on droplet size. An asterisk (*) indicates that the value is significantly different at $p < 0.05$ compared to the other group.

2.3.1. Physicochemical characterization of Nanoemulgel

The various nanoemulgel formulations were prepared, and their mean particle sizes were determined (Table 3 and Figure 6) [17]. The results revealed that the developed formulation (i.e., FD) showed a mean droplet size of 139 nm and a lower PDI. The value of the PDI should lie between 0.1 and 0.5. The developed formulation showed a PDI of 0.309, which shows a good size distribution for the droplets.

Table 3. Different nanoemulgel formulations with different polymer and drug concentrations.

Formulation Code	Drug Concentration		Carbopol 940 (% *w*/*v*) Concentration	Particle Size (nm)	Resultant PDI	ZOI (mm)
	Timur Oil (%)	Rosemary Oil (%)				
FA	1	2	0.5	255	0.590	9
FB	0.5	0.5	1.0	195	0.521	10
FC	1.0	1.0	1.5	166	0.412	11
FD	2.0	1.0	1.8	139	0.309	15
FE	1.0	0.5	2.0	172	0.389	10

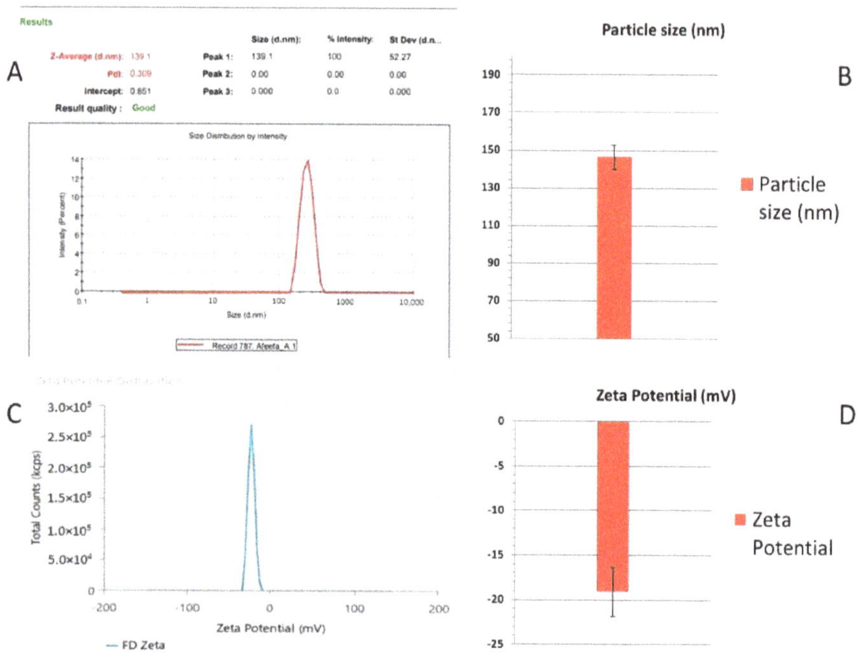

Figure 6. (**A**) Particle size of developed nanoemulgel. (**B**) Bar diagram for particle size of developed nanoemulgel. (**C**) Zeta potential for developed nanoemulgel. (**D**) Bar diagram for zeta potential of developed nanoemulgel.

This technique determines the charge on the globule surface of the dispersed phase. The zeta potential value of a nanoemulgel typically ranges from +100 mV to −100 mV [27,28,30,31]. The zeta potential of the developed nanoemulgel formulation was found to be −21.99 mV, −16.54 mV, and −18.85 mV, respectively. The mean zeta potential of the developed nanoemulgel formulation was observed as −19.12 ± 2.73 mV and we have shown these results in the bar chart (Figure 6D). We have considered a zeta potential value of −21.99 mV of the developed nanoemugel formulation (FD), indicating good formulation stability, as shown in Figure 6C.

2.3.2. Mechanical Strength of NEG

Patient acceptability depends on creating a gel with the necessary physical characteristics, such as excellent pourability, spreadability, and acceptable hardness and viscosity. A texture analyzer was used to investigate the mechanical characteristics of Tim–Ros–NEG. The findings showed that the characteristics varied with the concentration of Carbopol 940 (0.5%, 1%, 1.5%, and 1.8%). With an increase in hydroxypropyl methylcellulose content, gel cohesiveness was observed to rise by Karavana et al. [32] and the same observation was recorded by Cevher et al. [33] for Carbopol gels. The developed nanoemulgel's hardness was found to be 783.01 g force. Firmness or hardness refers to the formulation's deformational flexibility under stress, while cohesiveness refers to the number of cross-links between gel molecules and their ability to maintain shape. The higher the value, the better the gel strength [23]. The findings demonstrated Tim–Ros–NEG's resistance to deformation. However, our spreadability results demonstrated that Tim–Ros–NEG is spreadable even though it is more resistant to deformation. Typically, increased hardness results in less spreading. The developed Tim–Ros–NEG was determined to have a consistency of 898.34 g.s. The term "consistency" refers to a product's viscosity and characterizes its texture and hardness. A high consistency score suggests that Tim–Ros–NEG was consistent, according to the investigation. Cohesiveness is just a measure of how effectively a product resists a second deformation in comparison to how well it resists a first deformation. The greatest force represents a sample's stickiness or adhesiveness. According to the findings of our research, Tim–Ros–NEG is more cohesive or sticky than other materials since it demonstrated high cohesiveness (−459.81 g force) and an index of viscosity of 688.84 g/s. Higher Tim–Ros–NEG adhesion translates into a longer skin contact duration, which might result in less frequent nanoemulgel application. Osmałek et al. reported that Opokan®, a commercial product, had the lowest textural characteristics with a hardness of more than twice that of other gels [34]. Mohapatra et al. reported that Black Cohosh loaded ethosomal gel for the treatment of menopause. They have used Carbopol® 971P NF as a gelling agent. They have considered firmness, consistency, cohesiveness, and work of cohesion as evaluation parameters [35]. Hasan et al. developed combined therapy of 5-fluorouracil (5-FU) and cannabidiol (CBD)-loaded nanostructured lipid carrier gel (FU-CBD-NLCs gel) for the effective treatment of non-melanoma skin cancer. In this study, they have reported textural properties of 5-fluorouracil (5-FU) and cannabidiol (CBD)-loaded nanostructured lipid carrier gel (FU-CBD-NLCs gel). They have used Carbopol 934 as a gelling agent. The results of our investigation are in accordance with the published research article [36]. The findings of FD's texture analysis (Table 4 and Figure 7) indicated that the formulation was sufficiently firm and cohesive, showing that it would be user-friendly during application. Both formulations showed thermosensitivity, as indicated by a negative index of viscosity values.

Table 4. Texture analysis of developed formulation.

Test ID	Batch		Firmness (g) Force 1	Consistency (g. s) Area F-T 1:2	Cohesiveness (g) Force 2	Index of Viscosity (g. s) Area F-T 2:3
Start Batch Force	Force					
Force 1	Force		783.01	898.34	−459.81	−688.84
End Batch Force	Force					
Average:	Force (F)	AVERAGE ("BATCH")	783.01	898.34	−459.81	−688.84
S. D.	Force (F)	STDEVP ("BATCH")	0.00	0.00	0.00	0.00
C. V.	Force (F)	STDEVP ("BATCH")/AVERAGE ("BATCH") ×100	0.00	0.00	0.00	0.00
End of Test Data						

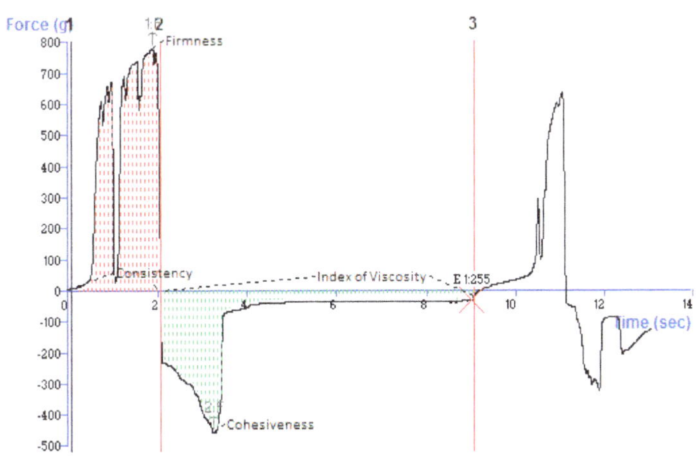

Figure 7. Texture Analysis of developed nanoemulgel showing the resultant force versus time.

2.3.3. In Vitro Release Study

The cumulative % release experiments of the Timur oil from developed nanoemulgel formulation were determined using phosphate buffer (with pH 5.5) and compared with that of the prepared nanoemulsion and pure Timur oil. Due to the essential oil's encapsulation in the lipid portion of the nanoemulgel, the essential oil from the developed nanoemulgel was released slowly over time. Within 24 h, the developed nanoemulgel formulation showed a considerable release (70%) while the developed nanoemulsion formulation exhibited an 85% release (Figure 8). A 100% release was seen in the case of pure Timur oil within 4 h [3]

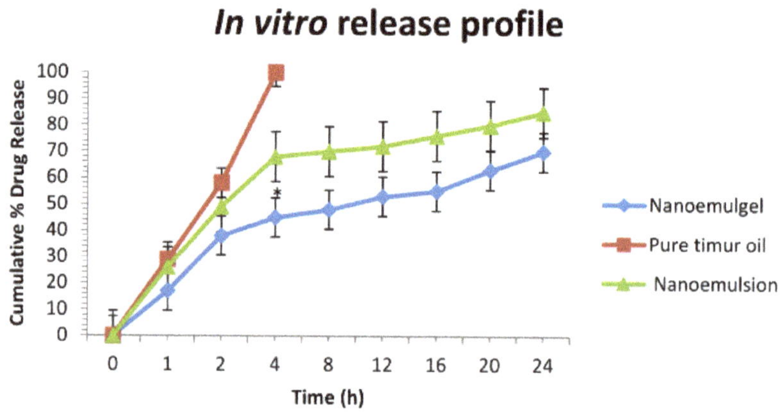

Figure 8. Comparison of in vitro release profile between pure Timur oil, nanoemulsion, and nanoemulgel. An asterisk (*) indicates that the value is significantly different at $p < 0.05$ compared to the other group.

Any drug's therapeutic effectiveness depends on how readily it is released from its pharmaceutical formulation [37]. The viscosity, surfactant content, polymer, and drug concentration are only a few of the variables that affect how quickly a topical pharmaceutical formulation releases a drug [38]. The development of a viscous formulation due to the synthetic polymer Carbopol 940′s large molecular weight delays the release profile of pharmaceuticals from topical preparations like nanoemulgel [39]. Additionally, it has been stated previously that emulgel serves as a reservoir for the drug, releasing it first from the internal phase to the exterior phase before entering the skin [40]. In contrast to nanoemulsion, which requires the oil droplets to pass through a hydrophilic phase transition, nanoemulgel allows the oil droplets to first be released into the gel matrix and then they can pass directly into the skin's subcutaneous layer [41]. Additionally, among all the formulations, the nanoemulgel showed a superior sustained release effect, as shown in Figure 8 [42]. Therefore, based on this study, it was concluded that the Timur oil release with the developed formulation having a polymer concentration of 1.8% w/v was considerably higher, with a sustained release, compared to the pure Timur oil. Therefore, based on this study, it was concluded that among all the prepared nanoemulgel formulations, (FD) was the optimal formulation and can be considered for further in vivo studies. The antimicrobial action is due to the oil components and the drug delivery system, we used. The developed nanoemulgel formulation showed a significantly ($p < 0.05$) higher release of Timur oil compared to the in vitro release of pure Timur oil.

2.3.4. Release Kinetics

The release data were examined to determine the goodness of fit in various kinetic models in order to establish a suitable Tim–Ros–NEG release pattern. The release data were run through various release kinetic models, including zero-order, first-order, Higuchi matrix, and Korsmeyer–Peppas models, and the best-fitting model was chosen based on the regression coefficient (R^2) value. The highest R^2 value was 0.9807, while the lowest R^2 value was 0.8879. The Higuchi matrix model was the best-fitting model for the nanoemulgel, having an R^2 value of 0.9807 (Figure 9) [43]. The value of n was found below 0.5 (i.e., 0.1561), showing that Timur oil release from the developed nanoemulgel formulation follows Fickian diffusion [44].

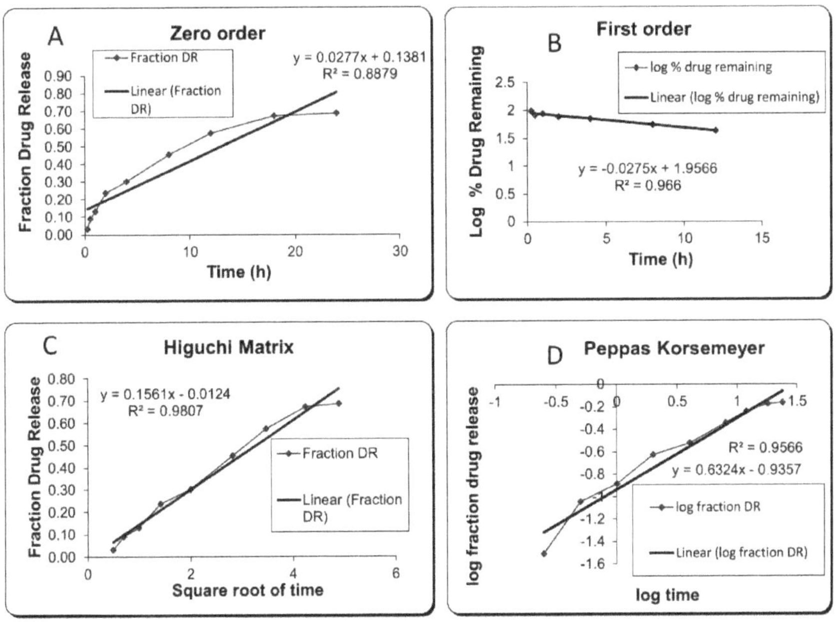

Figure 9. Various release kinetic models of the nanoemulgel: (**A**) zero-order kinetics; (**B**) first-order kinetics; (**C**) Higuchi matrix; (**D**) Korsmeyer–Peppas.

2.3.5. Drug Deposition Study

The skin retention properties of the nanoemulgel were determined using a drug deposition study. The skin retention of the Timur oil from the fourth nanoemulgel formulation (with the formulation code FD) was found to be 56.45 μg of the dose applied, whereas pure Timur oil was found to be 34.56 μg of the dose applied on the skin. The gel used for topical application has to have the right formulation properties to make it easy to apply and remain in contact with the skin for a long time. Tim–Ros–NEG at 1.8% had a greater viscosity and remained stuck to the skin for longer compared to Tim–Ros–NEG at 0.5%, 1.0%, and 1.5%. This nanoemulgel showed greater drug localization in the skin. Therefore, the nanoemulgel (FD) formulation resulted in more medication being retained in the skin, as demonstrated in Table 5.

Table 5. Drug retention of pure Timur oil and nanoemulgel on rat skin.

The Dose Applied to the Skin	Drug Deposited (μg)
1 g of nanoemulgel	56.45
1 mL of pure Timur oil	34.56

2.3.6. Morphological Studies/Internal Composition Study

TEM is the most effective characterization technique that uses high magnification to anticipate morphological characteristics and examine samples' surfaces [31,45,46]. TEM was used to study the morphology of the globules of the developed nanoemulgel. The globule size was in accordance with the results obtained using the Zetasizer. TEM images of the developed formulation (having the formulation code FD) indicated spherical globes with slight roughness, and no aggregation was observed among the globules (Figure 10).

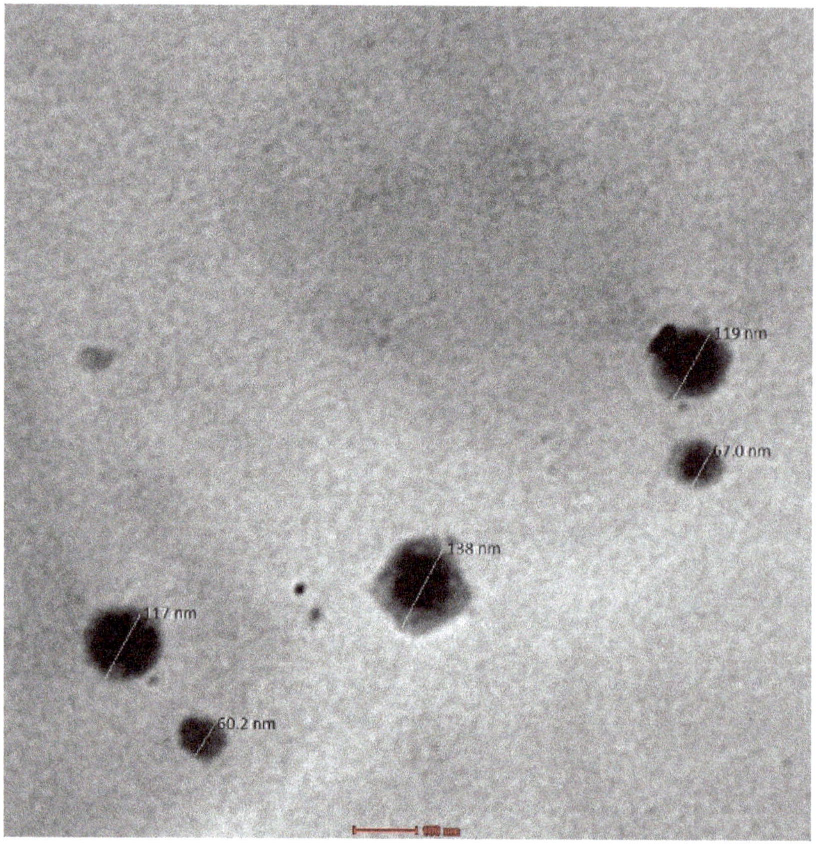

Figure 10. TEM of developed nanoemulgel (scale bar = 100 nm; 92,000× frame).

2.3.7. Confocal Laser Scanning Microscopy

The CLSM investigation evaluated the depth and intensity of permeation through the skin of the rats treated with a marker-loaded commercial gel, the developed nanoemulsion, and the developed nanoemulgel. The outcomes were contrasted with those for the skin from the group that received the control treatment. The breadth and depth of the carrier system's penetration into the skin were similar to the intensity and depth of the marker's penetration. Figure 11A–D display the photomicrographs of rat skin treated with the control, commercial gel, developed nanoemulsion, and optimal nanoemulgel formulations. Compared to the control group, the treatment group treated with nanoemulgel-based formulations showed dramatically enhanced depth and intensity of penetration into the skin. This showed that formulations based on developed carriers have more potential for skin penetration.

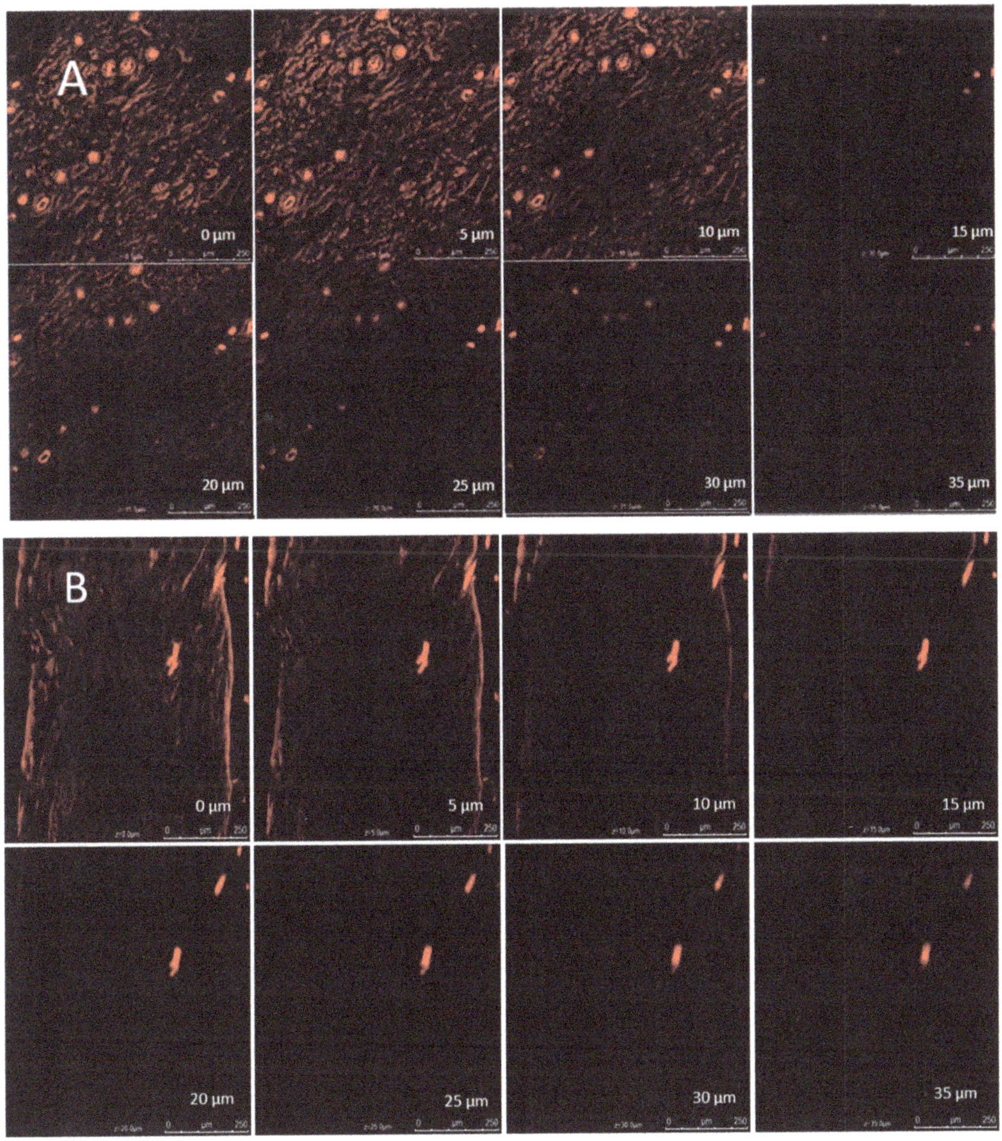

Figure 11. *Cont.*

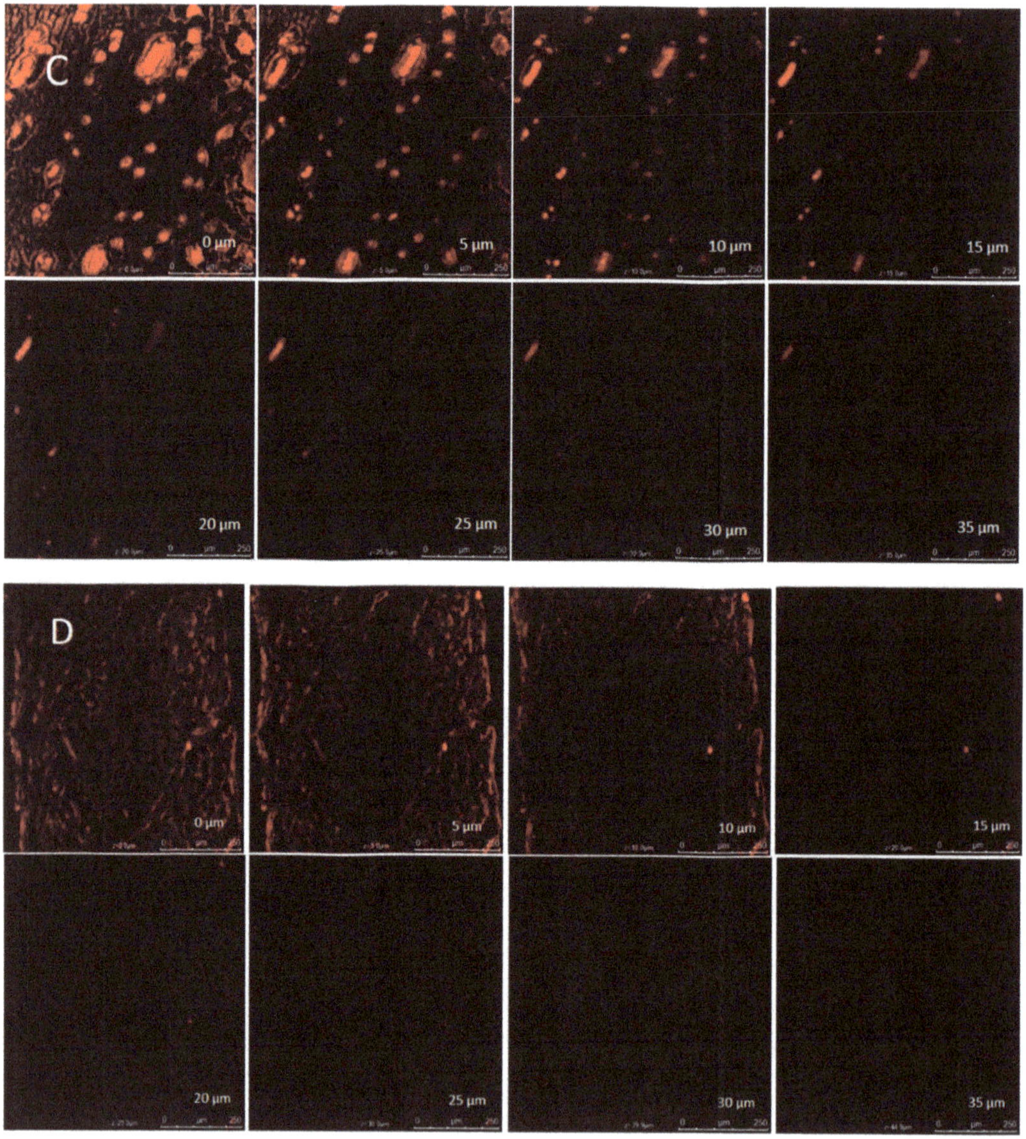

Figure 11. Confocal laser scanning microscopy images of excised rat skin treated with (**A**) plain dye solution, (**B**) dye-loaded marketed gel, (**C**) dye-loaded developed nanoemulsion, and (**D**) dye-loaded developed nanoemulgel formulation.

2.3.8. Ex Vivo Skin Permeation Study

In order to compare the permeability of the nanoemulgel with that of a pure Timur oil, ex vivo permeation tests were carried out. Franz diffusion cells and rat skin were used in the study. Figure 12 shows the outcomes for the ex vivo permeation of Timur oil from the Tim–Ros–NEG and pure Timur oil. The Tim–Ros permeation from the two formulations is compared, and it is shown that the permeation of Timur oil from the created NEG formulation (2265.01 ± 0.01 µg/cm^2) was considerably greater ($p < 0.05$) than that from the pure Timur oil (12.96 ± 0.33 µg/cm^2) after 24 h (Figure 12). For the Tim–Ros–NEG, the %

drug permeation for the nanoemulgel was found to be 21.7% at 0.5 h, 26.42% at 2 h, 44.73% at 6 h, 57.89% at 12 h, and 79.11% at 24 h, whereas the % DP of the pure Timur oil was 4.08% at 0.5 h, 5.71% at 2 h, 5.89% at 6 h, 12% at 12 h, and 15.03% at 24 h (Figure 12). The nanosized oil globules possessing Timur oil enhance the penetration, and may accelerate drug permeation through the skin's lipophilic layers [47–49]. As the permeation parameters were examined simultaneously, it was discovered that the permeation flux was considerably increased ($p < 0.05$) with Tim–Ros–NEG (484.12 ± 0.065 μg cm^{-2} h^{-1}) compared to the pure Timur oil (3.21 ± 0.074 μg cm^{-2} h^{-1}). The nanoemulgel showed a better permeability profile, as the flux was 94.947 μg/cm^2/h, whereas the flux of the pure Timur oil was 24.504 μg/cm^2/h, much lower than that of the nanoemulgel. The nanoemulgel permeability was four times better than that of the pure Timur oil. When comparing the apparent penetration and flux of the two formulations, it was discovered that there was a four-fold augmentation ratio between the Tim–Ros–NEG and the pure Timur oil. This could be because the skin permeability increased due to Carbopol-940 [50], and the nanometric lipophilic globules also made this possible.

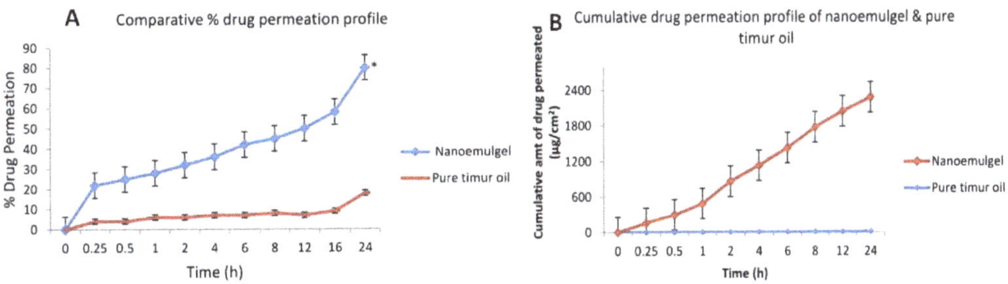

Figure 12. (**A**) Comparative % drug permeation profiles of the drug nanoemulgel and pure Timur oil. (**B**) Cumulative drug permeation profile of the nanoemulgel and pure Timur oil. An asterisk (*) indicates that the value is significantly different at $p < 0.05$ compared to the other group.

2.3.9. Skin Irritation Study

A skin irritancy test was performed to verify the safety of the gel formulation. The formulation exhibited no erythemal or edematous scores even after 72 h. The average skin irritation score was 0 (no erythema in a rat), which is less than 5, demonstrating that the gel does not irritate Wistar rat skin when applied to it (Figure 13) [20]. Table 6 presents the results of this study. According to Draize et al. (33), a primary irritancy index (PII) value of less than two (2) denotes that the applied formulation is not irritating to human skin. As a result, because the PII for optimised nanoemulgel and optimised nanoemulsion was less than 2, they were considered to be non-irritants. Developing an appropriate nanoemulgel formulation depends heavily on the choice of surfactant. To compare ionic and nonionic surfactants, Tween 80, Tween 20, Span 80, polyethylene glycol 400, and PEG 200 (a nonionic surfactant) were chosen to develop the Tim–Ros–NEG. These nonionic surfactants generate homogeneous, superior droplets with a low critical micelle concentration, which aids in the quick absorption and release of the nanoemulgel. They also have a minimal risk of causing irritation and show low toxicity compared to other materials [19]. Therefore, it was determined that all the excipients included in the formulation were non-irritating and safe for topical usage.

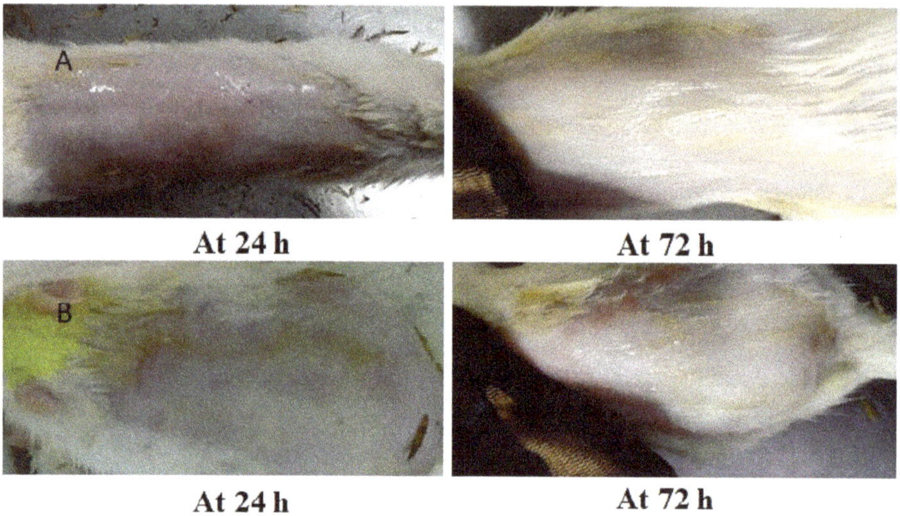

Figure 13. (**A**) Irritation study on nude rat skin conducted by applying the developed nanoemulsion. (**B**) Irritation study on nude rat skin conducted by applying the developed nanoemulgel.

Table 6. Mean erythema scores for various formulations obtained at the end of 24, 48 and 72 h.

Formulations	Erythyma Score (Total)		
	24 h	48 h	72 h
Control (Group I)	0	0	0
Aqueous formalin solution (Group II)	3	4	4
Developed nanoemulgel (1% rosemary oil + 2% Timur oil) (Group III)	0	0	0
Developed nanoemulsion (1% rosemary oil + 2% Timur oil) (Group IV)	0	0	0
Marketed gel (Group V)	1	2	2

2.3.10. Antifungal Activity

The results for the in vitro antifungal efficacy of Smix, Timur oil, lavender oil, rosemary oil, ketoconazole (standard), Timur:rosemary (2:1), the developed nanoemulsion, and a 1.8% nanoemulgel obtained using the cylinder plate method are presented in Figures 14 and 15, and in Table 7. The antifungal activity of each sample or formulation was assessed at a concentration of 10 μg/mL (Table 7). It was found that the developed (Tim–Ros–NEG) formulations exhibited promising antifungal activity against *C. albicans*. It was revealed that the Tim–Ros–NEG significantly suppressed the growth of *C. albicans* compared to the other test preparations.

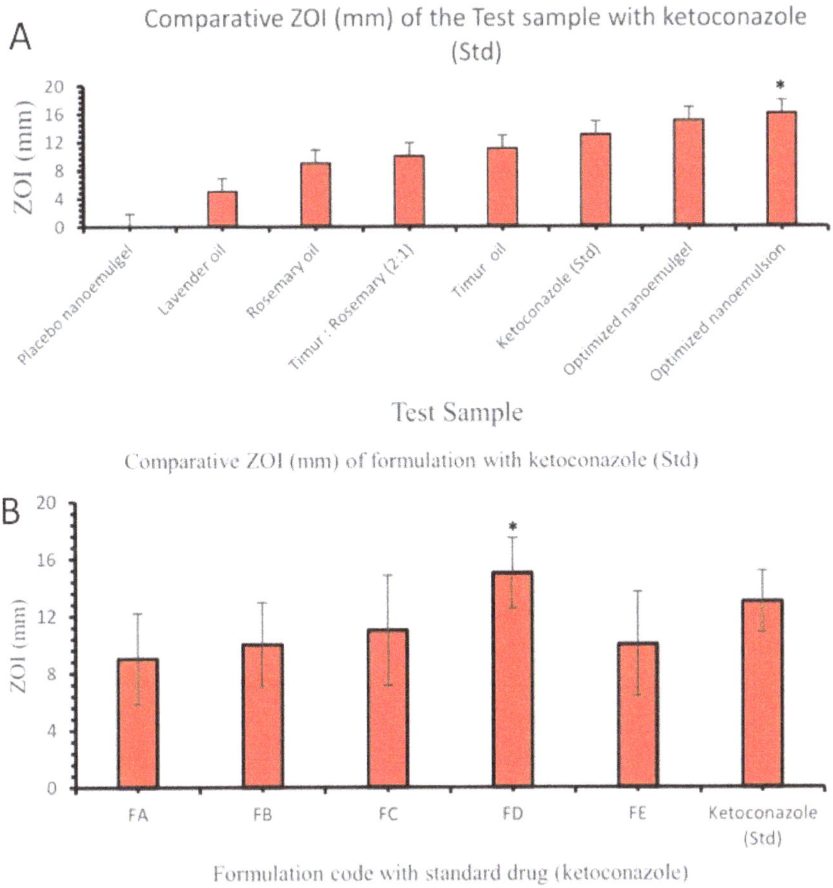

Figure 14. (**A**) Comparison of the ZOIs (mm) of the test samples with the ZOI of ketoconazole (Standard drug). (**B**) Comparison of the ZOI (mm) of the nanoemulgel formulation with that of ketoconazole (Standard drug). An asterisk (*) indicates that the value is significantly different at $p < 0.05$ compared to the other group.

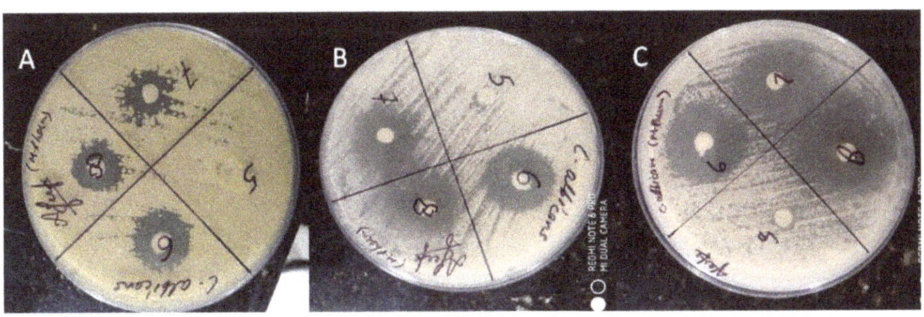

Figure 15. (**A**) ZOIs (mm) of the developed Smix (5), Timur oil (6), lavender oil (7), and rosemary oil (8) against *C. albicans*. (**B**) ZOIs (mm) of the placebo nanoemulgel (5), Timur (2): rosemary (1) (6), Timur oil (7), and developed nanoemulsion (8). (**C**) ZOIs (mm) of the placebo nanoemulsion (5), ketoconole standard (6), Timur (2): rosemary (1) (7), and 1.8% nanoemulgel (8) against *C. albicans*.

Table 7. Assessment of antifungal activity of samples against *C. Albicans*.

S. No.	Disk Code	Formulations/Samples	Concentration (μg/mL)	ZOI (mm)
1.	5 [a]	Smix	10	00 ± 00
2.	6 [a]	Timur oil	10	11 ± 0.7
3.	7 [a]	Lavender oil	10	5 ± 0.9
4.	8 [a]	Rosemary oil	10	9 ± 0.5
5.	5 [b]	Placebo nanoemulgel	10	0 ± 0.0
6.	6 [b]	Timur:rosemary (2:1)	10	10 ± 0.4
7.	7 [b]	Timur oil	10	11 ± 0.7
8.	8 [b]	Developed Nanoemulsion	10	16 ± 1.01 *
9.	5 [c]	Placebo nanoemulsion	10	0 ± 0.0
10.	6 [c]	Ketoconazole (standard)	10	13 ± 0.8
11.	7 [c]	Timur:rosemary (2:1)	10	10 ± 0.4
12.	8 [c]	1.8% nanoemulgel	10	15 ± 2.9

[a] Figure 15A; [b] Figure 15B; [c] Figure 15C. An asterisk (*) indicates that the value is significantly different at $p < 0.05$ compared to the other formulations.

The developed nanoemulsion formulation was selected for the fabrication of the nanoemulgel with different Carbopol Ultrez 21 and Carbopol 940 concentrations (viz., 0.5%, 1%, 1.5%, 1.8%, and 2% gels). The developed Tim–Ros–NEG formulation was notably effective as an antifungal agent.

Timur oil exhibited an inherent antifungal effect reflected by the zone of inhibition for the developed nanoemulsion. Due to the synergistic effects of Timur oil and rosemary oil when delivered together using nanoemulsion technology, which permitted the intensive diffusion of the drug-containing oil globules, Tim–Ros–NEG's action was enhanced. Timur oil, lavender oil, Timur:rosemary (2:1), and the developed nanoemulsion produced results comparable to those for the known antifungal essential oil and the reference drug (ketoconazole). The Tim–Ros nanoemulsion showed statistically significant growth suppression [51]. As a result, it was discovered that the nanoemulsion platform using Timur oil as the oil core had superior efficacy against *C. albicans* species. When both Timur oil and rosemary oil were used in combination, better antifungal activity than ketoconazole against *C. albicans* was observed. Timur oil and rosemary oil were combined in different ratios, and they gave comparable results when compared to the reference drug (ketoconazole). Both Timur oil and rosemary oil have antifungal activity themselves. When we used a 2:1 ratio of Timur:rosemary oil, the oils showed more synergistic activity compared with the other ratios of Timur oil and rosemary oil, as described in Table 7 and Figure 14. It was apparent that Tim–Ros nanoemulsion and Tim–Ros–NEG were active against *C. albicans*, and showed a significantly greater zone of inhibition compared to that resulting from placebo NEG and the marketed formulation ($p < 0.05$).

2.3.11. Histopathological Study

Histopathological analysis was carried out to evaluate the toxicity profile of the developed formulation (Tim–Ros–NEG). The animal groups were divided into four groups: normal control, formalin solution treatment (positive control), blank NEG, and Tim–Ros–NEG. The rat skin was observed in positive control, blank NEG, and Tim–Ros–NEG groups and was compared with the control group. A histopathological examination of skin from the rat's dorsal skin was performed to look for any indications of inflammatory reactions. Pathological alterations, including a thicker, deteriorated epidermis (E), intercellular edema, and inflammatory cell infiltrate were visible in the formalin-treated group (Figure 16). The Tim–Ros–NEG-treated groups did not exhibit aberrant alterations in the treated rat skin tissue compared to controls, other than a slightly thicker epidermis. Overall, the findings showed that the Tim–Ros–NEG was safe for topical administration and within the skin's tolerance limit. No macrophages or lymphocytes were observed on the rat's dorsal skin, indicating that there was no major inflammation in the surrounding tissues (Figure 16). This suggests that Tim–Ros–NEG might be safely administered topically using the formulated nanoemulgel.

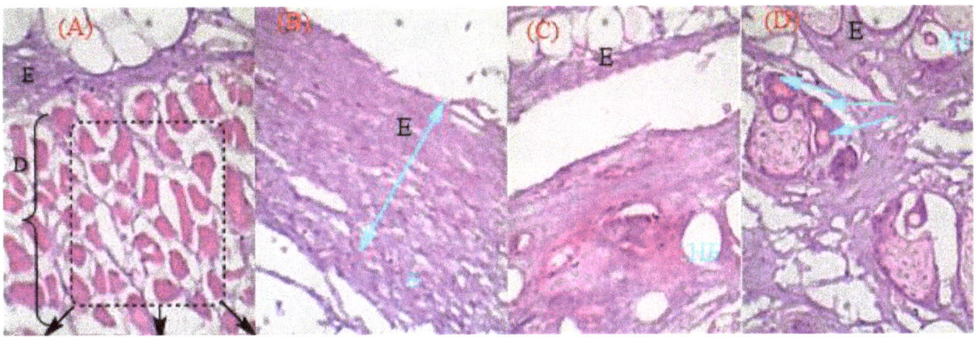

Figure 16. Histological micrographs of hematoxylin-eosin-stained skin section showing the normal epidermis, dermis tissues of (**A**) untreated rat skin, (**B**) formalin-exposed rat skin showed thickened degenerated epidermis E and inflammatory cells infiltrate(+). (**C**) Rat skin treated with conventional oil (Rosemary oil). (**D**) Rat skin treated with nanoemulgel enriched with Timur oil and Rosemary oil. The florescent blue arrow indicates blood vessel. E: Epidermis; D: Dermis; HF: Hair follicle.

3. Conclusions

In the present study, the topical application of a Timur–rosemary oil nanoemulgel (Tim–Ros–NEG) showed antifungal activity against *C. Albicans* count, improving its effect compared to the action of pure Timur oil. Moreover, we observed a 15 ± 2.9 mm ZOI from the Timur oil:rosemary oil (2:1) nanoemulgel in comparison to the pure Timur oil (11 ± 0.7 mm) and standard drug ketoconazole (13 ± 0.8 mm). The Timur oil:rosemary oil (2:1) nanoemulgel exhibited a synergistic effect. We believe that, if we used it in combination with ketoconazole, it would also have a synergistic effect. Therefore, we could reduce the therapeutic dose of ketoconazole, thereby reducing the toxicity caused by ketoconazole.

Natural substances such as Timur oil and rosemary oil have been considered to be suitable options for topical and even oral treatments because synthetic drugs can have undesirable effects and some infections caused by *Candida* are resistant to them. It was possible to create a nanoemulgel by adding Carpobol 940, a hydrogel material, at a concentration of 1.8% to a nanoemulsion containing Timur oil, rosemary oil, Tween 80, and Transcutol P. This method allows for greater penetration through the skin, improving the topical bioavailability and increasing the retention time.

When Timur oil and rosemary oil were added to the mixture, the tested microbial strains were suppressed more effectively due to better penetration of the oil globules from the NEG. The developed formulation was confirmed to be safe when tested on the skin of Wistar rats. The animals treated with Tim–Ros–NEG showed no indications of erythema. Additionally, the histopathological investigation showed no toxicity on the skin of the Wistar rats, indicating that the formulation is safe and effective for topical use. Consequently, the data suggest that the use of Tim–Ros–NEG might be a promising approach for the safe, highly effective, targeted delivery of Timur oil.

4. Materials and Methods

4.1. Materials

Tween 80, Carbopol 940, Carbopol Ultrez 21, acetone, chloroform, propylene glycol, disodium hydrogen phosphate, and Span 80 were obtained from S D Fine Chemicals

Ltd., Mumbai, India. Methanol was purchased from Merck, Mumbai, India. Mueller–Hinton agar (which is produced by Becton, Dickinson, and company, Franklin lakes, NJ, United States) was used in the culture media. Marketed Timur oil was purchased from Moksha lifestyle, New Delhi (India). The rosemary oil and lavender oil were obtained from Pharmacos, Faridabad, Haryana, India. The Indian variety of Timur seeds and Nepali variety of Timur seeds were obtained from Hari Gokul, Khari Baoli, Old Delhi, India, and Khopra, Nepal, respectively. Both Indian and Nepali varieties of Timur seeds, authenticated by Prof. Javed Ahmad, Ex-head Department of Botany, School of Chemical & Life Sciences, Jamia Hamdard, New Delhi, has been deposited in the Herbarium of the herbal garden, Jamia Hamdard, New Delhi.

4.2. *Extraction of Timur Oil from Timur Seeds*

A total of 200 g of Timur seeds were weighed and crushed using a mortar and pestle. Then, the crushed Timur seeds were transferred to the clean round-bottomed flask of the Clevenger apparatus setup. A hydro-distillation process lasting 6–7 h was conducted. After completion of the distillation process, the Timur oil was collected. The collected Timur oil was dried using sodium sulfate to absorb the excess water. Then, it was filtered and the obtained oil was stored at 2 °C to 7 °C [52].

4.2.1. Identification of Timur Oil

Fourier Transform Infrared (FTIR) Spectroscopy

A Perkin–Elmer 591B spectrophotometer was used to determine the FTIR spectra of the Timur oil and Rosemary oil with KBr pellets or Nujol films, and scanning was performed at wavelengths of 4000 to 400 (cm^{-1}) to obtain the characteristic spectra.

HPTLC Analysis for Qualitative Estimation of linalool in Timur Oil (Indian Variety and Nepali Variety)

The separate applications of each oil were made on 5 cm × 10 cm chromatographic precoated silica gel plates (Merck, TLC grade) that served as the stationary phase. Toluene and ethyl acetate (95:5 v/v) were used as the mobile phase in a twin-trough glass chamber where the TLC plates were prepared. The plates were taken off after the solvent front had moved 15 cm away from the initial extract site, and they were then left to dry. When the spots on the produced plates had dried, they were examined under visible (white), short UV (254 nm), and long UV (366 nm) light.

GC Analysis of Timur Oil, Marketed Timur Oil, and Nepali Timur Oil for Quantification of Linalool Content

The oils were subjected to GC analysis on a Shimadzu 15A gas chromatograph with a split/splitless injector (250 °C). The DB-5 capillary column (30 m 0.25 mm, film thickness 0.32 μm) was used, and nitrogen was utilized as the carrier gas (1 mL min^{-1}). After maintaining a temperature of 60 °C for three minutes, the column was heated to 220 °C at a rate of 5 °C min^{-1} and maintained at this temperature for five minutes. The relative percentage quantity was estimated using a Shimadzu C-R4A Chromatopac from the peak region [53].

4.3. *Screening of Components for Preparation of Nanoemulsion*

The miscibility of the chosen oil served as the criterion for choosing the surfactant and co-surfactant. To make the stable nanoemulsion, the combination of a high HLB value with a low HLB value at the optimum temperature was used. Miscibility studies with oil were carried out by mixing Smix in a 5:5 ratio and vortexing for 5 min. Then, the mixtures were kept at room temperature for 24 h. After 24 h, an evaluation was performed based on the color change and phase separation [54].

4.4. Construction of Pseudoternary Phase Diagram

After the selection of the surfactant and co-surfactant (Smix) based on the miscibility with the oil, Smix was mixed in different volume ratios (1:0, 1:1, 1:2, 1:3, 2:1, 3:1, and 4:1). Based on the titration chart, different ratios of Smix were developed. Based on the points obtained from the titration chart, the phase diagram was made. The phase diagram was studied in detail for the formulation of the nanoemulsion [54].

4.5. Method for Preparation of the Phase Diagram (Aqueous Titration Method)

For the preparation of the phase diagram, the oil and different volume ratios of Smix were mixed. The ratios ranged from 1:9 to 9:1, and these were mixed to make 16 different combinations, which were 1:9, 1:8, 1:7, 1:6, 1:5, 1:4, 1:3.5, 1:3, 3:7, 1:2, 4:6, 5:5, 6:4, 7:3, 8:2, and 9:1. The pseudoternary diagrams were developed based on the aqueous titration method. Each oil and Smix underwent a slow titration with the aqueous phase while being moderately stirred. Water was added in varying amounts, ranging from 5 to 95% of the total volume. The formulation was visually observed after the addition of the water to the volume [19].

4.6. Preparation of the Nanoemulsion

Using Timur–rosemary oil as the oil phase and distilled water as the aqueous phase, the nanoemulsion was formulated. Under continuous stirring at high speed using a vortex mixer at room temperature, rosemary oil was dissolved into the oil phase at room temperature. After the drug had been fully dissolved in the oil phase, Smix (Tween 80 and Transcutol P) was introduced. Under continuous stirring, the water was added drop by drop into the oil phase. The mixture was then subjected to gentle agitation using the vortex for 3 min. The ideal formulation was selected based on the polydispersity index (PDI) and droplet size. Before evaluating each formulation's droplet size, polydispersity index (PDI), and physical characteristics, each formulation was self-emulsified in distilled water with gentle agitation [55,56].

Physicochemical Characterization of Nanoemulsion

The developed formulations were characterized for their physicochemical parameters, droplet sizes, and PDIs. Equipment from Malvern Instruments was utilized to determine the particle size and PDI. The Timur–rosemary oil was prepared as a nanoemulsion using a titration process [57].

4.7. Fabrication of Timur Oil Nanoemulgel Formulation

The best formulation was chosen to be the nanoemulsion with the greatest concentration of Timur and rosemary oil, the smallest particle size, and the lowest PDI. The developed nanoemulsion formulation was selected for the fabrication in the nanoemulgel with different Carbopol Ultrez 21 and Carbopol 940 concentrations (viz., 0.5%, 1%, 1.5%, 1.8%, and 2% gels). First, Carbopol hydrogels were prepared using Carbopol 940 and Carbopol Ultrez 21 as thickening agents by dispersing Carbopol in purified water and left-over night for swelling, then the pH of the hydrogel was neutralized using 0.05% of triethanolamine (TEA), and DMDM was added as a preservative. Then, the hydrogel matrix was mixed with the developed nanoemulsion (5 mL) with 1% rosemary oil and 2% Timur oil at 100 rpm until the nanoemulgel was formed. In order to achieve uniformity, each formulation was thoroughly blended [58].

4.7.1. Physical Characterization of Timur Oil Nanoemulgel Formulation

We visually assessed a wide range of physical characteristics, including consistency, spreadability, homogeneity, phase separation, and visual appearance, when developing the nanoemulgel. Using a pH meter, the pH values were determined (Mettler Toledo Inc., Columbus, OH, USA).

Zeta Potential

Equipment from Malvern Instruments was used to determine the zeta potential to predict the dispersion stability and surface charge of the particles. The zeta potentials were measured in triplicate [59–63].

Transmission Electron Microscopy (TEM)

Transmission electron microscopy (TEM) is a high-resolution technology that can be used to investigate specimens at the nanoscale. The developed formulation was studied for its surface morphology with the help of TEM, which was operated at 20–120 kV (Thermo Scientific Talos L120C G2 (S)TEM Microscope, Waltham, MA, USA) and carried out at Jamia Hamdard, New Delhi, India, enabling point-to-point resolution. A droplet of the sample was applied to a 300 mesh grid of copper-coated carbon that had been adequately saturated with water (1:1000), dyed with phosphotungstic acid (2% *w*/*v*), and allowed to air dry for about one minute before evaluation [54].

Mechanical Properties of NEG

To determine the gel's mechanical attributes, including its hardness, cohesiveness, adhesiveness, gumminess, chewiness, and elasticity, a texture study was conducted. The texture analysis was performed using a texture analyzer (TA-XT Plus, Stable Microsystems, Godalming, UK). Using a probe that dips into the formulation at a certain speed and with a predetermined amount of force, data were collected and analyzed. A beaker free of air bubbles was filled with 100 g of the gel. A disc of Perspex (diameter 40 mm) was attached to the handle, and speeds of 1.0 mm/s before the test, 2.0 mm/s during the test, and 10.0 mm/s after the test were set. A 25 mm diameter cylinder probe was used for the penetration test, with a 5 mm penetration depth and a test speed of 2 mm/s in compression mode. Data computation was performed in Texture Exponent Software. All the tests were performed at room temperature using a triple (mean $\pm$ SD) set [62].

4.7.2. In Vitro Release Study

The dialysis bag (Sigma, St. Louis, MO, USA) was kept in running water for 3–4 h for the removal of glycerine and then treated with a 0.3% *w*/*v* sodium sulfide solution in water at 80 °C for 1 min for the removal of sulfur compounds. It was then washed with hot water at 60 °C for 2 min. The procured dialysis bag was acidified with 0.2% *v*/*v* H_2SO_4 in distilled water. It was then rinsed with hot water to remove the acid and stored in the dissolution medium in the refrigerator to keep the pores open [63]. The essential oil release from antifungal nanoemulgel was determined in phosphate buffer at pH 5.5 using the dialysis bag method [54,64].

A freshly prepared formulation (0.5 gm of gel) was put in the dialysis bag (MWCO 12KD, Sigma), pre-soaked in double-distilled water for 12 h before use, and sealed. The dialysis bags were put in a beaker containing 100 mL of phosphate buffer at pH 5.5, which acted as a receptor compartment; it was maintained at 37 °C and stirred at a speed of 150 rpm. At an appropriate time interval (0.5, 1, 2, 4, 6, up to 24 h), a 3 mL aliquot of the dissolution medium was withdrawn, and the same volume of fresh dissolution medium was then added. A good sink condition was maintained throughout the test. The amount of the drug was evaluated using a UV spectrophotometer at 272 nm for the developed formulation [63].

$$\%\text{Cumulative drug release} = \text{concentration}(\mu g/mL) \times \text{volume of dissolution media}(mL) \times \text{Dilution Factor} \times 100/\text{Initial amount of drug}.$$

Release Kinetics

The data from the drug release studies were fitted with different kinetic models, such as the zero-order model, first-order model, Higuchi matrix model, and Korsmeyer–Peppas model. The correlation coefficient (R^2) for each model was calculated using the following formula. The model resulting in the R^2 closest to 1 was chosen as the best fit for drug release [65].

$$\text{Zero order model } Q = kt \text{ First order model } \log Q = kt/2.303 \text{ Higuchi model } Q = k\sqrt{t} \text{ Korsmeyer-Peppas model} = kt^n$$

4.7.3. Antifungal Activity of Formulations

The antifungal activity of the rosemary oil, Timur oil, lavender oil (standard), ketoconazole (standard), developed nanoemulsion, and developed nanoemulgel with regard to *C. albicans* was determined following the method demonstrated by Kadimi et al. [66]. A McFarland standard was used for preparing the suspension of the organism. The organism was first grown on sabouraud dextrose agar medium (pH 6.2). The medium was prepared and sterilized in an autoclave for 20 min at 121 °C. A sterile borer was used to create wells that were 6 mm in diameter. After 24 h of incubation, the inoculum was firmly swept over the agar plate using a sterile cotton swab to make uniform culture lawns. Quantities of 0.1 gm of different gels (0.5, 1%, 1.5%, 1.8%, and 2%) with different concentrations of Timur oil and rosemary oil were poured into wells with the help of a sterile spatula and incubated for 48 h, and these plates were assessed for clear zones around the wells. The zones of inhibition were assessed by measuring the diameters [67].

Calculation

Based on the ZOIs of the control and test samples, the % inhibition was calculated using the formula below:

$$\%\ \text{inhibition} = \text{AIC} - \text{AIT}/\text{AIC} \times 100$$

where AIC is the area of inhibition for the control and AIT is the area of inhibition for the extracts.

4.7.4. Ex Vivo Permeation Study

Male Wistar rats (weighing between 150 and 200 g) were acquired from Institutional Animal Ethical Committee constituted by Jamia Hamdard, under protocol number 1575, which was utilized to prepare the skin for permeation testing. The rats were euthanized by CO_2 inhalation. The skin was removed from the area of the dorsal. With the use of "veet cream", the hairs were plucked. Using isopropyl alcohol and a cotton swab, the fat was eliminated [68]. The skin was then cleaned with distilled water and kept at −21 °C for scientific investigation. The skin penetration study was conducted on the dorsal skin of a rat used as the model skin for the Franz diffusion cell. The sample skin was positioned between the donor compartment and the receptor compartment, with the stratum corneum towards the donor compartment. The sample formulation served as the donor medium, while phosphate-buffered saline served as the receptor (pH 5.5). After 15 min, 30 min, 1 h, 2 h, 4 h, 6 h, and 24 h, the receptor medium (1 mL) was removed and replaced with a new medium. After the dilutions, a 0.45 µm membrane filter was used to filter the solutions, and a UV spectrophotometer was used to measure the quantity of drug in the receptor medium [69].

Data Analysis

After the study, the flux was plotted against time on a graph. The quantity of the drug that permeated was determined using the formula:

$$\text{Amount of drug permeated } (\mu g/cm^2) = \text{concentration } (\mu g/mL) \times \text{volume of dissolution media}/\text{Area } (cm^2)$$

The volume of dissolution media was equal to 35 mL, and the area of the diffusion cell was equal to 1.966 cm^2.

The flux was equal to the slope of the steady-state portion of the plot between the amount of drug permeated per cm^2 vs. time in $\mu g/cm^2/h$.

The permeability coefficient (Kp) was equal to flux/initial drug concentration (C_0).

4.7.5. Drug Deposition Study

The quantity of drug held in the skin sample used in permeation testing was used to measure the capacity of the globules to help keep the drug inside the skin layers. The skin placed on the diffusion cell was taken off once the permeation experiment was finished. In order to remove any adherent product, the skin was cleansed with cotton soaked in ordinary saline solution and wiped with tissue paper. In order to extract the medication, the skin sample was then homogenized with 10 mL of phosphate buffer at pH 5.5. The resulting homogenate solution was filtered through a membrane filter (0.45 μm), and its drug content was measured using a UV spectrophotometer at 272 nm. All the readings were collected three times for accuracy [70].

4.7.6. Confocal Laser Scanning Microscopy (CLSM)

For confocal microscopy, gels were applied to the dorsal skin of the rats. Formulations containing the vesicles loaded with the fluorescent dye rhodamine were prepared by adding the dye (as a marker) to the nanoemulgel. Confocal pictures were acquired as XY-planes to show the fluorescence distribution (parallel to the plane of the skin surface). The brightest fluorescence with a morphology of stratum corneum surface resembles the skin surface in the imaging plane. The process of creating an XZ-section involves drawing a horizontal line across the area of interest in the Z = 0 μm XY-plane. This line is then "optically sliced" across the digital image data of the subsequent XY sections, yielding (XA-planar) optical cross-sections. All the photos were produced using an identical optical aperture, lens, and scan speed, and were averages of the triplet scans. The CLSM was used to examine the extent and mechanism of the rhodamine-loaded nanoemulgel's skin permeability [71,72].

4.7.7. Skin Irritation Studies

This study was carried out on healthy Wistar rats weighing 150–200 gm. By performing a Draize patch test, the irritability potential of the antifungal nanoemulgel was compared with that of the commercialized antifungal cream [73–75]. The protocol (1575) was approved by the Institutional Animal Ethical Committee constituted by Jamia Hamdard for such a purpose. Throughout the whole experimental process, animals were cared for and handled in accordance with CPCSEA criteria. Experimental Wistar rats were used to assess the irritability of the developed nanoemulgel (Tim–Ros–NEG) [76]. With care taken not to harm the skin's top layer, the hair from the dorsal side of the acclimatized animals was removed 24 h before the experiment. The animals were subsequently split into five groups (n = 3), with the first group (Group I; the control group) receiving no treatment (no drug treatment) and Group II received 0.8% (v/v) aqueous formalin solution as a standard irritant. Group III, Group IV, and Group V received the developed nanoemulgel (1% rosemary oil + 2% Timur oil), developed nanoemulsion (1% rosemary oil + 2% Timur oil), and marketed antifungal preparation, respectively. In the drug-free treatment group, an equal amount of blank NEG was applied to the cleansed skin across a 1 cm^2 area instead of the formulations. At 24, 48 and 72 h, the skin of the test animals was examined for any dermal responses, namely erythema or edema [73].

4.7.8. Histopathology of Skin

Histopathology studies were performed for the evaluation of the rat skin for any damage resulting from the application of the various drugs on the skin surface. The animals were sacrificed after one week, and the skin was removed from the interscapular region using scissors and forceps. Samples were prepared and sectioned by utilizing a microtome. Furthermore, hematoxylin and eosin dye were used to stain the sectioned samples. The specimens were then examined using a high-power microscope to assess their integrity and perform stratum corneum and epidermal examinations. The results were compared with those of the rats in the control group [77].

4.7.9. Statistical Assessment

The data were acquired in triplicate for each of the established experiments, and the values were presented as mean ± S.D. Statistical significance was considered when the p-value was less than 0.05.

Author Contributions: Conceptualization, A.N. and K.K.; methodology, A.N.; validation, K.K. and M.S.M.A.; formal analysis, S.J. and T.W.S.; investigation, A.N.; resources, K.K.; writing—original draft preparation, A.N.; writing—review and editing, S.J.; supervision, K.K. All authors have read and agreed to the published version of the manuscript.

Funding: This research received no external funding.

Institutional Review Board Statement: This study was conducted after approval from CAHF, Jamia Hamdard, under protocol number 1575.

Informed Consent Statement: Not applicable.

Data Availability Statement: The data presented in this study are available on request from the corresponding author.

Acknowledgments: The authors are thankful to Jamia Millia Islamia, New Delhi, India, for providing the FTIR facilities. The authors are also thankful to AIIMS, New Delhi, India, for providing the zeta potential and Zetasizer facilities. We thank Jamia Hamdard, New Delhi, for providing the TEM facilities.

Conflicts of Interest: The authors declare no conflict of interest.

References

1. Garg, A.; Sharma, G.S.; Goyal, A.K.; Ghosh, G.; Si, S.C.; Rath, G. Recent Advances in Topical Carriers of Anti-Fungal Agents. *Heliyon* **2020**, *6*, e04663. [CrossRef] [PubMed]
2. Kyle, A.A.; Dahl, M.V. Topical Therapy for Fungal Infections. *Am. J. Clin. Dermatol.* **2004**, *5*, 443–451. [CrossRef] [PubMed]
3. Ullah, N.; Amin, A.; Farid, A.; Selim, S.; Rashid, S.A.; Aziz, M.I.; Kamran, S.H.; Khan, M.A.; Rahim Khan, N.; Mashal, S.; et al. Development and Evaluation of Essential Oil-Based Nanoemulgel Formulation for the Treatment of Oral Bacterial Infections. *Gels* **2023**, *9*, 252. [CrossRef] [PubMed]
4. Ojha, P. Antifungal Property of Essential Oil Extracted from Zanthoxylum Armatum (Timur). *J. Nutr. Health Food Eng.* **2015**, *3*, 266–270. [CrossRef]
5. Das, N.G.; Nath, D.R.; Baruah, I.; Talukdar, P.K.; Das, S.C. Field Evaluation of Herbal Mosquito Repellents. *J. Commun. Dis.* **1999**, *31*, 241–245.
6. Tiwary, M.; Naik, S.N.; Tewary, D.K.; Mittal, P.K.; Yadav, S. Chemical Composition and Larvicidal Activities of the Essential Oil of Zanthoxylum Armatum DC (Rutaceae) against Three Mosquito Vectors. *J. Vector Borne Dis.* **2007**, *44*, 198–204.
7. Soares, B.V.; Morais, S.M.; Dos Santos Fontenelle, R.O.; Queiroz, V.A.; Vila-Nova, N.S.; Pereira, C.M.C.; Brito, E.S.; Neto, M.A.S.; Brito, E.H.S.; Cavalcante, C.S.P.; et al. Antifungal Activity, Toxicity and Chemical Composition of the Essential Oil of *Coriandrum sativum* L. Fruits. *Molecules* **2012**, *17*, 8439–8448. [CrossRef]
8. Prakash, B.; Singh, P.; Mishra, P.K.; Dubey, N.K. Safety Assessment of Zanthoxylum Alatum Roxb. Essential Oil, Its Antifungal, Antiaflatoxin, Antioxidant Activity and Efficacy as Antimicrobial in Preservation of *Piper nigrum* L. Fruits. *Int. J. Food Microbiol.* **2012**, *153*, 183–191. [CrossRef]
9. Guleria, S.; Tiku, A.K.; Koul, A.; Gupta, S.; Singh, G.; Razdan, V.K. Antioxidant and Antimicrobial Properties of the Essential Oil and Extracts of Zanthoxylum Alatum Grown in North-Western Himalaya. *Sci. World J.* **2013**, *2013*, 790580. [CrossRef]
10. Jiang, Y.; Wu, N.; Fu, Y.J.; Wang, W.; Luo, M.; Zhao, C.J.; Zu, Y.G.; Liu, X.L. Chemical Composition and Antimicrobial Activity of the Essential Oil of Rosemary. *Environ. Toxicol. Pharmacol.* **2011**, *32*, 63–68. [CrossRef]
11. Cuvelier, M.; Richard, H.; Berset, C. Antioxidative Activity and Phenolic Composition of Pilot-Plant and Commercial Extracts of Sage and Rosemary. *J. Am. Oil Chem. Soc.* **1996**, *73*, 645–652. [CrossRef]
12. Erkan, N.; Ayranci, G.; Ayranci, E. Antioxidant Activities of Rosemary (*Rosmarinus officinalis* L.) Extract, Blackseed (*Nigella sativa* L.) Essential Oil, Carnosic Acid, Rosmarinic Acid and Sesamol. *Food Chem.* **2008**, *110*, 76–82. [CrossRef]

13. Xavier, C.P.R.; Lima, C.F.; Fernandes-Ferreira, M.; Cristina, M.; Pereira-Wilson, C. Salvia Fruticosa, Salvia Officinalis and Rosmarinic Acid Induce Apoptosis and Inhibit Proliferation of Human Colorectal Cell Lines: The Role in MAPK/ERK Pathway *Nutr. Cancer Int. J.* **2009**, *61*, 564–571. [CrossRef]
14. Almeida, R.; Maggi, F. Antioxidant and Antimicrobial Activities of Rosemary Extracts Linked to Their Polyphenol Composition Related Papers. *Free. Radic. Res.* **2006**, *40*, 223–231. [CrossRef]
15. Radtke, O.A.; Yeap, L.; Lu, Y.; Kiderlen, A.F. Evaluation of Sage Phenolics for Their Antileishmanial Activity and Modulatory Effects on Interleukin-6, Interferon and Tumour Necrosis Factor- α -Release in RAW 264.7 Cells. *Z. Naturforsch. C* **2003**, *58*, 395–400. [CrossRef]
16. Nejati, H.; Farahpour, M.R.; Nagadehi, M.N. Topical Rosemary Officinalis Essential Oil Improves Wound Healing against Disseminated Candida Albicans Infection in Rat Model. *Comp. Clin. Path.* **2015**, *24*, 1377–1383. [CrossRef]
17. Eid, A.M.; Issa, L.; Al-Kharouf, O.; Jaber, R.; Hreash, F. Development of Coriandrum Sativum Oil Nanoemulgel and Evaluation of Its Antimicrobial and Anticancer Activity. *BioMed Res. Int.* **2021**, *2021*, 5247816. [CrossRef]
18. Krishnamoorthy, R.; Gassem, M.A.; Athinarayanan, J.; Periyasamy, V.S.; Prasad, S.; Alshatwi, A.A. Antifungal Activity of Nanoemulsion from Cleome Viscosa Essential Oil against Food-Borne Pathogenic Candida Albicans. *Saudi J. Biol. Sci.* **2021**, *28*, 286–293. [CrossRef]
19. Shafiq-un-Nabi, S.; Shakeel, F.; Talegaonkar, S.; Ali, J.; Baboota, S.; Ahuja, A.; Khar, R.K.; Ali, M. Formulation Development and Optimization Using Nanoemulsion Technique: A Technical Mote. *AAPS PharmSciTech* **2007**, *8*, 1–6. [CrossRef]
20. Algahtani, M.S.; Ahmad, M.Z.; Ahmad, J. Nanoemulgel for Improved Topical Delivery of Retinyl Palmitate: Formulation Design and Stability Evaluation. *Nanomaterials* **2020**, *10*, 848. [CrossRef]
21. Buya, A.B.; Ucakar, B.; Beloqui, A.; Memvanga, P.B.; Préat, V. Design and Evaluation of Self-Nanoemulsifying Drug Delivery Systems (SNEDDSs) for Senicapoc. *Int. J. Pharm.* **2020**, *580*, 119180. [CrossRef] [PubMed]
22. Rajput, R.L.; Narkhede, J.S.; Mujumdar, A.; Naik, J.B. Synthesis and Evaluation of Luliconazole Loaded Biodegradable Nanogels Prepared by PH-Responsive Poly (Acrylic Acid) Grafted Sodium Carboxymethyl Cellulose Using Amine Based Cross Linker for Topical Targeting: In Vitro and Ex Vivo Assessment. *Polym. Technol. Mater.* **2020**, *59*, 1654–1666. [CrossRef]
23. Kaur, R.; Sharma, N.; Tikoo, K.; Sinha, V.R. Development of Mirtazapine Loaded Solid Lipid Nanoparticles for Topical Delivery: Optimization, Characterization and Cytotoxicity Evaluation. *Int. J. Pharm.* **2020**, *586*, 119439. [CrossRef] [PubMed]
24. Aggarwal, G.; Dhawan, B.; Harikumar, S. Enhanced Transdermal Permeability of Piroxicam through Novel Nanoemulgel Formulation. *Int. J. Pharm. Investig.* **2014**, *4*, 65. [CrossRef] [PubMed]
25. Mao, Y.; Chen, X.; Xu, B.; Shen, Y.; Ye, Z.; Chaurasiya, B.; Liu, L.; Li, Y.; Xing, X.; Chen, D. Eprinomectin Nanoemulgel for Transdermal Delivery against Endoparasites and Ectoparasites: Preparation, in Vitro and in Vivo Evaluation. *Drug Deliv.* **2019**, *26*, 1104–1114. [CrossRef]
26. Aman, R.M.; Hashim, I.I.A.; Meshali, M.M. Novel Clove Essential Oil Nanoemulgel Tailored by Taguchi's Model and Scaffold-Based Nanofibers: Phytopharmaceuticals with Promising Potential as Cyclooxygenase-2 Inhibitors in External Inflammation. *J. Nanomed.* **2020**, *15*, 2171–2195. [CrossRef]
27. Goel, S.; Sachdeva, M.; Agarwal, V. Nanosuspension Technology: Recent Patents on Drug Delivery and Their Characterizations. *Recent Pat. Drug Deliv. Formul.* **2019**, *13*, 91–104. [CrossRef]
28. Goel, S.; Agarwal, V.; Sachdeva, M. Development and Characterization of Oral Nanosuspension Using Esomeprazole Magnesium Trihydrate. *Nanosci. Nanotechnol.-Asia* **2019**, *10*, 909–917. [CrossRef]
29. Barhoum, A.; García-Betancourt, M.L.; Rahier, H.; Van Assche, G. Physicochemical Characterization of Nanomaterials: Polymorph, Composition, Wettability, and Thermal Stability. In *Emerging Applications of Nanoparticles and Architectural Nanostructures: Current Prospects and Future Trends*; Elsevier: Amsterdam, The Netherlands, 2018; pp. 255–278, ISBN 9780128135167.
30. Talegaonkar, S.; Dutta, D.; Chaudhary, N.; Goel, S.; Singh, R. Metal Nanoparticles as a Surrogate Carrier in Drug Delivery and Diagnostics. In *Micro- and Nanotechnologies-Based Product Development*; CRC Press: Boca Raton, FL, USA, 2021; pp. 139–161. [CrossRef]
31. Goel, S.; Agarwal, V.; Sachdeva, M. Nanocrystals: A Strategic Approach for the Development of Futuristic Nano-Formulations and Their Patenting Applicability. *Curr. Nanosci.* **2021**, *18*, 291–303. [CrossRef]
32. Yaprak Karavana, S.; Güneri, P.; Ertan, G. Benzydamine Hydrochloride Buccal Bioadhesive Gels Designed for Oral Ulcers: Preparation, Rheological, Textural, Mucoadhesive and Release Properties. *Pharm. Dev. Technol.* **2009**, *14*, 623–631. [CrossRef]
33. Cevher, E.; Taha, M.A.M.; Orlu, M.; Araman, A. Evaluation of Mechanical and Mucoadhesive Properties of Clomiphene Citrate Gel Formulations Containing Carbomers and Their Thiolated Derivatives. *Drug Deliv.* **2008**, *15*, 57–67. [CrossRef]
34. Osmałek, T.; Milanowski, B.; Froelich, A.; Górska, S.; Białas, W.; Szybowicz, M.; Kapela, M. Novel Organogels for Topical Delivery of Naproxen: Design, Physicochemical Characteristics and in Vitro Drug Permeation. *Pharm. Dev. Technol.* **2017**, *22*, 521–536 [CrossRef]
35. Mohapatra, S.; Mirza, M.A.; Ahmad, S.; Farooq, U.; Ansari, M.J.; Kohli, K.; Iqbal, Z. Quality by Design Assisted Optimization and Risk Assessment of Black Cohosh Loaded Ethosomal Gel for Menopause: Investigating Different Formulation and Process Variables. *Pharmaceutics* **2023**, *15*, 465. [CrossRef]
36. Hasan, N.; Imran, M.; Nadeem, M.; Jain, D.; Haider, K.; Moshahid Alam Rizvi, M.; Sheikh, A.; Kesharwani, P.; Kumar jain, G.; Jalees Ahmad, F. Formulation and Development of Novel Lipid-Based Combinatorial Advanced Nanoformulation for Effective Treatment of Non-Melanoma Skin Cancer. *Int. J. Pharm.* **2023**, *632*, 122580. [CrossRef]

37. Deng, Y.; Liu, Z.; Geng, Y. Anti-Allergic Effect of Artemisia Extract in Rats. *Exp. Ther. Med.* **2016**, *12*, 1130–1134. [CrossRef]
38. Islam, N.; Irfan, M.; Zahoor, A.F.; Iqbal, M.S.; Syed, H.K.; Khan, I.U.; Rasul, A.; Khan, S.U.D.; Alqahtani, A.M.; Ikram, M.; et al. Improved Bioavailability of Ebastine through Development of Transfersomal Oral Films. *Pharmaceutics* **2021**, *13*, 1315. [CrossRef]
39. Ojha, B.; Jain, V.K.; Gupta, S.; Talegaonkar, S.; Jain, K. Nanoemulgel: A Promising Novel Formulation for Treatment of Skin Ailments. *Polym. Bull.* **2022**, *79*, 4441–4465. [CrossRef]
40. Sengupta, P.; Chatterjee, B. Potential and Future Scope of Nanoemulgel Formulation for Topical Delivery of Lipophilic Drugs. *Int. J. Pharm.* **2017**, *526*, 353–365. [CrossRef]
41. Ma, Q.; Zhang, J.; Lu, B.; Lin, H.; Sarkar, R.; Wu, T.; Li, X. Nanoemulgel for Improved Topical Delivery of Desonide: Formulation Design and Characterization. *AAPS PharmSciTech* **2021**, *22*, 163. [CrossRef]
42. Cecchini, M.E.; Paoloni, C.; Campra, N.; Picco, N.; Grosso, M.C.; Soriano Perez, M.L.; Alustiza, F.; Cariddi, N.; Bellingeri, R. Nanoemulsion of Minthostachys Verticillata Essential Oil. In-Vitro Evaluation of Its Antibacterial Activity. *Heliyon* **2021**, *7*, e05896. [CrossRef]
43. Yeo, E.; Yew Chieng, C.J.; Choudhury, H.; Pandey, M.; Gorain, B. Tocotrienols-Rich Naringenin Nanoemulgel for the Management of Diabetic Wound: Fabrication, Characterization and Comparative in Vitro Evaluations. *Curr. Res. Pharmacol. Drug Discov.* **2021**, *2*, 100019. [CrossRef] [PubMed]
44. Taleuzzaman, M.; Sartaj, A.; Kumar Gupta, D.; Gilani, S.J.; Mirza, M.A. Phytosomal Gel of Manjistha Extract (MJE) Formulated and Optimized with Central Composite Design of Quality by Design (QbD). *J. Dispers. Sci. Technol.* **2023**, *44*, 236–244. [CrossRef]
45. Gao, M.; Shen, X.; Mao, S. Factors Influencing Drug Deposition in the Nasal Cavity upon Delivery via Nasal Sprays. *J. Pharm. Investig.* **2020**, *50*, 251–259. [CrossRef]
46. Agarwal, V.; Bajpai, M. Stability Issues Related to Nanosuspensions: A Review: Ingenta Connect. *Pharm. Nanotech.* **2013**, *1*, 85–92. [CrossRef]
47. Godwin, D.A.; Michniak, B.B.; Creek, K.E. Evaluation of Transdermal Penetration Enhancers Using a Novel Skin Alternative. *J. Pharm. Sci.* **1997**, *86*, 1001–1005. [CrossRef]
48. Shin, S.C.; Choi, J.S. Enhanced Efficacy of Triprolidine by Transdermal Application of the EVA Matrix System in Rabbits and Rats. *Eur. J. Pharm. Biopharm.* **2005**, *61*, 14–19. [CrossRef]
49. Datta, D.; Panchal, D.S.; Venuganti, V.V.K. Transdermal Delivery of Vancomycin Hydrochloride: Influence of Chemical and Physical Permeation Enhancers. *Int. J. Pharm.* **2021**, *602*, 120663. [CrossRef]
50. Zhang, Y.; Lane, M.E.; Moore, D.J. An Investigation of the Influence of PEG 400 and PEG-6-Caprylic/Capric Glycerides on Dermal Delivery of Niacinamide. *Polymers* **2020**, *12*, 2907. [CrossRef]
51. Sampathi, S.; Mankala, S.K.; Wankar, J.; Dodoala, S. Nanoemulsion Based Hydrogels of Itraconazole for Transdermal Drug Delivery. *J. Sci. Ind. Res.* **2015**, *74*, 88–92.
52. Al Mentaji, H.N. *Official Methods of Analysis of AOAC International*, 18th ed.; Horwitz, D.W.L., Latimer, D.G.W., Eds.; AOAC International: Rockville, MD, USA, 2005; ISBN 0935584773.
53. Sourmaghi, M.H.S.; Kiaee, G.; Golfakhrabadi, F.; Jamalifar, H.; Khanavi, M. Comparison of Essential Oil Composition and Antimicrobial Activity of *Coriandrum sativum* L. Extracted by Hydrodistillation and Microwave-Assisted Hydrodistillation. *J. Food Sci. Technol.* **2015**, *52*, 2452–2457. [CrossRef]
54. Shafiq, S.; Shakeel, F.; Talegaonkar, S.; Ahmad, F.J.; Khar, R.K.; Ali, M. Development and Bioavailability Assessment of Ramipril Nanoemulsion Formulation. *Eur. J. Pharm. Biopharm.* **2007**, *66*, 227–243. [CrossRef]
55. Eid, A.M.; Jaradat, N.A.; Al-Masri, M.; Issa, L.; Zubidat, F.; Asrawi, H.; Ahmad, S. Development and Antimicrobial Evaluation of Eruca Sativa Oil Nanoemulgel with Determination of the Oil Antioxidant, Sun Protection Factor and Elastase Inhibition. *Curr. Pharm. Biotechnol.* **2020**, *21*, 244–255. [CrossRef]
56. Zakir, F.; Ahmad, A.; Farooq, U.; Mirza, M.A.; Tripathi, A.; Singh, D.; Shakeel, F.; Mohapatra, S.; Ahmad, F.J.; Kohli, K. Design and Development of a Commercially Viable in Situ Nanoemulgel for the Treatment of Postmenopausal Osteoporosis. *Nanomedicine* **2020**, *15*, 1167–1187. [CrossRef]
57. Eid, A.M.; Istateyeh, I.; Salhi, N.; Istateyeh, T. Antibacterial Activity of Fusidic Acid and Sodium Fusidate Nanoparticles Incorporated in Pine Oil Nanoemulgel. *Int. J. Nanomed.* **2019**, *14*, 9411–9421. [CrossRef]
58. Dordević, S.M.; Cekić, N.D.; Savić, M.M.; Isailović, T.M.; Randelović, D.V.; Marković, B.D.; Savić, S.R.; Stamenić, T.T.; Daniels, R.; Savić, S.D. Parenteral Nanoemulsions as Promising Carriers for Brain Delivery of Risperidone: Design, Characterization and in Vivo Pharmacokinetic Evaluation. *Int. J. Pharm.* **2015**, *493*, 40–54. [CrossRef]
59. Vivo, E.; Assessment, I.V.; Barkat, A.; Ansari, M.J. Development of Natural Polysaccharid–Based Nanoparticles of Berberine to Enhance Oral Bioavailability: Formulation, Optimization, Ex Vivo, and In Vivo Assessment. *Polymers* **2021**, *13*, 3844.
60. Mangla, B. Lipid-Nanopotentiated Combinatorial Delivery of Tamoxifen and Sulforaphane: Ex Vivo, in Vivo and Toxicity Studies. *Nanomedicine* **2020**, *15*, 2563–2583. [CrossRef]
61. Singh, A.; Neupane, Y.R.; Panda, B.P.; Kohli, K. Lipid Based Nanoformulation of Lycopene Improves Oral Delivery: Formulation Optimization, Ex Vivo Assessment and Its Efficacy against Breast Cancer. *J Microencapsul.* **2017**, *34*, 416–429. [CrossRef]
62. Hurler, J.; Engesland, A.; Kermany, B.P.; Skalko-Basnet, N. Improved Texture Analysis for Hydrogel Characterization: Gel Cohesiveness, Adhesiveness, and Hardness. *J. Appl. Polym. Sci.* **2012**, *125*, 180–188. [CrossRef]
63. Jain, P.; Mirza, M.A.; Talegaonkar, S.; Nandy, S.; Dudeja, M.; Sharma, N.; Anwer, K.; Alshahrani, S.M.; Iqbal, Z. Design and in Vitro/in Vivo Evaluations of a Multiple-Drug-Containing Gingiva Disc For. *RSC Adv.* **2020**, *10*, 8530–8538. [CrossRef]

64. Indian Pharmacopoeia Commission. *Indian Pharmacopeia;* IPC: Raknpa, India, 2007; Volume 2, ISBN 81-903436-0-3.
65. Jain, P.; Taleuzzaman, M.; Kala, C.; Gupta, D.K.; Ali, A.; Aslam, M. Quality by Design (Qbd) Assisted Development of Phytosomal Gel of Aloe-Vera Extract for Topical Delivery. *J. Liposome Res.* **2021**, *31*, 381–388. [CrossRef]
66. Sankar Kadimi, U.; Balasubramanian, D.R.; Ganni, U.R.; Balaraman, M.; Govindarajulu, V. In Vitro Studies on Liposomal Amphotericin B Obtained by Supercritical Carbon Dioxide-Mediated Process. *Nanomed. Nanotechnol. Biol. Med.* **2007**, *3*, 273–280. [CrossRef] [PubMed]
67. Banjara, R.A.; Jadhav, S.K.; Bhoite, S.A. Antibacterial Activity of Di-2-Ethylaniline Phosphate Screened by Paper Disc Diffusion Method. *J. Appl. Pharm. Sci.* **2012**, *2*, 230–233. [CrossRef]
68. Pillai, O.; Hamad, M.O.; Crooks, P.A.; Stinchcomb, A.L. Physicochemical Evaluation, in Vitro Human Skin Diffusion, and Concurrent Biotransformation of 3-O-Alkyl Carbonate Prodrugs of Naltrexone. *Pharm. Res.* **2004**, *21*, 1146–1152. [CrossRef] [PubMed]
69. Elsayed, M.M.A.; Abdallah, O.Y.; Naggar, V.F.; Khalafallah, N.M. Lipid Vesicles for Skin Delivery of Drugs: Reviewing Three Decades of Research. *Int. J. Pharm.* **2007**, *332*, 1–16. [CrossRef] [PubMed]
70. Dave, N.; Chan, M.Y.; Huang, P.J.J.; Smith, B.D.; Liu, J. Regenerable DNA-Functionalized Hydrogels for Ultrasensitive, Instrument-Free Mercury(II) Detection and Removal in Water. *J. Am. Chem. Soc.* **2010**, *132*, 12668–12673. [CrossRef]
71. Hussain, T.; Tan, B.; Yin, Y.; Blachier, F.; Tossou, M.C.B.; Rahu, N. Oxidative Stress and Inflammation: What Polyphenols Can Do for Us? *Oxid. Med. Cell. Longev.* **2016**, *2016*, 7432797. [CrossRef]
72. Suntharalingam, M.; Alcázar-Román, A.R.; Wente, S.R. Nuclear Export of the Yeast MRNA-Binding Protein Nab2 Is Linked to a Direct Interaction with Gfd1 and to Gle1 Function. *J. Biol. Chem.* **2004**, *279*, 35384–35391. [CrossRef]
73. Draize, J.H.; Nelson, A.A.; Calvery, H.O. The Percutaneous Adsorption of DDT (2, 2,-Bis (p-Chlorophenyl) 1,1,1-Triehlorethane) to Laboratory Animals. *J. Pharmacol. Exp. Ther.* **1944**, *82*, 159–166.
74. Patel, M.R.; Patel, R.B.; Parikh, J.R.; Patel, B.G. Novel Isotretinoin Microemulsion-Based Gel for Targeted Topical Therapy of Acne: Formulation Consideration, Skin Retention and Skin Irritation Studies. *Appl. Nanosci.* **2016**, *6*, 539–553. [CrossRef]
75. Joshi, M.; Patravale, V. Formulation and Evaluation of Nanostructured Lipid Carrier (NLC)-Based Gel of Valdecoxib. *Drug Dev. Ind. Pharm.* **2006**, *32*, 911–918. [CrossRef]
76. Pople, P.V.; Singh, K.K. Development and Evaluation of Topical Formulation Containing Solid Lipid Nanoparticles of Vitamin A. *AAPS PharmSciTech* **2006**, *7*, E63–E69. [CrossRef]
77. Dammu, R.K.M.; Saileela, K. Clinico-Microbiological Study of Dermatophytosis in the Tertiary Care Hospital, KMC & H, Chinakondrupadu, Guntur. *Indian J. Microbiol. Res.* **2020**, *7*, 258–264. [CrossRef]

gels

Article

New Biopharmaceutical Characteristics of In Situ Systems Based on Poloxamer 407

Elena O. Bakhrushina [1], Elizaveta V. Novozhilova [1,2], Marina M. Shumkova [1,*], Victor S. Pyzhov [1], Maria S. Nikonenko [1], Alexander I. Bardakov [1], Natalia B. Demina [1], Ivan I. Krasnyuk, Jr. [3] and Ivan I. Krasnyuk [1]

1 Department of Pharmaceutical Technology, A.P. Nelyubin Institute of Pharmacy, I.M. Sechenov First Moscow State Medical University (Sechenov University), Moscow 119048, Russia; bakhrushina_e_o@staff.sechenov.ru (E.O.B.); elizaveta.novozhilova@stud.unifi.it (E.V.N.); pyzhov_v_s@student.sechenov.ru (V.S.P.); nikonenko_m_s@student.sechenov.ru (M.S.N.); bardakov_a_i@staff.sechenov.ru (A.I.B.); demina_n_b@staff.sechenov.ru (N.B.D.); krasnyuk_i_i@staff.sechenov.ru (I.I.K.J.)

2 Department of Chemistry "Ugo Schiff", Università di Firenze, Via Della Lastruccia 3-13, 50019 Sesto Fiorentino, Italy

3 Department of Analytical, Physical and Colloidal Chemistry A.P. Nelyubin Institute of Pharmacy, I.M. Sechenov First Moscow State Medical University (Sechenov University), Moscow 119048, Russia; krasnyuk_i_i_1@staff.sechenov.ru

* Correspondence: shumkovamm@gmail.com

Abstract: Thermosensitive systems based on poloxamer 407 are widely used in targeted drug delivery; however, the stability of the phase transition temperature remains insufficiently studied. This article presents the results of a study on the effect of adding polyethylene glycols (PEG) with different molecular weights and some classical gel-forming polymers on the gelation temperature of thermoreversible compositions based on poloxamer 407 in a long-term experiment. The study showed a positive effect of PEG addition with average molecular weights at concentrations of 1.5–2.0%, as well as gelling agents at a concentration below the critical gelation concentration. The proposed rheological test for studying the samples' adhesion can give an indirect forecast of the composition adhesive rate. Based on the conducted studies, three experimental binary systems based on poloxamer 407 were selected, with the addition of HPMC 0.5%, sodium alginate 0.5%, and PEG 1500 1.5%. These systems are the most promising for the further development of in situ targeted drug delivery systems.

Keywords: in situ systems; Kolliphor; poloxamer; polyethylene glycol; gelation temperature; adhesion

check for updates

Citation: Bakhrushina, E.O.; Novozhilova, E.V.; Shumkova, M.M.; Pyzhov, V.S.; Nikonenko, M.S.; Bardakov, A.I.; Demina, N.B.; Krasnyuk, I.I., Jr.; Krasnyuk, I.I. New Biopharmaceutical Characteristics of In Situ Systems Based on Poloxamer 407. *Gels* **2023**, *9*, 508. https://doi.org/10.3390/gels9070508

Academic Editor: Shige Wang

Received: 31 May 2023
Revised: 17 June 2023
Accepted: 20 June 2023
Published: 21 June 2023

1. Introduction

Thermosensitive in situ systems based on poloxamers have become one of the most popular products for targeted therapy in the last decade. Based on thermosensitive in situ systems, drugs for parenteral, intranasal, vaginal, rectal, and ophthalmological usage are being developed [1–4].

In comparison to many other directionally synthesized polymers, poloxamers are commercially available, the technology of obtaining targeted delivery systems based on them is highly cost-effective, and patients' medical adherence to therapy with their use is high. In addition to this, poloxamers, undergoing sol–gel conversion and back, do not change their rheological and technological characteristics [5].

Despite the significant advantages, monocomponent aqueous solutions of poloxamers have several disadvantages: a relatively low gelation temperature (26–27 °C for poloxamer 407), average adhesive ability, and low adsorption activity [6–8]. Due to the narrow range of concentrations used to produce thermoreversible dosage forms (18.0–20.0%), it is possible to vary the technological characteristics of the finished dosage form by introducing an additional polymer into the composition. The resulting binary or non-binary polycomplexes could either improve thermoreversible properties or result in their loss [9].

At the moment, many studies have proven the feasibility of creating polycomplexes based on poloxamers, since these binary and non-binary systems not only have improved technological characteristics and are more relevant to the physiological conditions of the sol-gel transition, but also have improved biopharmaceutical characteristics [7,10–13]. Thus, Ricci et al. in their studies demonstrated that the introduction of polyethylene glycol 400 or inorganic salts into a poloxamer solution causes an increase in the active pharmaceutical ingredient (API) release and diffusion coefficients [14,15]. In Wang P. et al., the effect of the multi-gel core of poloxamer 407 with polyethylene glycol on the completeness of the Exenatide release profile from microspheres was shown [16].

The standard methods of investigation for in situ systems based on poloxamers are the determination of the gelation temperature, gelation time, and gel strength [17–19]. Additionally, adhesive (bioadhesive) properties of pharmaceutical compositions are often studied, and structural and mechanical analyses are performed on a viscometer and a texture analyzer [18,20,21].

Rheological analysis is an in vitro method for determining adhesion, which makes it possible to predict the behavior of polymer matrices in vivo, as well as to investigate their structural interactions by interpreting changes following physicochemical interactions [22]. The mucoadhesion process is a phenomenon that combines different types of interactions, and a rheological method is used to evaluate these interactions between mucin and a gel-forming polymer [23,24].

For the first time, the use of rheological indicators to determine the degree of adhesion to the mucosa was proposed by Hassan and Gallo in their 1990 paper [25]. Today, some studies demonstrate a correlation between the adhesion parameters of polymer matrices measured using classical methods (such as the flow method, separation force from the mucin layer, and the use of texture analyzers) and rheological measurements of polymer matrices at low shear rates (up to 20 s^{-1}) [26].

The range of shear rates chosen for the developed technique was from 0 to 10 s^{-1}. In vivo, the forces counteracting adhesion are influenced by gravity and the slope of the drug application surface. Consequently, the shear rate magnitude formed during the flow and separation of the adhesive sample from the mucous membrane is relatively small.

In the study conducted by Bakhrushina et al., the potential of utilizing the rheological method without the addition of a mucin solution was demonstrated. This approach allows the obtention of primary data suitable for the screening stages of promising formulations. The experiment involved a diverse range of known polymers, and formulations both with and without the addition of mucin exhibited similar characteristics in terms of the nature of the rheological curve and shear stresses within the measured range [27].

Thus, it was concluded that the screening of samples can be carried out through a general visual assessment of the yield curve type, including its shape, smoothness, peaks, plateau zone, and shear stress range.

At the same time, the gelation temperature parameter, which is the main screening parameter when choosing the pool of optimal compositions, is not studied by most researchers of dynamics and, as a rule, is tracked only for several weeks from the moment of sample production. At the same time, the loss of the initial gelation temperature due to thermosensitive complexes is the main problem that prevents the widespread introduction of thermosensitive in situ systems into clinical practice. Stability is crucial in thermosensitive systems for targeted drug delivery during storage as well as after administration, as it can protect the encapsulated drugs from degradation and dilution upon administration. It also improves the expected treatment effect by releasing a specific load and controlling the systemic distribution of therapeutic drugs [28]. Studies by Grela et al. [20] show that in monocomponent placebo formulations based on poloxamer 407, significant changes in the main rheological parameters are observed that exceed the established criteria ($\pm$10% of each initial value). In addition, a complete loss of thermal reversibility was observed in samples stored at 40 °C.

In the study by Ivanova et al. [29], the problem of the stability of monocomponent compositions based on poloxamer 407 is proposed to be solved by adding various glycols, including propylene glycol and polyethylene glycol. The positive effect of glycols of different molecular weights on the stability of the liquid crystal gel phases of the poloxamer solution was established. Thus, the creation of poloxamer 407 polycomplexes with various excipients may also be advisable to increase the stability of the thermosensitive system.

This study aimed to investigate various polycomponent complexes containing poloxamer 407 for a determination of the stability of the gelation temperature in a long-term experiment, as well as the use of a rheological method to study the main parameters of thermosensitive in situ systems.

2. Results and Discussion

2.1. Results

2.1.1. Gelation Temperature Screening Test

After preparation, all experimental samples were stored for 7 days for structuring. The structuring of the samples was necessary because, during the first 7 days, the temperature of the sol–gel transition of copolymers with PEGs was above 45 °C. Then, it rapidly decreased and reached a plateau by the seventh day. Because of this, the temperature values of the first 7 days were not taken into account since many samples subsequently showed stability.

After that, samples with aggregate stability were selected for further study, showing no signs of delamination, coagulation, or syneresis. All the selected compositions were transparent homogeneous liquids at the storage temperature. As a result of the initial selection, 12 experimental samples were selected for the experiment on long-term storage and the study of gelation temperature stability, in addition to the thermoreversible component (poloxamer 407, 18.0%) containing additional gelling agents in the following concentrations: PEG 400 1.5%; PEG 1500 1.5%, 1.0%, 2.5%; PEG 3400 1.5%, 2.0%, 2.5%; PEG 4000 1.5%, 2.0% 2.5%; PEG 6000 1.5%; HPMC 0.5%; sodium alginate 0.5%. A monocomponent composition of poloxamer 407 in a concentration of 18.0% and water was used as a reference sample. Figure 1 shows the dynamics of change in the analyzed indicator observed over 12 weeks.

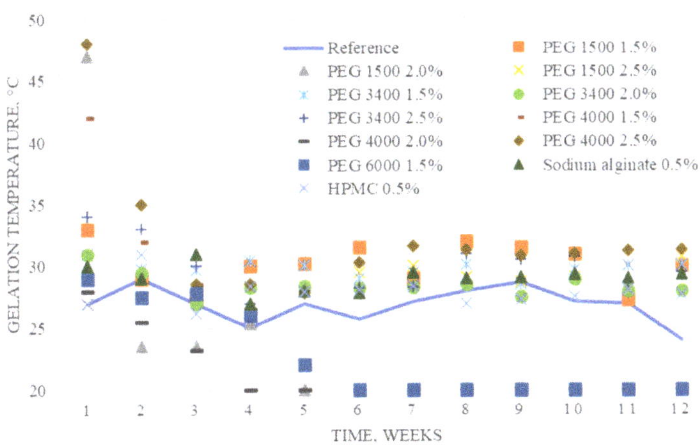

Figure 1. Dynamics of the gelation temperature change over 12 weeks for experimental samples.

Statistical analysis was used to determine the statistical significance of differences in gelation temperatures and the gelation time of samples (Table 1). It was shown that the introduction of additional gel-forming agents (PEG 1500, PEG 3400, sodium alginate, and

HPMC) not only increased the average gelation temperature but also made this indicator stable during the experiment.

Table 1. Average temperature measurement results of gelation time for experimental samples.

Polymer Excipients n = 3	The Average Value of the Gelation Temperature, °C	Standard Deviation	Sample Variance	Internal Scatter
Reference composition	26.68	1.832	3.358	[22.0; 29.0]
PEG 1500 1.5%	29.85	1.699	2.888	[27.5; 33.0]
PEG 1500 2.0%	27.05	9.923	98.475	[20.0; 47.0]
PEG 1500 2.5%	28.92	1.294	1.674	[27.0; 31.5]
PEG 3400 1.5%	30.13	1.400	1.960	[28.0; 33.0]
PEG 3400 2.0%	28.64	1.036	1.074	[27.0; 31.0]
PEG 3400 2.5%	30.62	4.160	17.307	[27.0; 42.0]
PEG 4000 1.5%	29.80	8.098	65.575	[20.0; 42.0]
PEG 4000 2.0%	27.74	11.055	122.218	[20.0; 48.0]
PEG 4000 2.5%	33.72	7.380	54.470	[28.0; 50.0]
PEG 6000 1.5%	29.05	6.792	46.135	[22.0; 42.0]
Sodium alginate 0.5%	29.09	1.241	1.541	[27.0; 31.0]
HPMC 0.5%	28.50	1.780	3.167	[26.0; 31.5]

The gelation time of the studied samples was not exposed to significant fluctuations during the entire storage period and varied in the range from 40 s (poloxamer 407 18.0%, PEG 4000 2.0%) to 1 min 51 s on average (poloxamer 407 18.0%, PEG 4000 2.5%).

2.1.2. Rheological Properties of Selected Samples

At the next stage, the characteristics of the five selected stable samples were investigated in comparison with the reference composition using a rotational viscosity test. Figure 2 shows rheograms based on the results of measurements in the temperature range from 20 to 50 °C. All rheograms graphically confirm the presence of thermoreversible gel formation in both the reference and five multicomponent compositions. On the rheograms, a growth zone, a decline zone, as well as a plateau zone of dynamic viscosity (measured at a shear rate of 100 s^{-1}) are distinguished.

The longest plateau zone was observed in the reference sample (Figure 2a), as well as in the sample containing HPMC in a concentration of 0.5% as an additional gel-forming agent (Figure 2b). At the same time, it should be noted that the stable high dynamic viscosity (2.5 ± 1.0 Pa·S) of the sample with HPMC was maintained at a temperature of 40 °C. Thus, the temperature of the reverse gel–sol transition for the studied complex was significantly higher than that for the other samples.

In the sample with alginate (Figure 2c), a stable high viscosity was observed at a temperature of 32 °C (2.17 Pa·S) but at a temperature of 35 °C, the value of the dynamic viscosity began to decrease, which indicates an opposite gel–sol transition in this temperature range. Thus, the plateau zone of a sample containing sodium alginate is short, which may limit the use of this complex in vivo, in cases where phase transition temperatures exceeding 35° C are required or pathologically caused temperature fluctuations at the application site are possible.

A similar pattern is observed on the rheograms of samples PEG 1500 2.5% and PEG 3400 1.5% (Figure 2e,f). For the sample, which is a binary mixture of poloxamer 407 (18.0%) and PEG 1500 (2.5%), the sol–gels phase transition based on rheological studies is fixed at a temperature of 31 °C, and the reverse gel–sol transition is fixed at 34 °C. For a composition using PEG 3400 (1.5%) as an additional gel-forming agent, a constant viscosity (0.6 ± 0.5 Pa·S) is set in the temperature range of 30–33 °C, which makes this composition the least promising of the studied pool for use in medicine. In addition, the compositions containing PEG 1500 (2.5%) and PEG 3400 (1.5%) are characterized by the lowest values of dynamic viscosity compared to those of other analyzed compositions.

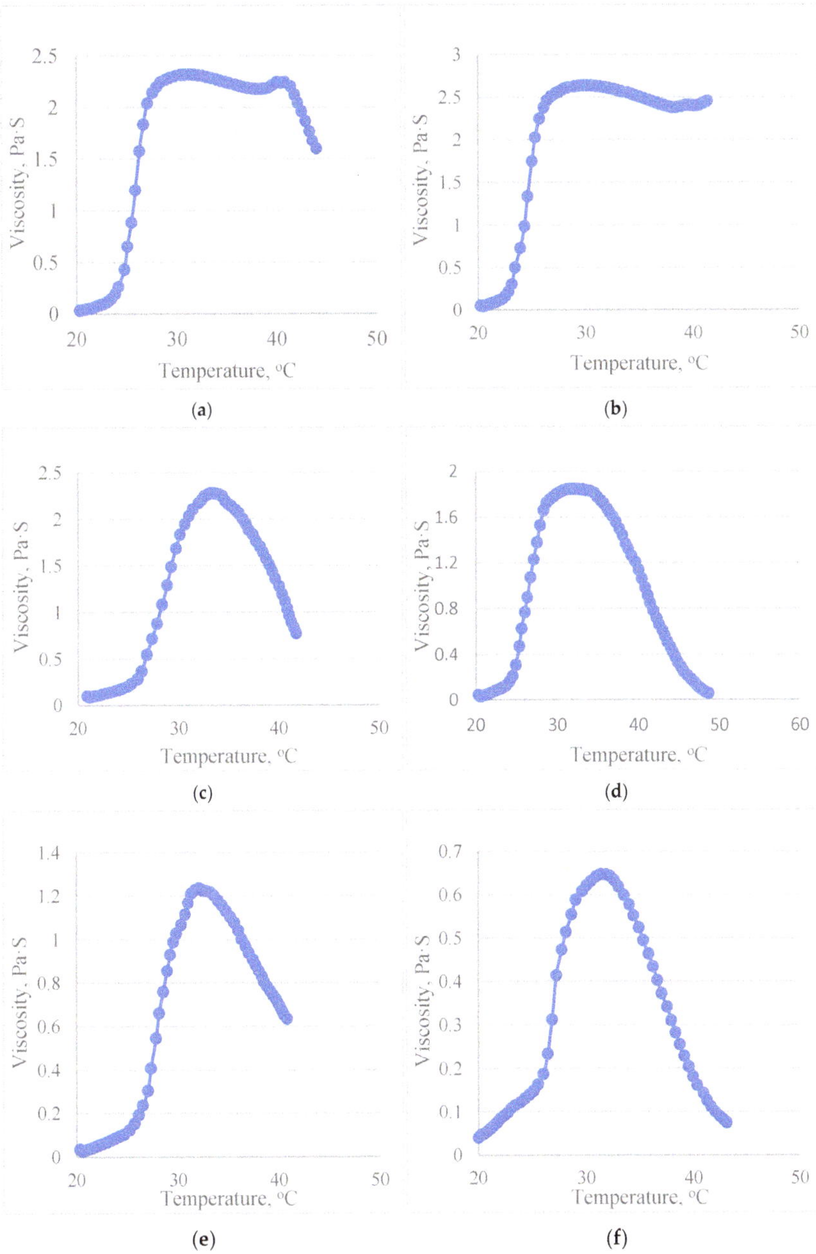

Figure 2. Rheograms of the studied samples: (**a**) poloxamer 407 18.0%; (**b**) poloxamer 407 18.0%, HPMC 0.5%; (**c**) poloxamer 407 18.0%, sodium alginate 0.5%; (**d**) poloxamer 407 18.0%, PEG 1500 1.5%; (**e**) poloxamer 407 18.0%, PEG 1500 2.5%; (**f**) poloxamer 407 18.0%, PEG 3400 1.5%.

On the rheogram of the sample of the binary complex of poloxamer 407 (18.0%) and PEG 1500 (1.5%) (Figure 2d), the plateau zone is fixed in the temperature range of 30–35 °C The dynamic viscosity of the complex in the gel-like state is 1.75 ± 0.5 Pa·S, which is the best result among complexes with PEG and is comparable to the indicators of the complex with sodium alginate and the reference sample.

2.1.3. Adhesive Properties of Selected Samples

The next stage was the study of the adhesive properties of the pool of experimental samples. The measurement results are shown on rheograms (Figure 3). The measurements in this study were carried out at a variable shear rate in the small shift–large shift–small shift mode when the rotation speed of the internal geometry of the viscometer first increased sequentially, remained at the highest point for several seconds (10 s^{-1}) and then gradually decreased until the rotation stopped completely. The shear stress on the above rheograms gradually increased for all analyzed samples with an increase in the shear rate from a small shift to a large shift (from 0 to 5 s^{-1}), then a clearly pronounced peak of the shear stress values (Figure 3d–f) or an exit to a plateau (Figure 3a–c) was recorded. When the rotation speed decreased from a large shift to small shift (from 10 to 0 s^{-1}), the shear stress remained constant or decreased slightly, without returning to the initial values.

The shear stress in this experiment recorded the presence of temporary connections between the sample and the material of the internal rotating geometry.

The suitability of this method for studying the adhesive strength of samples was determined by comparing the values of the sample viscosities and the shear stress formed in the experiment in response to the slow rotation of the geometry. Table 2 shows the average values of the dynamic viscosity of the samples at a temperature of 32 °C, which corresponds to the conditions of the experiment.

Table 2. Results of rheological measurements of experimental samples.

Composition of the Analyzed Sample n = 3	Average Dynamic Viscosity (100 s^{-1}) at a Temperature of 32 °C	Standard Deviation	Sample Variance	Ranges of Average Values of the Shear Stress in the Large Shift–Small Shift Mode (from 10 to 0 s^{-1})
Reference sample	2.3 Pa·s	0.187	0.035	116–140 Pa
HPMC 0.5%	2.6 Pa·s	0.107	0.011	164–189 Pa
Sodium alginate 0.5%	2.2 Pa·s	0.135	0.018	158–194 Pa
PEG 1500 1.5%	1.8 Pa·s	0.164	0.027	140–164 Pa
PEG 1500 2.5%	1.2 Pa·s	0.072	0.005	97–114 Pa
PEG 3400 1.5%	0.6 Pa·s	0.207	0.043	52–70 Pa

It is shown that the reference sample and the composition using sodium alginate (0.5%) have similar values of dynamic viscosity (2.3 Pa·S and 2.2 Pa·S, respectively). At the same time, the maximum values of shear stresses in the reference sample, arising in response to the rotation of the geometry of the viscometer, are in the range from 116 to 140 Pa, and those in the analyzed sample with the addition of sodium alginate were in the range from 158 to 194 Pa. Thus, the viscosity of the gel did not have a decisive influence on the magnitude and nature of the change in the shear stress determined in the experiment, and the results obtained may indirectly indicate differences in the adhesive ability of the samples.

The character of the adhesive ability of the samples was judged according to the determined range of values of shear stresses when the geometry operated in the large shift–small shift mode (Table 2). The expression of the adhesive ability of poloxamer 407 was positively affected by the addition of polymers such as HPMC, sodium alginate, and PEG 1500 (1.5%) to the mixture.

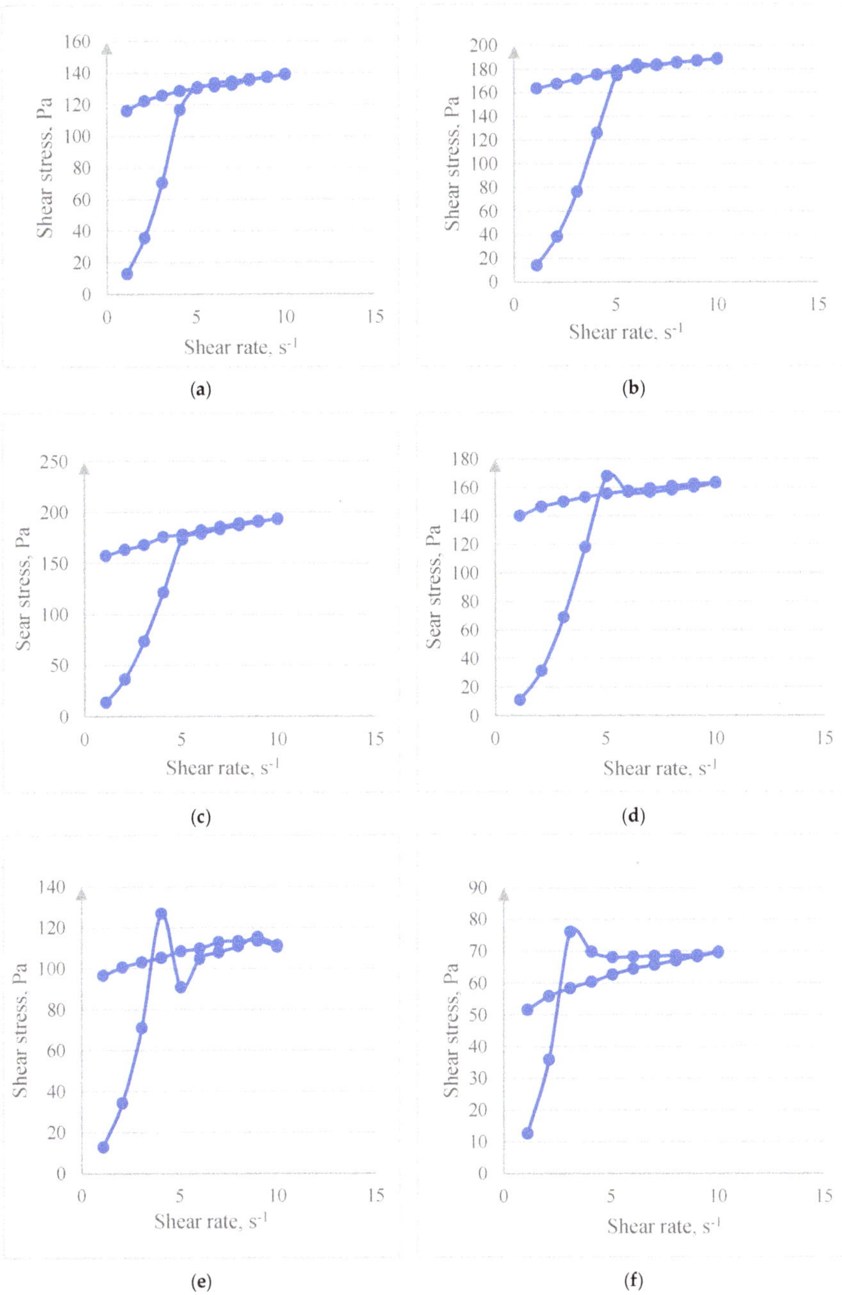

Figure 3. Graphs of the dependence of the shear stress on the shear rate of experimental samples, determined in the range of low speeds (from 0 to 10 s^{-1}): (**a**) poloxamer 407 18.0%; (**b**) poloxamer 407 18.0%, HPMC 0.5%; (**c**) poloxamer 407 18.0%, sodium alginate 0.5%; (**d**) poloxamer 407 18.0%, PEG 1500 1.5%; (**e**) poloxamer 407 18.0%, PEG 1500 2.5%; (**f**) poloxamer 407 18.0%, PEG 3400 1.5%.

2.2. Discussions

To reproduce experimental data on the instability of thermosensitive solutions of poloxamer 407 obtained in the studies of Grela et al. [20], an 18.0% composition based on Kolliphor® P 407 (BASF) was used as a reference sample in the experiment. The purpose of the long-term experiment was to evaluate the stability of the gelation temperature for various compositions with poloxamer 407. This was necessary because a significant variability of values has been observed over time. The ultimate goal was to ensure that the properties of the potential in situ dosage form remain unchanged during storage.

First of all, polyethylene glycols with different molecular weights (from 1500 to 6000) were considered the introduced polymers, since their influence on the stability of compositions with poloxamer 407 was shown in studies [29]. In addition, PEGylation, the coating of nanoparticles with polyethylene glycol (PEG), is a widely employed strategy in drug and gene delivery to enhance targeting efficiency, as PEG is considered a safe and FDA-approved "stealth" polymer in the field [30]. Additionally, classical gel-forming agents (HPMC and sodium alginate) were considered auxiliary substances in concentrations preceding their critical concentration of gel formation. Hence, the three-dimensional gel structure they formed did not interfere with the reverse gel–sol transition of the composition [31,32].

It is known that poloxamer 407 is a hydrophilic nonionic surfactant, a three-block copolymer consisting of two hydrophilic blocks (from 54 to 60 polyethylene oxide units) separated by a hydrophobic block (from 95 to 105 polypropylene oxide units). Due to the presence of a hydrophilic and hydrophobic block, poloxamers have amphiphilic properties and are able to form micelles [33].

There are various hypotheses of the mechanism of micelle formation. The most plausible one is that there is an initial phase separation stage with areas that are very rich in poloxamer and areas that are more saturated with water. Then, the further process of desolvation of a part of the PPO leads to the aggregation of unimers and the formation of micelles close to each other, with the formation of clusters. As for the sol–gel transition during gel formation, the partial destruction of the PEO chains in the micellar mantle leads to the formation of a denser structure [34].

Micelles can have different shapes, namely spherical, cylindrical, or lamellar, depending on the length of the PO and EO blocks, and the core always consists of hydrophobic blocks, while hydrophilic blocks form the outer shell [35].

Thus, when polyethylene oxides with different molecular weights are added to the poloxamer solution, the micellar mantle of the poloxamer is probably strengthened, which can lead to both an improvement in the thermosensitive properties (an increase in the phase transition temperature) and a complete loss of them.

When studying mixtures of poloxamer 407 with various PEGs, the atypical behavior of gels was observed. The first feature was a strong increase in the gelation temperature compared to that of other excipients (in some cases up to 48 °C). The second feature was the long duration of gel structuring (up to 7 days, compared to 1–2 days for samples that do not contain PEG), during which the gel formation temperature dropped rapidly (up to 10–15 points), after which it reached a plateau, and the system became relatively stable.

Additionally, for some compositions containing high-molecular PEG (4000 and 6000) in high concentrations (2.0–2.5%), a very long sol–gel transition was determined. The samples remained liquid when heated in a water bath to temperatures above 50 °C; however, when placed in a thermostat in a climate chamber (constant temperature of 38 °C), a dense gel formed for several hours.

According to the experimental data obtained, the main role in the stabilization of the thermodynamic parameters of the system belongs to the molecular weight and concentration of PEG. It was noted that when using polymers with an average molecular weight (PEG 1500 and PEG 3400) at a minimum experimental concentration (1.5%), it was possible to stabilize the thermosensitive system for the entire period of the study (12 weeks).

During the conduction of the long-term studies, the following phenomenon was identified. When PEG 1500 was added to the poloxamer 407 at a concentration of 2.0%, in contrast to the introduction of PEG 3400 in a similar concentration, the system was not stabilized until the fifth week of the experiment, and the gelation temperature of the composition was kept at 23.5 °C, decreasing after this to 20 °C and becoming stable. The thermal reversibility of the composition was preserved; however, the low temperature of the sol–gel transition made the use of the composition unpromising.

It should be noted that the composition with PEG 1500 in a concentration of 2.5% demonstrated excellent stability and a high average gelation temperature (about 29 °C). This phenomenon requires further detailed study using modern analytical methods. The use of sodium alginate gelling agents and HPMC as stabilizers of the thermoreversible system at a concentration of 0.5% was considered highly promising since it provided a relatively high average temperature of gelling (29 °C and 27.8 °C, respectively), which remained stable throughout the entire period of the experiment. The gelation times for these compositions were also average among those for the entire analyzed pool.

The rheological methods of studying the indicators of thermosensitive in situ systems used in this study also showed their prospects. Rotary viscometers with a temperature-controlled system are examples of affordable laboratory and analytical equipment, and the methods produced with their help are highly reproducible and can be validated. The temperature test for analytical samples at a constant shear rate made it possible to conduct a reasonable and fast screening of compositions based on the indicator of the duration of the plateau zone—a high dynamic viscosity with increasing temperature. The results reflected in the rheograms (Figure 2) are visual and can be used for the effective selection of promising samples.

The proposed rheological test for studying the adhesion of samples gives an indirect forecast of the degree of adhesion of a composition. However, as it was noted earlier, it is easily reproducible and available in most laboratories, unlike the well-known mucoadhesive screenings, and can be used as an additional factor in choosing the optimal composition, along with the main rheological test.

Future advances in this area may be supported by linking nanostructural characterization techniques to spectroscopic tools to probe into the interplay between changes in structure and inter-macromolecular interactions and provide a rational basis for additive selection [36].

3. Conclusions

In the course of the conducted studies, the effect of the introduced additional polymers on the stability of the thermosensitive compositions of poloxamer 407 was shown. The best results of stability studied during 12 weeks of storage (at a temperature of +5/+8 °C), demonstrated the following compositions: poloxamer 407 18.0%, HPMC 0.5%; poloxamer 407 18.0%, sodium alginate 0.5%; poloxamer 407 18.0%, PEG 1500 1.5%; poloxamer 407 18.0%, PEG 1500 2.5%; poloxamer 407 18.0%, PEG 3400 1.5%; poloxamer 407 18.0%, PEG 3400 2.0%. Thus, the positive effect of the addition of medium-molecular-weight PEGs in the minimum operating concentrations (1.5–2.0%), as well as that of gel-forming agents (HPMC, sodium alginate) in a concentration lower than the critical concentration of gel formation (0.5%) was shown.

According to the data obtained after rheological tests, three experimental binary systems based on poloxamer 407 (18.0%) can be distinguished as the most promising for creating targeted drug delivery systems based on them, according to the parameters of gelation temperature, gelation time, and stability, as well as mucoadhesion with the addition of HPMC 0.5%, sodium alginate 0.5% and PEG 1500 1.5%.

Poloxamer 407 is a thermoreversible gel with a low viscosity at room temperature that rapidly transforms into a viscous gel at body temperature (approximately 37 °C). It can be conveniently administered using a syringe or applicator, facilitating in situ gelation at the desired location [36]. Additionally, polymer micelles have proven to be effective carriers for compounds with poor solubility, unfavorable pharmacokinetics, and low stability in the physiological environment [37]. These promising formulations, exhibiting improved pharmaceutical properties, can be selected for further development as dosage forms with active substances. Moreover, the gelation temperature can be adjusted to meet the specific requirements of the study, ranging from room temperature to body surface temperature, and even to pathological temperatures associated with inflammation.

Studies of the issues of stabilization of specific indicators of thermosensitive systems will be continued in the future using other analytical methods as well as by creating non-binary systems.

4. Materials and Methods

To develop thermoreversible compositions, the poloxamer 407 Kolliphor® P 407 (BASF SE, Ludwigshafen, Germany) was used in the experiment, as well as the following excipients for the formation of stable polycomplexes: PEG 1500-Pluriol® E 1500; PEG 3400-Pluriol® E 3400; PEG 4000-Pluriol® E 4000; PEG 6000-Pluriol® E 6000 (BASF SE, Ludwigshafen, Germany); hydroxypropylmethylcellulose Benecel® K100M PHARM; Sodium Alginate Protanal® CR 8133 (FMC, Philadelphia, PA, USA). Gelling systems were prepared using 'Cold Methods'. All compositions were binary mixtures of poloxamer 407 and an additional gel-forming agent. The concentration of poloxamer 407 in all experimental samples was constant and amounted to 18.0%. The polymer concentrations ranged from 0.5 to 2.5% depending on weight. The weighed sample of the poloxamer was dispersed in purified water, mixed with an additional polymer of a given weight, and cooled to −15/−18 °C for 5–7 min until it was completely dissolved.

4.1. Long-Term Screening Test

Temperature and gelation time in a long-term experiment, as well as viscosity using a rotational temperature test and adhesion using the rotational viscosity test, were determined and proposed as new screening parameters for evaluating thermoreversible systems.

The gelation temperature was measured in the samples at a volume of 20 mL +/−0.5 mL that were stored for 12 weeks of the experiment in a refrigerator at a temperature of +5/+8 °C, by heating them in a water bath at a constant temperature of 50 °C with an immersion measuring thermosonde with a periodicity of 7 days.

The gelation time was measured from the moment the sample was immersed in a polymer container in a water bath at 50 ± 1 °C. The values determined by increasing the dynamic viscosity of the sample during its mixing were taken as an indicator of the gelation temperature and time.

To process the obtained data, statistical methods were used to calculate the average value (gelation temperature) and the variance of a random variable, which was taken as a measure of the stability of the samples.

4.2. Rheological Property Test

The MS-DIN 33 measuring system (Lamy, Rhône-Alpes, France) was utilized for the test. It consists of an internal rotating cylinder with a cone-shaped tip and an internal static cylinder. The recommended sample loading volume is approximately 17 mL.

For the rheological tests assessing the thermal sensitivity of the developed systems, the dependency of dynamic viscosity (Pa·S) at a constant shear rate of 100 s^{-1} on temperature was determined. The measuring geometry of the device was placed in THERMACELL PACKAGE RT-1 Plus (La, Paris, France). The sample temperature gradually increased from 20 to 50 °C during the test, and the measurements were carried out in three repetitions with an interval of 5 min.

4.3. Adhesion Property Test

The adhesion value was measured using the rheological method in the range of shear rates from 0 to 10 s^{-1} at a constant temperature of 32 ± 0.5 °C, which simulates the conditions of external and topical application of the system by the patient. The nature and magnitude of adhesion were judged by the shape of the rheological curve, as well as by the maximum values of the shear stress that occurred during the experiment.

4.4. Statistical Analysis

In the statistical study, Microsoft Excel 2018 was utilized for data analysis. The formulas employed included standard deviation (STDEV) and sample variance (VAR). The STDEV formula was applied to calculate the standard deviation for a sample of a population, while the VAR formula was used to determine the sample variance. The level of statistical significance was defined at $p < 0.05$.

Author Contributions: Conceptualization, E.O.B., N.B.D. and I.I.K.; methodology, E.V.N., E.O.B. and N.B.D.; validation, I.I.K.J., V.S.P. and A.I.B.; formal analysis, M.S.N.; investigation, E.V.N. and M.S.N.; resources, E.V.N. and M.M.S.; data curation, E.V.N., E.O.B. and V.S.P.; writing—original draft preparation, E.V.N. and M.M.S.; writing—review and editing, E.O.B. and A.I.B.; visualization, V.S.P. and M.S.N.; supervision, N.B.D., I.I.K.J. and I.I.K.; project administration, E.O.B., E.V.N. and M.M.S. All authors have read and agreed to the published version of the manuscript.

Funding: This research received no external funding.

Institutional Review Board Statement: Not applicable.

Informed Consent Statement: Not applicable.

Data Availability Statement: The authors can provide any necessary information on the study upon request.

Conflicts of Interest: The authors declare no conflict of interest.

References

1. Agrawal, M.; Saraf, S.; Saraf, S.; Dubey, S.K.; Puri, A.; Gupta, U.; Kesharwani, P.; Ravichandiran, V.; Kumar, P.; Naidu, V.G.M. Stimuli-responsive in situ gelling system for nose-to-brain drug delivery. *J. Control. Release* **2020**, *327*, 235–265. [CrossRef] [PubMed]
2. Vigani, B.; Rossi, S.; Sandri, G.; Bonferoni, M.C.; Caramella, C.M.; Ferrari, F. Recent Advances in the Development of In Situ Gelling Drug Delivery Systems for Non-Parenteral Administration Routes. *Pharmaceutics* **2020**, *12*, 859. [CrossRef] [PubMed]
3. Zhang, L.; Pang, L.; Zhu, S.; Ma, J.; Li, R.; Liu, Y.; Zhu, L.; Zhuang, X.; Zhi, W.; Yu, X. Intranasal tetrandrine temperature-sensitive in situ hydrogels for the treatment of microwave-induced brain injury. *Int. J. Pharm.* **2020**, *583*, 119384. [CrossRef]
4. Demina, N.B.; Anurova, M.N. Suppositories: Aspects of dosage form modernization. *Drug Dev. Regist.* **2016**, *4*, 92–97.
5. Shaikhullina, M.; Khaliullina, A.; Gimatdinov, R.; Butakov, A.; Chernov, V.; Filippov, A. NMR relaxation and self-diffusion in aqueous micellar gels of pluronic F-127. *J. Mol. Liq.* **2020**, *306*, 112898. [CrossRef]
6. Almeida, H.; Amaral, M.H.; Lobão, P.; Sousa Lobo, J.M. Applications of poloxamers in ophthalmic pharmaceutical formulations: An overview. *Expert Opin. Drug Deliv.* **2013**, *10*, 1223–1237. [CrossRef]
7. Anurova, M.N.; Bakhrushina, E.O.; Demina, N.B. Review of contemporary gel-forming agents in the technology of dosage forms. *Pharm. Chem. J.* **2015**, *49*, 627–634. [CrossRef]
8. Cabana, A.; Aït-Kadi, A.; Juhász, J. Study of the gelation process of polyethylene oxide$_a$–polypropylene oxide$_b$–polyethylene oxide$_a$copolymer (poloxamer 407) aqueous solutions. *J. Colloid Interface Sci.* **1997**, *190*, 307–312. [CrossRef]
9. Dumortier, G.; Grossiord, J.L.; Agnely, F.; Chaumeil, J.C. A review of poloxamer 407 pharmaceutical and pharmacological characteristics. *Pharm. Res.* **2006**, *23*, 2709–2728. [CrossRef]
10. Escobar-Chávez, J.J. Applications of thermo-reversible pluronic F-127 gels in pharmaceutical formulations. *J. Pharm. Pharm. Sci.* **2006**, *9*, 339–358.
11. Ullah, K.H.; Raza, F.; Munawar, S.M.; Sohail, M.; Zafar, H.; Zafar, M.I.; Ur-Rehman, T. Poloxamer 407 Based Gel Formulations for Transungual Delivery of Hydrophobic Drugs: Selection and Optimization of Potential Additives. *Polymers* **2021**, *13*, 3376. [CrossRef] [PubMed]
12. Seo, H.J.; Kim, J.-C. Effects of additives on phase transitions of Poloxamer 407/Poloxamer 188 mixture and release property of monoolein cubic phase containing the poloxamers. *J. Ind. Eng. Chem.* **2012**, *18*, 88–91. [CrossRef]
13. Giuliano, E.; Paolino, D.; Fresta, M.; Cosco, D. Mucosal Applications of Poloxamer 407-Based Hydrogels: An Overview. *Pharmaceutics* **2018**, *10*, 159. [CrossRef] [PubMed]
14. Fakhari, A.; Corcoran, M.; Schwarz, A. Thermogelling properties of purified poloxamer 407. *Heliyon* **2017**, *3*, e00390. [CrossRef] [PubMed]
15. Ricci, E.J.; Bentley, M.V.L.B.; Farah, M.; Bretas, R.E.S.; Marchetti, J.M. Rheological characterization of poloxamer 407 lidocaine hydrochloride gels. *Eur. J. Pharm. Sci.* **2002**, *17*, 161–167. [CrossRef]
16. Wang, P.; Wang, Q.; Ren, T.; Gong, H.; Gou, J.; Zhang, Y.; Cai, C.; Tang, X. Effects of pluronic f127-peg multi-gel-core on the release profile and pharmacodynamics of exenatide loaded in PLGA microspheres. *Colloids Surf. B* **2016**, *147*, 360–367. [CrossRef]
17. Hu, K.-L.; Mei, N.; Feng, L.; Jiang, X.-G. Hydrophilic nasal gel of lidocaine hydrochloride. *Arzneimittelforschung* **2009**, *59*, 635–640.

18. Mohamed, S.; Nasr, M.; Salama, A.; Refai, H. Novel lipid–polymer hybrid nanoparticles incorporated in thermosensitive in situ gel for intranasal delivery of terbutaline sulphate. *J. Microencapsul.* **2020**, *37*, 577–594. [CrossRef]
19. Salem, H.F.; Kharshoum, R.M.; Abou-Taleb, H.A.; Naguib, D.M. Nanosized transferosome-based intranasal in situ gel for brain targeting of resveratrol: Formulation, optimization, in vitro evaluation, and in vivo pharmacokinetic study. *AAPS PharmSciTech* **2019**, *20*, 181. [CrossRef]
20. Grela, K.; Marciniak, D.; Pluta, J. Stability evaluation of thermosensitive drug carrier systems based on Pluronic F-127 polymer. *Acta Pol. Pharm.* **2014**, *71*, 167–180.
21. Paul, A.; Fathima, K.M.; Nair, S.C. Intra nasal in situ gelling system of lamotrigine using ion activated mucoadhesive polymer. *Open Med. Chem. J.* **2017**, *11*, 222–244. [CrossRef]
22. Mackie, A.R.; Goycoolea, F.M.; Menchicchi, B.; Caramella, C.M.; Saporito, F.; Lee, S.; Stephansen, K.; Chronakis, I.S.; Hiorth, M.; Adamczak, M. Innovative Methods and Applications in Mucoadhesion Research. *Macromol. Biosci.* **2017**, *17*, 1600534. [CrossRef] [PubMed]
23. Bassi da Silva, J.; Ferreira, S.; Reis, A.; Cook, M.; Bruschi, M. Assessing Mucoadhesion in Polymer Gels: The Effect of Method Type and Instrument Variables. *Polymers* **2018**, *10*, 254. [CrossRef] [PubMed]
24. Barbosa de Souza, A.; Vinícius Chaud, M.; Francine Alves, T.; Ferreira de Souza, J.; Andrade Santana, M. Hyaluronic Acid in the Intestinal Tract: Influence of Structure, Rheology, and Mucoadhesion on the Intestinal Uptake in Rats. *Biomolecules* **2020**, *10*, 1422. [CrossRef] [PubMed]
25. Hassan, E.E.; Gallo, J.M. A Simple Rheological Method for the in Vitro Assessment of Mucin-Polymer Bioadhesive Bond Strength. *Pharm. Res.* **1990**, *7*, 491–495. [CrossRef] [PubMed]
26. Sun, Y.; Perez, A.F.; Cardoza, I.L.; Baluyot-Reyes, N.; Ba, Y. Mucoadhesive and Rheological Studies on the Co-Hydrogel Systems of Poly(Ethylene Glycol) Copolymers with Fluoroalkyl and Poly(Acrylic Acid). *Polymers* **2021**, *13*, 1956. [CrossRef]
27. Bakhrushina, E.O.; Kondratyeva, V.M.; Khodenok, A.I.; Demina, N.B.; Krasnyuk, I.I. Study of adhesive characteristics of mucoadhesive polymer systems by rotational viscometry. *Khimiko-Farmatsevtichesky Zhurnal* **2023**, *57*, 35–44. [CrossRef]
28. Abuwatfa, W.H.; Awad, N.S.; Pitt, W.G.; Husseini, G.A. Thermosensitive Polymers and Thermo-Responsive Liposomal Drug Delivery Systems. *Polymers* **2022**, *14*, 925. [CrossRef]
29. Ivanova, R.; Lindman, B.; Alexandridis, P. Effect of pharmaceutically acceptable glycols on the stability of the liquid crystalline gels formed by Poloxamer 407 in water. *J. Colloid Interface Sci.* **2002**, *252*, 226–235. [CrossRef]
30. Suk, J.S.; Xu, Q.; Kim, N.; Hanes, J.; Ensign, L.M. PEGylation as a strategy for improving nanoparticle-based drug and gene delivery. *Adv. Drug Deliv. Rev.* **2016**, *99*, 28–51. [CrossRef]
31. Bakhrushina, E.O.; Novozhilova, E.V.; Kashperko, A.S.; Sokolova, A.V.; Demina, N.B.; Krasnyuk, I.I. Biopharmaceutical study of binary poloxamer systems as in situ drug delivery systems poloxamer polycomplexes: The study. *Int. J. Appl. Pharm.* **2022**, *14*, 162–165. [CrossRef]
32. Kurniawansyah, I.S.; Rusdiana, T.; Sopyan, I.; Subarnas, A. A review on poloxamer and hydroxy propyl methyl cellulose combination as thermoresponsive polymers in novel ophthalmic in situ gel formulation and their characterization. *Int. J. Appl. Pharm.* **2021**, *13*, 27–31. [CrossRef]
33. Erdodi, G.; Kennedy, J.P. Amphiphilic conetworks: Definition, synthesis, applications. *Prog. Polym. Sci.* **2006**, *31*, 1–18. [CrossRef]
34. Alexandridis, P.; Alan Hatton, T. Poly(ethylene oxide)-poly(propylene oxide)-poly(ethylene oxide) block copolymer surfactants in aqueous solutions and at interfaces: Thermodynamics, structure, dynamics, and modeling. *Colloids Surf.* **1995**, *96*, 1–46. [CrossRef]
35. Bonacucina, G.; Cespi, M.; Mencarelli, G.; Giorgioni, G.; Palmieri, G.F. Thermosensitive self-assembling block copolymers as drug delivery systems. *Polymers* **2011**, *3*, 779–811. [CrossRef]
36. Cook, M.T.; Haddow, P.; Kirton, S.B.; McAuley, W.J. Polymers Exhibiting Lower Critical Solution Temperatures as a Route to Thermoreversible Gelators for Healthcare. *Adv. Funct. Mater.* **2021**, *31*, 2008123. [CrossRef]
37. Kabanov, A.V.; Batrakova, E.V.; Alakhov, V.Y. Pluronic® block copolymers as novel polymer therapeutics for drug and gene delivery. *J. Control. Release* **2002**, *82*, 189–212. [CrossRef]

gels

MDPI

Article

Innovative Wound Healing Hydrogel Containing Chicken Feather Keratin and Soy Isoflavone Genistein: In Vivo Studies

Nilesh M. Mahajan [1], Kalyani Wanaskar [1], Nemat Ali [2,*], Debarshi Kar Mahapatra [3], Muzaffar Iqbal [4], Abid R. Bhat [5] and Mohammed Kaleem [6,*]

[1] Department of Pharmaceutics, Dadasaheb Balpande College of Pharmacy, Rashtrasant Tukadoji Maharaj Nagpur University, Nagpur 440037, Maharashtra, India; nmmahajan78@gmail.com (N.M.M.); wanaskarkalyani@gmail.com (K.W.)
[2] Department of Pharmacology & Toxicology, College of Pharmacy, King Saud University, P.O. Box 55760, Riyadh 11451, Saudi Arabia
[3] Department of Pharmaceutical Chemistry, Dadasaheb Balpande College of Pharmacy, Nagpur 440037, Maharashtra, India; dkmbsp@gmail.com
[4] Department of Pharmaceutical Chemistry, College of Pharmacy, King Saud University, Riyadh 11451, Saudi Arabia; muziqbal@ksu.edu.sa
[5] Department of Emergency Medicine, University of Maryland School of Medicine, Baltimore, MD 21201, USA
[6] Department of Pharmacology, Dadasaheb Balpande College of Pharmacy, Rashtrasant Tukadoji Maharaj Nagpur University, Nagpur 440037, Maharashtra, India
* Correspondence: nali1@ksu.edu.sa (N.A.); kaleemmubin88@gmail.com (M.K.)

Abstract: The current study was performed to isolate keratin from chicken feathers with an intention to develop a keratin–genistein wound-healing hydrogel, along with its in vivo analysis. Pre-formulation aspects were analysed by using FTIR; SEM; HPTLC, while gel was characterized for gel strength, viscosity, spreadability, drug content, etc. Additionally, an in vivo study along with biochemical factors against pro-inflammatory factors and histopathological studies were conducted to determine possible wound-healing and anti-inflammatory effects. Pre-formulation studies revealed the presence of amide bonds with region of dense fibrous keratin and an internal porous network in extracted keratin, which corresponds with standard keratin. Evaluation of optimised keratin–genistein hydrogel indicated the development of neutral, non-sticky hydrogel which spread evenly on the skin. In vivo studies in rats indicate higher degrees of wound-healing in combined hydrogel (94.65%) for a duration of 14 days as compared to an individual hydrogel formulation with the development of the epidermis and excessive proliferation of fibrous connective tissue indicating wound repair. Furthermore, the hydrogel inhibited the overexpression of IL-6 gene along with other pro-inflammatory factors, indicating its anti-inflammatory effects. In order to find out the possibility of closure of wounds and anti-inflammatory properties of the novel product, an in vivo investigation into the healing of wounds in laboratory animals was carried out through biochemical (ELISA and qRT-PCR) analyses against inflammatory markers (IL-2, IL-6, IL-1, IL-10, and COX-2) and histopathological (liver, skin, and the kidneys) investigations. Based on the results, we conclude that keratin–genistein hydrogel is a promising therapeutic molecule for the management of wound repair.

Keywords: chicken feather; keratin; genistein; hydrogel; wound healing; ELISA; qRT-PCR

Citation: Mahajan, N.M.; Wanaskar, K.; Ali, N.; Mahapatra, D.K.; Iqbal, M.; Bhat, A.R.; Kaleem, M. Innovative Wound Healing Hydrogel Containing Chicken Feather Keratin and Soy Isoflavone Genistein: In Vivo Studies. *Gels* **2023**, *9*, 462. https://doi.org/10.3390/gels9060462

Academic Editor: Shige Wang

Received: 17 May 2023
Revised: 30 May 2023
Accepted: 31 May 2023
Published: 5 June 2023

1. Introduction

For healthcare systems, skin wounds represent a significant burden. Their high rates of morbidity chronification, and relapse, pose a significant strain on healthcare-related aspects of living, elevating both individual and societal expenses as well as posing a problem for healthcare systems across the globe [1–3]. Based on the ageing of the population and a rise in the rates of diabetes, overweight, and heart disease, skin wounds and their potential to become chronic will likely grow more common [2,4]. As a result, managing wounds is a major issue that is becoming more and more prevalent globally [4,5]. As such, it is crucial to develop new products that can accelerate the healing process.

When the integrity of any tissue is damaged, a wound develops (such as when the skin or a muscle becomes damaged, a bone is fractured, or burns occur) [6]. A wound could occur from a fall, surgery, an infection, circumstances, or another event [7]. Several antioxidant genes, including catalase (CAT) and glutathione peroxide (GPx), are expressed more frequently during the natural healing of wounds, indicating a possible function for free-radical-scavenging enzymes [8].

In order to promote cellular growth and cell-guided tissue development, the protein layer has a distinct three-dimensional architecture. Hair, nails, and other areas of the skin contain large amounts of the naturally occurring protein keratin [9]. In comparison with the other proteins, it contains greater levels of cysteine [10]. The protein is hard, resilient, and lighter because the sulphur atoms in the remnants of cysteine frequently interconnect with each other [11]. Chicken feathers can serve as a good source of protein due to their elevated keratin protein level [12].

Genistein, which is extracted from soybean, is used as an inhibitor of epidermal growth factor (EGF)-induced proliferation [11,13]. It promotes the synthesis of collagen, which is important for preserving the equilibrium of the skin as well as accelerating wound repair [12]. In human dermal fibroblasts, genistein has a direct influence on the signalling pathways which govern collagen production [14]. T-BHP (t-butylhydroperoxide) induces collagen production in the presence of oxidative stress [15]. Because of its antioxidant effects, this phytochemical guards human dermal fibroblasts from oxidative-stress-related production of collagen suppression [16]. Through changing the inflammatory reaction, it further modulates the healing of wounds [17,18]. Through enhancing antioxidant activity while regulating the expression of cytokines that are pro-inflammatory in the initial stages of wound repair, genistein supplements decrease the risk of oxidative stress [19]. Without proper antioxidant activities, wound healing might be delayed, or severe tissue damage can occur [20]. Isoflavone (genistein) in soybean has been reported to have potent antioxidant properties and also possess pharmaceutical potency with oestrogen-like activity [21].

The current study was performed to isolate keratin protein from chicken feathers with the intention to develop a keratin–genistein-combination-based wound-healing hydrogel. Since hydrogel polymer material can absorb and hold a greater amount of exudates, it was rationally chosen for this study. The pre-formulation aspects, such as protein content, drug–excipient interactions, etc., were determined through standard methods and sophisticated instruments (UV, FT-IR, SEM, and HPTLC). The prepared gel was evaluated comprehensively with respect to the physicochemical parameter (gel strength, viscosity, pH, spreadability, drug content, etc.). An in vivo wound-healing study in experimental animals followed by biochemical assays (ELISA and qRT-PCR) against pro-inflammatory factors (IL-2, IL-6, IL-1β, IL-10, and COX-2) and histopathological (liver, skin, and kidney) studies were performed to determine the possible wound closure and anti-inflammatory effects of the innovative product.

2. Results and Discussion

2.1. Characterisation of Keratin

2.1.1. Protein Content

The protein content as determined by the Kjeldahl method showed 86.89% *w*/*w* of protein content from 500 mg of keratin sample. This was found to be a good percentage of protein content available in the keratin sample, as it was expected to contain above 80% *w*/*w* [22].

2.1.2. Amino Acid Profiling

As per the literature, total amino acid contents must exceed 70% of the keratin obtained from the chicken feathers. The keratin sample displayed the presence of aspartic acid, alanine, glutamic acid, serine, threonine, and tyrosine in the range of 0.36 to 12.1 g/100 g.

These values corresponded with the reference range of standard keratin (Table 1). Serine, glutamic acid, and aspartic acid were found in higher concentrations than tyrosine, alanine, and threonine [16].

Table 1. Amino acid profiling of extracted keratin.

Test Parameter	Test Unit (g/100 g)	Reference Range (g/100 g)
1. Aspartic Acid	2. 3.15	3. 4.70
4. Threonine	5. 2.81	6. 4.10
7. Serine	8. 12.10	9. 13.57
10. Alanine	11. 2.41	12. 3.64
13. Glutamic Acid	14. 5.22	15. 9.10
16. Tyrosine	17. 0.36	18. 1.84

2.1.3. Fourier-Transform Infrared (FT-IR) Spectroscopy

The absorption bands that were observed in the keratin sample are mainly because of the amide bonds (-CONH-). Among nine amide bands, five were prominently observed, viz., amide-I, II, III, A and B. The signal at approximately 3300 cm^{-1} corresponds to amide A. The Amide B band has been reported and appeared within 3056 to 3075 cm^{-1}. The strong peaks at 1654 cm^{-1} were assigned to amide-I, which is closely related to the carbonyl groups. Amide–II was observed within 1480–1580 cm^{-1}, which is related to symmetric bending vibrations of –NH and –CN. The bands of approximately 1230–1240 cm^{-1} were observed and associated with β-sheet of amide-III. All the characteristic bands that were observed in the FT-IR spectrum of extracted keratin were indicative of its similar secondary structure as that of the standard [23].

The FT-IR spectrum of pure genistein shows a broad band at 3464.15 cm^{-1}, corresponding to the phenolic–OH group, and carbonyl stretching vibrations were observed at 1625.99 cm^{-1}. The C-H aromatic band was observed at 3041.74 cm^{-1}, whereas the band of carbonyl functionality C=O was observed at 1159.22, and that of the C-O stretching band was seen at 1022.27 cm^{-1}. Aromatic C=C stretching bands were observed in the range of 1600–1400 cm^{-1}. All the characteristic bands were observed in the FT-IR spectrum of pure genistein, confirming its identity and purity [24,25]

The FT-IR spectrum of the physical mixture of keratin and genistein with Carbopol 940 and other excipients was found to retain all major peaks shown in pure keratin and genistein. It revealed no interaction of the drug component with the excipients, and thus, the formulation is considered stable and compatible, as shown in Figure 1.

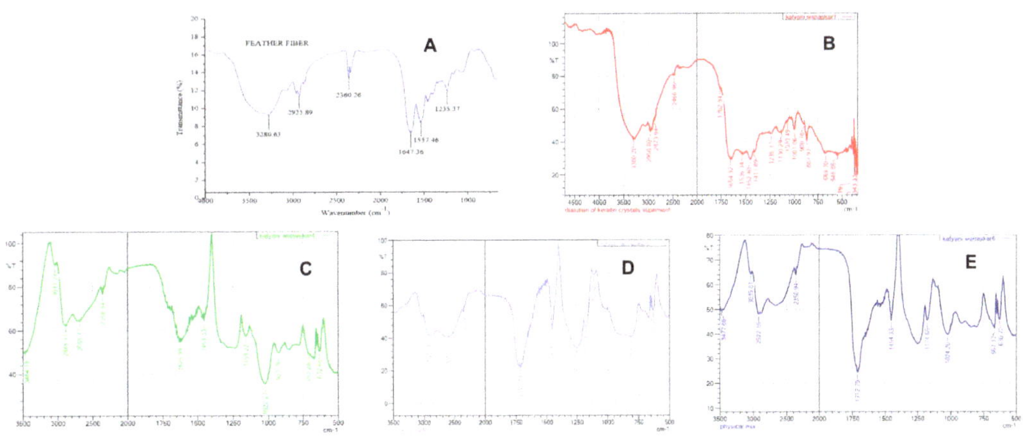

Figure 1. FT-IR spectra of (**A**) reference keratin; (**B**) extracted keratin; (**C**) genistein; (**D**) Carbopol 934; and (**E**) physical mixture.

2.1.4. High-Performance Thin Layer Chromatography (HPTLC)

The solvent system butanol: acetic acid: water in the ratio of 6.5:3.5:1 presented Rf value of 0.63 for the extracted keratin, which was also found in the keratin gel with an alike Rf value (Table 2). The chromatogram of the keratin gel produced the related characteristic peaks with a% corresponding area of 0.57 and 0.95, respectively (Figure 2). The most abundant constituents of keratin were found in the final formulation, which represents the intactness of the keratin in the gel.

Table 2. HPTLC interpretation of various gel formulations.

Peak	Start Position (Rf)	Start Height (AU)	Max Position (Rf)	Max Height (AU)	Max %	End Position (Rf)	End Height (AU)	Area (AU)	Area %
Keratin extract and keratin gel formulation									
3	0.62	6.5	0.63	12.8	2.52	0.64	0.4	90.3	0.57
7	0.62	7.7	0.63	18.9	3.54	0.65	6.4	159.5	0.95
Genistein extract and genistein gel formulation									
6	0.56	42.9	0.56	49.2	8.81	0.60	38.7	802.0	11.03
5	0.48	26.8	0.50	41.4	13.37	0.51	25.6	448.2	13.34
Keratin, genistein extract and keratin, genistein combination gel (Mobile Phase: butanol: acetic acid: water)									
10	0.56	3.5	0.58	14.3	2.33	0.60	9.8	175.0	1.46
9	0.49	5.9	0.51	13.0	2.35	0.55	1.2	135.9	1.78
Keratin, genistein extract and keratin, genistein combination gel (Mobile Phase: chloroform: methanol)									
4	0.73	21.3	0.74	27.7	6.09	0.75	19.5	279.2	3.68
8	0.68	15.2	0.70	25.4	7.02	0.72	14.2	271.8	4.87

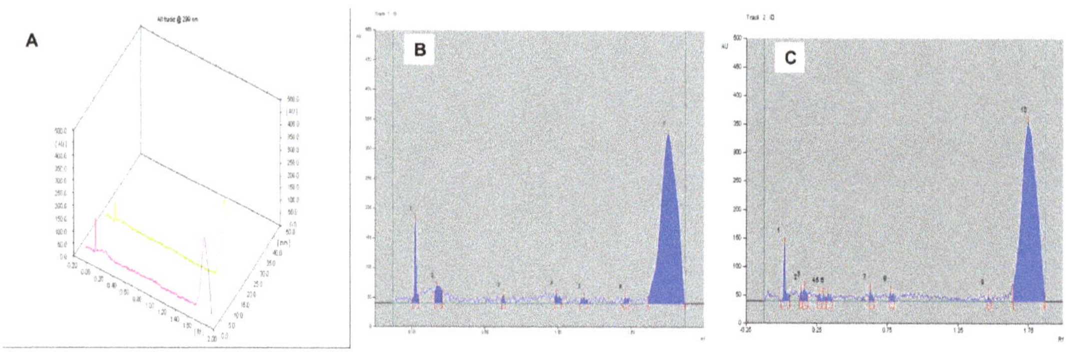

Figure 2. HPTLC of (**A**) Comparative chromatograms of extracted keratin and keratin gel; (**B**) extracted keratin; and (**C**) keratin gel.

2.1.5. Scanning Electron Microscopy (SEM)

The morphology of the keratin crystals at both 5000× (Figure 3B) and 1000× (Figure 3C) was investigated via SEM analysis. The microphotographs presented a region of dense fibrous keratin with an internal porous network (Figure 3A).

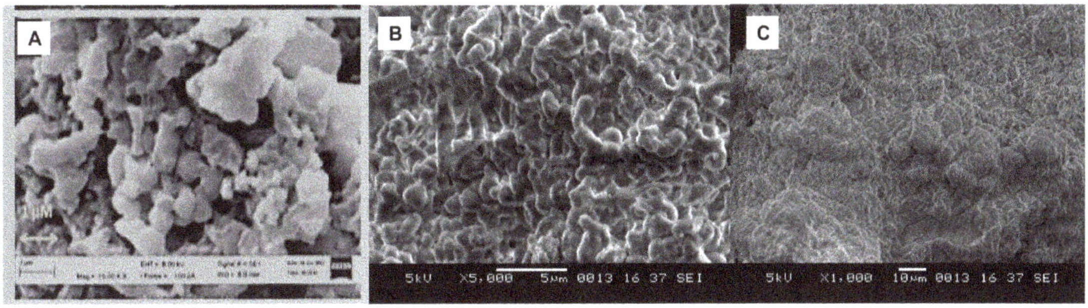

Figure 3. Microphotograph of (**A**) reference keratin; (**B**) extracted keratin at 5000×; and (**C**) extracted keratin at 1000×.

2.2. *Characterisation of Genistein*

High-Performance Thin Layer Chromatography (HPTLC)

The solvent system chloroform: methanol in the ratio of 10:1 presented an Rf value of 0.49, which was also found in genistein gel (Rf = 0.50) (Table 2). The chromatogram of the genistein gel produced the related characteristic peaks with a % corresponding area of 11.03 and 13.34, respectively (Figure 4). The most abundant constituents of genistein were found in the final formulation, which represents the intactness of the genistein in the gel.

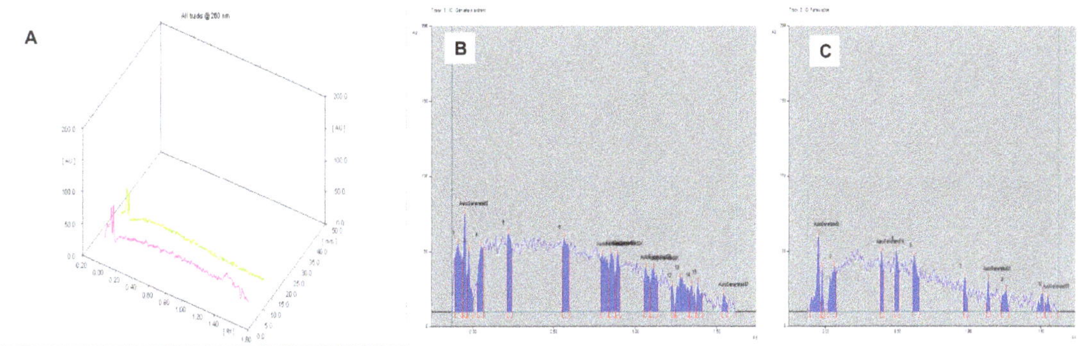

Figure 4. HPTLC of (**A**) comparative chromatograms of extracted genistein and genistein gel; (**B**) extracted genistein; and (**C**) genistein gel.

2.3. *Characterisation of Gel Formulations*

2.3.1. General Characterisation

The initial batches of the gel base were characterised for their consistency, pH, viscosity, and spreadability. Based on the study base formula, F1 was found to be difficult and not acceptable. The viscosity was also found to be quite high. In contrast, the F2 showed a slightly acidic nature and sticky content. The formulation F3 represented a non-sticky consistency, which favoured the ease of its application to the skin surface, and the pH was found to be neutral. Furthermore, the drug content study was performed to determine the content of the drug that was present in the optimised formulation. An optimised gel formulation was found to be a satisfactory drug content that ranges from 91 to 100% of keratin and genistein extract (Table 3).

Table 3. Post-stability characterisation of gel formulations.

Gel Type	Gel Strength (Sec)	Drug Content Mean ± SD	pH Mean ± SD		Viscosity (Cps) Mean ± SD		Spreadability (gm·cm/s) Mean ± SD	
			Pre-Stability	Post-Stability	Pre-Stability	Post-Stability	Pre-Stability	Post-Stability
Keratin gel	4	96.07 ± 0.7214	7.6 ± 0.15	6.8 ± 0.15	470,913 ± 1978.7	433,342 ± 9495.1	9.78 ± 1.0	8.4 ± 0.85
Genistein gel	3	91.89 ± 0.2134	6.8 ± 0.1	6.4 ± 0.20	738,063 ± 2411.4	706,920 ± 1126.8	14 ± 1.4	13.65 ± 0.63
Keratin–Genistein gel	4	93.21 ± 0.764 (Keratin) 92.79 ± 0.7543 (Genistein)	7.2 ± 0.2	6.5 ± 0.26	714,547.53 ± 3040.3	700,446 ± 6777.86	12.56 ± 0.92	10.83 ± 1.2

2.3.2. High-Performance Thin Layer Chromatography (HPTLC)

The most abundant constituents of keratin and genistein were found in the final formulation. The mobile phase system comprised butanol: acetic acid: water (6.5:3.5:1 $v/v/v$); the keratin extract showed an Rf value of 0.58, which was also found in the keratin–genistein gel (Rf value of 0.51) (Table 2). The most abundant constituents of keratin and genistein were found in the final formulation. The mobile phase system comprised chloroform: methanol (10:1 v/v); the genistein extract demonstrated an Rf value of 0.74, which was equivalent to the keratin–genistein gel (Rf value of 0.70) (Table 2). However, only genistein runs in this solvent system. The chromatogram of keratin, genistein gel revealed similar characteristic peaks to that of genistein, which represented the intactness of the genistein in the gel formulation. The corresponding % area was found to be about the same (3.68 and 4.87) (Figure 5).

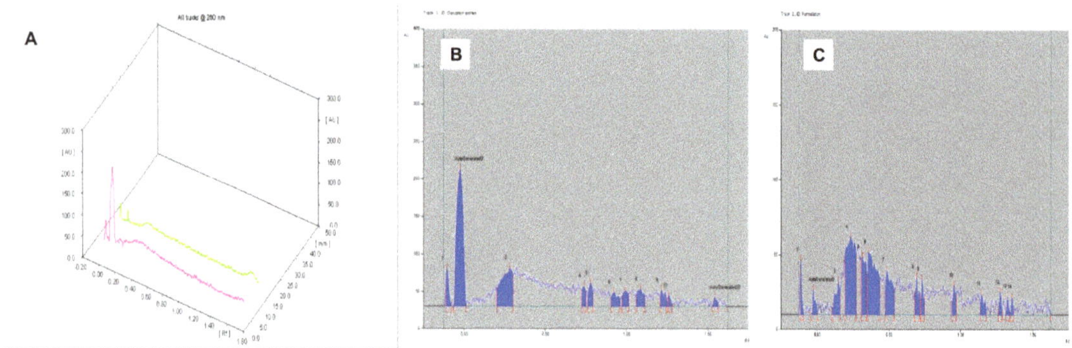

Figure 5. HPTLC of (**A**) comparative chromatograms of extracted keratin, genistein, and keratin–genistein combination gel; (**B**) extracted keratin, genistein; and (**C**) keratin–genistein gel under mobile phase system chloroform: methanol (10:1 v/v).

2.4. Stability Studies

The stability studies were carried out for the optimised formulation at 40 ± 2 °C temperature and 65 ± 5% RH for a period of 90 days [26]. The formulation presented good stability over the time regimen with no remarkable change in the pH, viscosity, and spreadability test profile (Table 4).

Table 4. Short-term (90 days) stability studies of gel formulations.

Formulation	pH Mean ± SD		Viscosity (cPs) Mean ± SD		Spreadability (gm·cm/s) Mean ± SD	
	0 Days	90 Days	0 Days	90 Days	0 Days	90 Days
Keratin gel	7.6 ± 0.15	6.8 ± 0.15	470,913 ± 19,787.5	433,342 ± 9495.15	9.78 ± 1.0	8.4 ± 0.85
Genistein gel	6.8 ± 0.1	6.4 ± 0.20	738,063 ± 24,114.7	706,920 ± 11,268.6	14 ± 1.4	13.65 ± 0.63
Keratin-Genistein gel	7.2 ± 0.2	6.5 ± 0.26	714,547.53 ± 30,403.9	700,446 ± 6777.86	12.56 ± 0.92	10.83 ± 1.2

2.5. Wound-Healing Activity

Keratin–genistein gel formulation showed higher degrees of wound healing (94.65%) for the duration of 14 days as compared to the keratin gel formulation (87.57%) and genistein (87.40%) gel formulation (Table 5). Furthermore, the keratin–genistein-combination gel formulation presented a comparable activity to that of the marketed hydrogel product (96.66%), as well as the control group (85.98%) (Figure 6). The stages of healing by the keratin–genistein-combination gel formulation in the experimental animal model are described in Figure 7. The present observations and potential roles of genistein in promoting wound healing were found to be comparable with the results stated in previous reports on early-stage cutaneous wound healing, refractory wound healing in type-1 diabetes, skin wound repair in an incision model, and wound healing in ovariectomised mice [27].

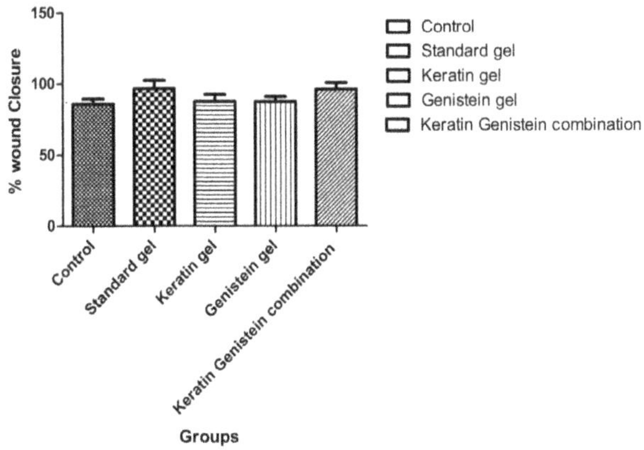

Figure 6. Percentage of wound closure after treatment with various gel formulations.

Table 5. Wound healing after 7 days and 14 days.

Groups	Day 7 (% Wound Healing) Mean ± SD	Day 14 (% Wound Healing) Mean ± SD
Keratin gel	54.29 ± 4.170	87.57 ± 5.00
Genistein gel	48.51 ± 7.88	87.40 ± 3.57
Keratin and genistein combination gel	51.18 ± 2.043	94.65 ± 4.64
Standard gel	54.99 ± 13.63	96.66 ± 5.77
Control	48.15 ± 9.32	85.98 ± 3.53

2.6. Histopathological Investigations

After 14 days, the wound-healing activity produces a normal architecture of the skin. As compared to the genistein gel and keratin gel, the keratin–genistein combination gel produced proper development of the epidermis, along with the proliferation of fibrous connective tissue and the growth of hair follicles. After 14 days, the normal functioning of the kidney and liver was studied, where wound healing does not produce any toxic effect. Genistein purposely solved the development of epidermal growth on the skin. The control sample of the skin showed failing to develop an epidermis and no proliferation of fibrous connective tissue. The keratin sample presented the development of epidermis and the proliferation of fibrous connective tissue, indicating wound healing. The genistein gel sample revealed the development of epidermis and excessive proliferation of fibrous connective tissue, demonstrating wound healing [23,28].

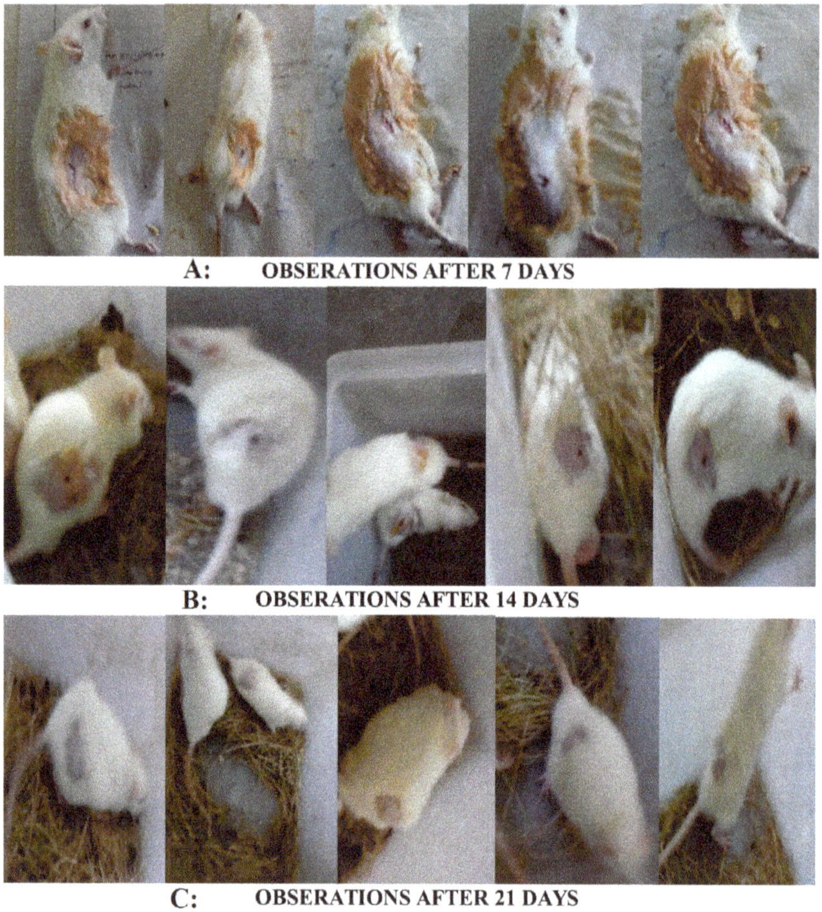

Figure 7. This figure indicates morphological studies of wound-healing activity of formulation after (**A**) 7 days; (**B**) 14 days; and (**C**) 21 days.

The combination of keratin and genistein showed proper development of epidermis, along with the proliferation of fibrous connective tissue and growth of hair follicles. The standard sample displayed the best wound healing, with development of epidermis and proliferation of fibroblastic cells and hair follicles. The control sample of the liver revealed vacuolar degenerative changes in hepatocytes around the central vein. The keratin, the genistein, the standard, and the combination sample showed the normal architecture of hepatocytes around the central vein. The control sample of the kidney revealed the inflamed architecture of glomeruli and tubules. In contrast, keratin, genistein, standard, and the combination sample showed the normal architecture of glomeruli and tubules (Figure 8).

In the case of vital organ toxicity studies, the histological section of liver shows normal architecture of hepatocytes around the central vein as compared to vacuolar degenerative changes in positive control group. Kidney histology shows the normal architecture of glomeruli and tubules. This result proved the safety of keratin and genistein for medicinal use.

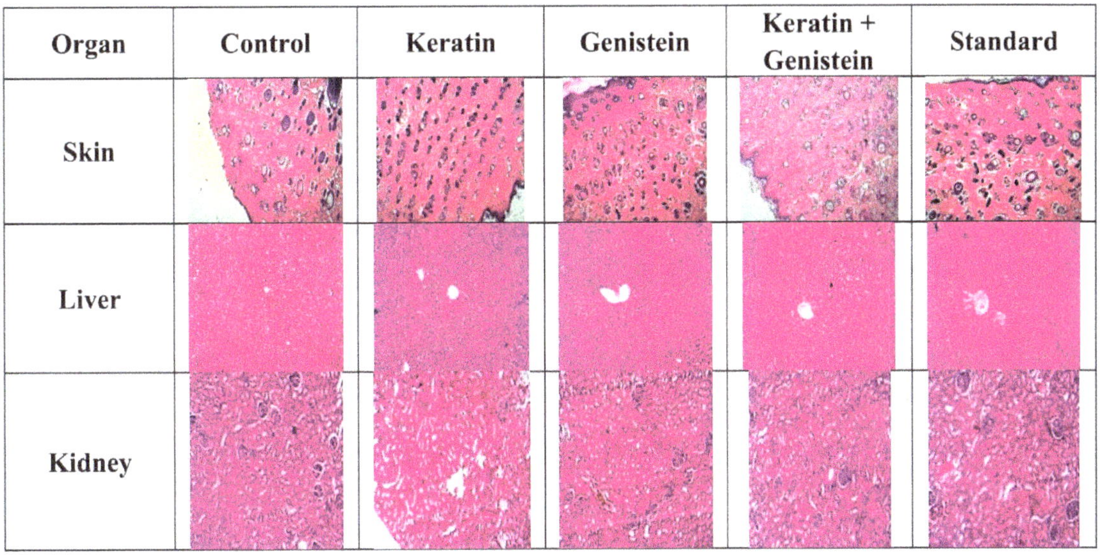

Figure 8. This figure shows that histopathological studies of skin, liver, and kidney.

2.7. Biochemical Investigations

The biochemical investigations through ELISA and qRT-PCR studies strongly advocated that the keratin–genistein combination gel-formulation-treated group produced a comparatively high impact over the IL-6 expression via the downregulation of the overexpressed IL-6 gene in contrast to the working control group (Figure 9). In comparison with other pro-inflammatory factors such as COX-2, IL-2, IL-10, and IL-1β expressed in the liver tissues, genistein and keratin, both in combination, have the capacity to regulate the higher expression of mRNA of IL-6 in inflamed liver tissues and, therefore, play an imperative role in the suppression of inflammation [29,30]. The exploration of the underlying mechanism(s) supported our hypothesis that the combination markedly normalised ($p < 0.001$) the upregulation of the IL-6 gene and exerted an anti-inflammatory effect by modulating the molecular targets (Table 6). The efficacy of the gel combination was observed to be quite comparable with the positive control (marketed standard formulation) at an analogous dose.

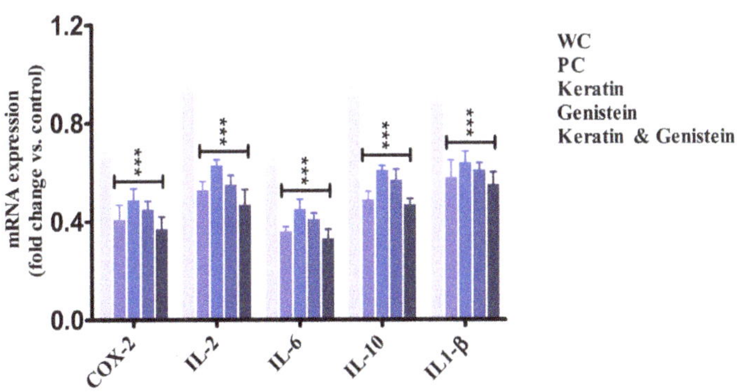

Figure 9. This figure shows mRNA expression levels of COX-2, IL-2, IL-6, IL-10, and IL1-β in the liver through qRT-PCR. [WC: Working Control; PC: Positive Control; *** represents the level of significance in comparision with working control].

Table 6. ELISA measurement of interleukins (IL-2, IL-6, IL-1β, and IL-10) and COX-2 levels in liver after the application of keratin, genistein, and keratin–genistein gel.

Parameters	NC	WC	PC	Keratin	Genistein	Keratin and Genistein
COX-2 (pg/mL)	188.78 ± 8.09	449.24 ± 12.95	210.84 ± 10.51 ***	240.97 ± 8.46 ***	223.79 ± 11.49 ***	196.25 ± 9.61 ***
IL-2 (pg/mL)	492.27 ± 12.03	1103.32 ± 19.37	594.38 ± 12.45 ***	620.31 ± 16.85 ***	601.76 ± 18.17 ***	519.37 ± 11.51 ***
IL-6 (pg/mL)	148.91 ± 4.61	327.94 ± 8.16	179.24 ± 5.34 ***	204.37 ± 9.17 ***	191.38 ± 7.42 ***	163.84 ± 7.19 ***
IL-10 (pg/mL)	630.94 ± 20.37	1294.01 ± 25.79	764.85 ± 23.14 ***	804.59 ± 22.67 ***	789.41 ± 11.64 ***	681.93 ± 9.43 ***
IL1-β (pg/mL)	552.37 ± 18.19	1109.86 ± 24.17	624.87 ± 12.37 ***	655.49 ± 11.27 ***	634.59 ± 9.73 ***	611.48 ± 10.70 ***

Legends: NC: Normal Control; WC: Working Control; PC: Positive Control. *** Represents the level of significance in comparision with normal control/working control.

The inflammation-reducing potential of genistein by reducing the levels of cytokines (IL-1β, IL-2, IL-6, and IL-10) was found to have complied with the previous reports on the protective actions of this natural product in gut inflammation, inflammation that is related to colitis, TNF-α-induced endothelial inflammation, and vascular inflammation [23,31].

3. Conclusions

A safe and effective keratin, genistein wound-healing combined gel formulation was successfully developed and comprehensively characterised through sophisticated analytical techniques. The major active components that were needed for the wound-healing activity were identified, along with other components, by using TLC, FTIR, and amino acid profiling techniques. The present investigation revealed that the fabrication of keratin gel showed excellent wound-healing properties, and with the combination of the inhibitor of EGF- induced proliferation, genistein showed noteworthy wound-closure activity. The fabricated formulations demonstrated good spreadability with a pH suitable for human skin application. The formulations also retained a good stability condition over a period of 90 days. The ability of the natural-component-containing gel product as an effective wound healer was confirmed via histopathological studies. From all the biochemical studies, it may be concluded that the novel keratin, genistein wound-healing gel effectually reduced the pro-inflammatory factors (IL-2, IL-6, IL-1β, IL-10, and COX-2), which led to acute lessening of inflammation. Comparing the gel to other commercially available formulations, its desirable formulation texture characteristics, in addition to its viscosity, pH, spreadability, and wound-healing function, contributed to the quality and stability attributes of the gel.

4. Material and Methods

4.1. Material

Keratin was successfully extracted from chicken feathers based on an existing method. Genistein was purchased from a local shop, namely, Green Heaven Lab Ltd., Nagpur, Maharashtra, India. Carbopol 934 was procured from Apex Drug House Ltd., Mumbai, India. LOBA Chemie Ltd., Mumbai, India, supplied analytical-grade methyl paraben and propyl paraben. Cremophor RH-40 was obtained from PIOMA Chemicals, Mumbai, India. The chemicals that were used in the study were of analytical grade and obtained in the highest pure form. Double-distilled water (Borosil, Mumbai, Maharashtra, India) was utilised for the experiment. The standard keratin was obtained from Sigma-Aldrich, St. Louis, MO, USA, with CAS No. 69430-36-0.

4.2. Extraction of Keratin from Chicken Feathers

Chicken feathers of a sufficient quantity were obtained from a nearby poultry farm and soaked in ether for a period of 24 h. The wet feathers were exposed to sunlight until they were completely dried. The dried feathers were further cut uniformly into small pieces and blended. The blended feathers were sealed carefully in a sealed plastic bag. In a 2 L conical flask, 0.5 M of sodium sulphide solution was prepared, and the weighed amounts of 50 g (Shimadzu BL-220H, Kyoto, Japan) of blended chicken feathers were added. The solution was stirred by a mechanical stirrer (Biotechniques BIPO5B, India) for the duration of 6 hrs at a temperature of 30 °C and the target pH range of 10–12. Furthermore, the solution was filtered off to acquire the supernatant liquid. The supernatant liquid was taken into a beaker and stirred. In a conical flask, ammonium sulphate solution (0.7 g/mL) was prepared in double-distilled water with continuous stirring until all particles were dissolved [32,33]. The solution was subsequently filtered, and the obtained filtrate was added drop wise to the previous breaker containing the feather filtrate solution in a ratio of 1:1 with fast stirring. The supernatant precipitates were collected separately. The collected solid particles were added to ethanol (100 mL), and a pH of 12 was maintained with a NaOH solution (0.1 N) with continuous stirring. Additionally, the solution was placed overnight on a mechanical shaker. Finally, the solution was filtered to procure the solid particles and the supernatant liquid. The solid particles were dissolved in 0.5 N NaOH solution. Ultimately, the liquid was collected, and the solid materials were discarded. The liquid sample was then dried, and the keratin crystals were collected [22].

4.3. *Characterisation of Extracted Keratin*

4.3.1. Protein Content

The protein content of the keratin sample was estimated via the Kjeldahl digestion flask method, as reported in the standard literature [34]. Protein content was determined by multiplying nitrogen contents by conventional factor 6.25, as the average nitrogen (N) content of most of the proteins was found to be about 16%. The % total nitrogen content in the sample was determined from the following formula [32]:

$$\begin{aligned}\text{\% Total Nitrogen} = {} & (\text{Blank titration} - \text{Sample titration}) \times 1.4\ (\text{constant factor for Nitrogen}) \\ & \times 2 \times \text{Exact normality of 0.1N NaOH solution}\end{aligned}$$

4.3.2. Amino Acid Analysis

The amino acids present in the feather keratin were estimated quantitatively by using the post-column derivatisation method with Ninhydrin applications followed by an amino acid analyser. By using 6 N HCl, the keratin samples (n = 4) were hydrolysed in a glass tube. The hydrolysed amino acids were then characterised via reverse-phase HPLC (RP-HPLC) [35].

4.3.3. Scanning Electron Microscopy (SEM)

The morphology of the extracted keratin sample was characterised by electron microscopy (Jeol JSM-6360A, Japan). The dried sample was sprinkled lightly over the aluminium stub (3–5 nm; 75 s; 40 W) connected with double adhesive tape. The sample underwent a gold coating to enhance the electrical conductance. The stub containing the gold-coated sample was scanned randomly, and at 10 kV acceleration voltage, the photomicrographs were recorded [32].

4.3.4. Fourier-Transform Infrared (FT-IR) Spectroscopy

The Fourier-transform infrared (FT-IR) spectroscopic (Shimadzu IR-Affinity-1, Kyoto, Japan) investigation of keratin, genistein, Carbopol 934, and a physical mixture in the range of 4000 cm^{-1} to 400 cm^{-1} was studied to determine the inherent stability and compatibility of the formulation through various chemical groups. Any such changes perceived through the spectra are a clear reflection of some possible interactions between the drug and the polymer [36].

4.3.5. High-Performance Thin Layer Chromatography (HPTLC)

(a) For Keratin

High-performance thin-layer chromatography (HPTLC) (Camag, North Carolina, USA) was performed for the keratin samples on 10 cm × 10 cm F_{254} HPTLC Si-60, diol, CN, NH2, and RP-18 plates. The sample and standard were spotted as bands of 5 mm width by using the automatic applicator Desaga AS 30 (Heidelberg, Germany). The samples were applied in the triplicate manner in the form of a band of 6 mm in the concentration of 10 µL. The distance between the tracks was kept to 10 mm at an application rate of 110 nL/s for the application of the bands. The plates were developed in a linear ascending mode at 85 mm in a chromatographic horizontal Teflon DS chamber (Chromdes, Lublin, Poland) that was previously saturated with some vapours of the mobile phase of butanol: acetic acid: water in the ratio of 4:1:1 *v*/*v*/*v*. The plates were developed after drying in the warm air streams and observed at 254 nm and 366 nm under a UV lamp. Densitometric scanning was performed by using the TLC scanner-III in the pre-absorbance/reflectance mode [37].

(b) For Genistein

In the present study, 10 cm × 10 cm F_{254} HPTLC Si 60, diol, and CN, NH2, and RP-18 plates were used for thin-layer chromatography (TLC). A Desaga AS 30 automatic applicator (Heidelberg, Germany) was used to identify standards and samples within 5 mm bands. In a chromatographic horizontal Teflon DS chamber (Chromdes, Lublin, Poland), which had been earlier saturated with mobile phase vapours, the plates were developed to an 85 mm thickness. The plates were examined at 254 nm and 366 nm under a UV light after being dried in a stream of hot air. On pre-coated TLC Plates (10 cm × 10 cm) and by using a concentration of 10 µL, the samples were used in triplicates in the shape of a band with a width of 6 mm via a Linomat-V sample applicator. A 110 nL/s rate of application was used to apply bands while maintaining a 10 mm space among the tracks. In a glass chamber with twin troughs that had been saturated with mobile phase, linear ascending progression was performed. By utilising the TLC scanner III in the pre-absorbance/reflectance mode, densitometric scanning was carried out. HPTLC-plates were used to hold the commercial and enzymatic genistein hydrolysates. Chloroform: methanol (10:1 *v*/*v*) was used for the development as long as the solvent front travelled to the highest point of the plate [33,37].

4.4. Characterisation of Genistein

A pre-formulation study of the drug moiety was performed to determine the authenticity of the drug and to estimate its associative properties that may influence formulation development. As per the methods that were comprehensively described in the previous section, the λmax of the compound through UV-Vis spectroscopy, FT-IR, and HPTLC were performed.

4.5. Formulation Development (Gels)

4.5.1. Optimisation of Gel Base

A lack of adequate activity of antioxidants, tissue injury, or healing of wounds may both be impaired. In-process quality control (IPQC) is a vital phase in the formulation of keratin protein wound-healing gel. In order to ensure that the final goods remain uniform from run to run, efficient over a period of time, and risk-free to be utilised, certain particular tests are carried out at different stages of the process of production. The first step that is performed is to verify the unprocessed products to determine if they agree with the standards that have already been established. As a result, the pH, viscosity, and amount of drug of the test formulation were examined. Certain distinct quality control criteria, such as texture profile, must be properly taken into account in the production of a gel (a semisolid formulation), with the aim of increasing stability, elegance, and subsequent approval from customers. The 1% gel formulation was prepared by taking 1 g of Carbopol 934, which was further added to 100 mL of distilled water and hydrolysed for duration of 24 h.

4.5.2. Formulation Development of 2% Keratin Gel

In total, 1 g of keratin was taken (powder form obtained through freeze-drying technique) and triturated with 3 mL of Cremophor RH-40 (as a solubiliser) in a mortar and pestle. Furthermore, 50 g of hydrated Carbopol 934 was added to the above content. The preservatives (propyl paraben and methyl paraben) were added to the required amount.

The pH of the formulation was maintained by adding Triethanolamine to form the gel product (Table 7).

Table 7. Composition of keratin gel, genistein gel, and keratin–genistein gel (50 g quantity).

2% KERATIN GEL		1% GENISTEIN GEL		KERATIN–GENISTEIN GEL	
Ingredients	**Quantity**	**Ingredients**	**Quantity**	**Ingredients**	**Quantity**
Carbopol 934	1% *w*/*v*	Carbopol 934	1% *w*/*v*	Carbopol 934	1% *w*/*v*
Keratin	1 g	Genistein	0.5 g	Keratin	1 g
Cremophor RH 40	3 mL	Methyl paraben	0.01 g	Genistein	0.5 g
Methyl Paraben	0.01 g	Propyl paraben	0.01 g	Cremophor RH 40	3 mL
Propyl Paraben	0.01 g	Triethanolamine	q.s.	Methyl paraben	0.01 g
Triethanolamine	q.s.			Propyl paraben	0.01 g
				Triethanolamine	q.s.

4.5.3. Formulation Development of 1% Genistein Gel

In total, 50 g of hydrated Carbopol 934 was taken, and 0.5 g of genistein was added. The content was mixed vigorously with a stirrer, and in required quantities, the preservatives (propyl paraben and methyl paraben) were added. The pH of the product was maintained by using triethanolamine (Table 7).

4.5.4. Formulation Development of Keratin–Genistein Combination Gel

In total, 0.5 g genistein was dissolved in 50 mL of hydrated Carbopol 934 (Sample 1). A total of 1 g keratin crystal was mixed with 3 mL of the solubiliser Cremophor RH-40 in a mortar and pestle (Sample 2). Sample 2 was added to sample 1, and preservatives (propyl paraben and methyl paraben) were added. The pH of the gel formulation was maintained by using Triethanolamine (Table 7).

4.6. Characterisation of the Gel Formulation

4.6.1. Physical Evaluation

Gel Strength

The gel strength was tested by the gel strength apparatus in accordance with the method. The 100 mL measuring cylinder was filled with the gel formulations, and then the gel was loaded with a 50 g piston. The gel strength was determined from the time (in seconds) required to move the piston 5 cm down through the gel. In this case, it took more than 5 min to drop the apparatus into the gel strength, and it was described by the minimum weights that pushed the apparatus 5 cm down through the gel.

pH

The pH of the gel formulations was determined by utilising the digital type of pH meter (Global-pH-DPH-507, Delhi, India) in the calibrated range of 4 and 7. In total, 1 g of the gel was placed into double-distilled water, and both the reference electrode and the glass electrode were dipped completely into the formulations to achieve the pH of the products. The experiment was executed in a triplicate form.

Viscosity

The viscosity of the products was determined by using the Brookfield Viscometer (DV-E1 model, Middleboro, MA, USA) at 0.5 rpm and using the no. 64 spindle in the temperature range of 25 $\pm$ 1 °C. The experiment was carried out in a triplicate manner.

Spreadability

On the principle of slip–drag, the spreadability of the gel formulations was determined. The required amount of the formulation was placed on the ground slide, and the test material was prepared such that it was sandwiched by another glass slide which comprised a hook system. A pressure of 1 kg of mass was applied to the slides to remove the air entrapped between the thin films. The excess content of the formulation was swiped off from the edges of the slide, and the top slide was dragged with a force equivalent to 50 g intensity. The time required for the top slide to cover 7.5 cm distance was estimated and concluded from the following formula:

$$S = M \times L/T$$

where S = Spreadability coefficient, M = weight applied, L = length moved by the glass slides, and T = time taken to separate the glass slides completely from each other [38].

4.6.2. Keratin and Genistein Contents

In each formulation, the drug content was estimated by taking 0.1 mL of sample in a 10 mL volumetric flask and making up the volume with double-distilled water to prepare a stock solution. Further dilutions were prepared and analysed spectrophotometrically at λmax 222 nm and 280 nm, respectively. The experiment was performed in a triplicate manner.

4.6.3. HPTLC

A previously discussed method, as stated above, with the mobile phase in butanol: acetic acid: water in the ratio of 4:1:1 *v/v/v* for keratin gel and chloroform: methanol in the ratio of 10:1 *v/v* for genistein gel was followed.

4.6.4. Short-Term Stability Study

The formulation was subjected to accelerated conditions of temperature (40 ± 2 °C) and humidity (75 ± 5%) for the duration of 90 days. The formulations were packed in aluminium foil and kept inside the PVC bottle. After the completion of the study, the essential physical parameters and drug content were determined at pre-determined intervals of 30 days, 60 days, and, finally, 90 days. The obtained observations/parameters were then compared with the pharmacopoeia guidelines.

4.7. Animals

Male albino rats [(n = 30) (age: 5 to 6 weeks; weight range: 200–300g)] were randomly divided into five groups. There were four hydrogel-treated experimental groups, including (1) feather keratin hydrogel, (2) genistein hydrogel, (3) keratin–genistein hydrogel, and (4) marketed preparation, while the fifth group was a non-treated control group (six rats in each group) after obtaining animal study permission from the institutional animal ethical committee (IAEC) as per the Committee for Control and Supervision of Experiments on Animals (CCSEA) guidelines and histopathological studies with approval numbers 1426/PO/Re/S/11/CPCSEA. The animals were kept in the animal house, provided free access to water, also fed standard food, and maintained under a controlled environment (temperature 22 ± 2 °C; humidity 50–60% RH; and 12/12 h light and dark conditions) with proper hygiene. All animals were housed under standard environmental conditions to prevent any significant antibacterial interference with the wound due to external factors [39–41].

4.8. Wound Healing Study

For evaluating the wound-healing properties of the feather keratin, genistein, and keratin–genistein hydrogel, and marketed preparation, a full-thickness excision model was applied for the evaluation. All animals were anaesthetised with an intra-peritoneal ketamine injection (100 mg/mL) and xylazine, ilium-xylazine-20 (20 mg/mL). For anaesthetic administration, 7.5 mL of ketamine and 5 mL of xylazine were diluted with 7.5 mL of double-distilled water. A dosage of 0.2 mL/100 g body weight was used for the induction of complete anaesthesia. The dorsal hair was removed with a hair-removing cream (Anne French, Ahmedabad, Gujarat, India). A partial-thickness skin wound of dimension 1.5 × 1.5 cm was prepared via excision of the dorsal skin of the animal by using surgical scissors and forceps. Subsequently, the excised wound of the experimental group of rats was covered with the tested hydrogel (applied 1 fingertip unit (FTU)), and the bare wound was kept as a negative control. The probable changes observed in the wound area with time progression were measured by using a method of transparency-based digital imaging. On the 7th and 14th postoperative days, the reconstructed skins of the wounds were excised and fixed in 4% paraformaldehyde for histological observations and collagen deposition determination. At the selected post-wound intervals, photographs were taken. The wound closure was estimated according to the following equation [42,43]:

$$\text{Wound closure (\%)} = [A_0 - A_t]/A_0 \times 100$$

where A_0 is the initial wound area and at is the wound area at the same time interval of "t" days.

4.9. Biochemical Analysis

4.9.1. Estimation of Cytokines by ELISA

Enzyme-Linked Immunosorbent Assay (ELISA) was performed to determine the cytokine levels of IL-1β, IL-2, IL-6, and IL-10 in liver tissue as per the manufacturer's instructions [44].

4.9.2. qRT-PCR Analysis

In total, 10 mg of tissue samples from all the groups was used to isolate total mRNA via the Triazole technique in order to examine the amount of mRNA that is expressed for the gene of interest. The RNeasy small kit was used to purify the mRNA as per the manufacturer's instructions. Finally, qRT-PCR (Real-Time Quantitative Reverse Transcription PCR) was performed via the standard instrument with the help of the PCR master mix. Denaturation of the cDNA was performed at 94 °C for 5 min, followed by annealing at 58 °C for 30 s and subsequent extension at 72 °C for 35 s. Forty times repetition of the cycle was set by using qRT-PCR, which helps in the detection of the amplified DNA in real-time. To normalise mRNA, GAPDH was used as a housekeeping reference. For all the treated groups, ΔCt values was normalised with untreated control samples (ΔCt = Ct gene of interest − Ct housekeeping gene). The relative changes in the expression level of a particular gene were measured in terms of 2−ΔΔCt (ΔΔCt = ΔCt test − ΔCt control) [45,46].

4.10. Histopathological Investigations

The repaired skin of the wounds was taken out and preserved in a 4% paraformaldehyde solution on the 7th and 14th day following surgery for histopathological examination and to determine where collagen was deposited. To check the biocompatibility of biological macromolecules, viz., keratin and genistein, major tissues such as the liver and kidney were resected and fixed in 10% formalin for histological observations with respect to major changes associated with the tissue toxicity study.

Via selective staining with the dye haematoxylin and eosin, the histological analyses of wound skin, liver, and kidney were performed. The 7 mm to 10 mm thick sections were cut and initially fixed in formal calcium and embedded in paraffin wax. By employing xylene, the cut sections were de-waxed, treated with decreasing amounts of alcohol for hydration, and stained with haematoxylin. Then, the sections were dehydrated with alcohol to 70% concentration and further stained with 1% alcoholic eosin solution. Subsequently, the sections were differentiated in 90% alcohol solution, cleaned with xylene, mounted cautiously, and observed under the microscope [47,48].

4.11. Digital Image Analysis

From the histological images of tissues, luminescence was identified and transformed into the heights of the final 3D, interactive surface plot, which recognises the pure stained areas adjacent to the adjoining sampling in a square dimension and represents the background stains, nuclear sites, and other characteristics. A computerised scoring was allocated by examining and determining the staining pattern by using the Image J® program.

4.12. Statistical Analysis

All values were provided as means with standard deviations (±SD), with n = 6; ** represented significance at $p < 0.01$ versus the control group, and * represented significance at $p < 0.05$ in comparison with the standard gel and keratin–genistein combination groups. By using Statistica® v.17.0 software, one-way ANOVA was used to analyse the data, and then Bonferroni's Multiple Comparison Test was performed.

Author Contributions: N.A. and N.M.M. analysed the data and prepared the article; K.W. carried out the experimental portion. The design, in vivo work processing, and tissue histopathological analysis interpretation were all performed with help from M.I., N.A. and M.K. The manuscript has been reviewed by N.M.M., M.K., D.K.M., N.A. and A.R.B. N.M.M. supervised all experimental work and provided valuable recommendations for troubleshooting throughout the project. All authors have read and agreed to the published version of the manuscript.

Funding: This research was funded through Researchers Supporting Project no. RSPD2023R734, King Saud University, Riyadh, Saudi Arabia.

Institutional Review Board Statement: The animal study protocol was recommended and approved by the Institutional Animal Ethical Committee of Dadasaheb Balpande College of Pharmacy, Besa, Nagpur, Maharashtra, India (Registration no. 1426/PO/Re/S/11/CPCSEA).

Informed Consent Statement: Not applicable.

Data Availability Statement: Not applicable.

Acknowledgments: The authors extend their appreciation to "Researchers Supporting Project number (RSPD2023R734),, King Saud University, Riyadh, Saudi Arabia" for their financial support. The authors are thankful to Dadasaheb Balpande College of Pharmacy for their support and for providing facilities.

Conflicts of Interest: The authors declare no conflict to interest.

Abbreviations

TLC	Thin-Layer Chromatography
HPTLC	High-Performance Thin-Layer Chromatography
FTIR	Fourier-Transform Infrared Spectroscopy
SEM	Scanning Electron Microscopy
ELISA	Enzyme-Linked Immunoassay
qRT-PCR	Real-Time Quantitative Reverse Transcription PCR
IAEC	Institutional Animal Ethical Committee
CCSEA	Committee for Control and Supervision of Experiments on Animals
T-BHP	t-butylhydroperoxide
COX-2	Cyclooxygenase-2
IPQC	In-Process Quality Control

References

1. Guest, J.F.; Vowden, K.; Vowden, P. The health economic burden that acute and chronic wounds impose on an average clinical commissioning group/health board in the UK. *J. Wound Care* **2017**, *26*, 292–303. [CrossRef]
2. Cheng, Q.; Gibb, M.; Graves, N.; Finlayson, K.; Pacella, R.E. Cost-effectiveness analysis of guideline-based optimal care for venous leg ulcers in Australia. *BMC Health Serv. Res.* **2018**, *18*, 421. [CrossRef]
3. Olsson, M.; Järbrink, K.; Divakar, U.; Bajpai, R.; Upton, Z.; Schmidtchen, A.; Car, J. The humanistic and economic burden of chronic wounds: A systematic review. *Wound Repair Regen.* **2019**, *27*, 114–125. [CrossRef]
4. Gray, T.A.; Rhodes, S.; Atkinson, R.; Rothwell, K.; Wilson, P.; Dumville, J.C.; Cullum, N. Opportunities for better value wound care: A multiservice, cross-sectional survey of complex wounds and their care in a UK community population. *BMJ Open* **2018**, *8*, e019440. [CrossRef]

5. Casado-Díaz, A.; La Torre, M.; Priego-Capote, F.; Verdú-Soriano, J.; Lázaro-Martínez, J.L.; Rodríguez-Mañas, L.; Pérez, M.B.; Tunez, I. EHO-85: A Multifunctional Amorphous Hydrogel for Wound Healing Containing *Olea europaea* Leaf Extract: Effects on Wound Microenvironment and Preclinical Evaluation. *J. Clin. Med.* **2022**, *11*, 1229. [CrossRef] [PubMed]
6. Pastar, I.; Stojadinovic, O.; Yin, N.C.; Ramirez, H.; Nusbaum, A.G.; Sawaya, A.; Patel, S.B.; Khalid, L.; Isseroff, R.R.; Tomic-Canic, M. Epithelialization in Wound Healing: A Comprehensive Review. *Adv. Wound Care* **2014**, *3*, 445–464. [CrossRef] [PubMed]
7. Boateng, J.S.; Matthews, K.H.; Stevens, H.N.; Eccleston, G.M. Wound Healing Dressings and Drug Delivery Systems: A Review. *J. Pharm. Sci.* **2008**, *97*, 2892–2923. [CrossRef] [PubMed]
8. Chattopadhyay, N.; Zastre, J.; Wong, H.-L.; Wu, X.Y.; Bendayan, R. Solid Lipid Nanoparticles Enhance the Delivery of the HIV Protease Inhibitor, Atazanavir, by a Human Brain Endothelial Cell Line. *Pharm. Res.* **2008**, *25*, 2262–2271. [CrossRef]
9. Rouse, J.G.; Van Dyke, M.E. A Review of Keratin-Based Biomaterials for Biomedical Applications. *Materials* **2010**, *3*, 999–1014. [CrossRef]
10. Khosa, M.A.; Ullah, A.J.J.F.P. A sustainable role of keratin biopolymer in green chemistry: A review. *J. Food Process. Beverages* **2013**, *1*, 8.
11. Hill, P.; Brantley, H.; Van Dyke, M. Some properties of keratin biomaterials: Kerateines. *Biomaterials* **2010**, *31*, 585–593. [CrossRef]
12. Kumaran, P.; Gupta, A.; Sharma, S. Synthesis of wound-healing keratin hydrogels using chicken feathers proteins and its properties. *Int. J. Pharm. Pharm. Sci.* **2017**, *9*, 171–178. [CrossRef]
13. Ar, B.; Choudhury, K. Study on the Effect of Genistein, a Soy Isoflavone in Insulin Tolerance in Albino Rat (*Rattus albicans*). *Sch. Acad. J. Biosci.* **2018**, *6*, 366–371.
14. Emmerson, E.; Campbell, L.; Ashcroft, G.S.; Hardman, M.J. The phytoestrogen genistein promotes wound healing by multiple independent mechanisms. *Mol. Cell. Endocrinol.* **2010**, *321*, 184–193. [CrossRef]
15. Duchnik, E.; Kruk, J.; Baranowska-Bosiacka, I.; Pilutin, A.; Maleszka, R.; Marchlewicz, M. Effects of the soy isoflavones, genistein and daidzein, on male rats' skin. *Adv. Dermatol. Allergol.* **2019**, *36*, 760–766. [CrossRef] [PubMed]
16. Marini, H.R.; Polito, F.; Altavilla, D.; Irrera, N.; Minutoli, L.; Calò, M.; Adamo, E.B.; Vaccaro, M.; Squadrito, F.; Bitto, A. Genistein aglycone improves skin repair in an incisional model of wound healing: A comparison with raloxifene and oestradiol in ovariectomized rats. *Br. J. Pharmacol.* **2010**, *160*, 1185–1194. [CrossRef] [PubMed]
17. Emmerson, E.; Hardman, M.J. The role of estrogen deficiency in skin ageing and wound healing. *Biogerontology* **2012**, *13*, 3–20. [CrossRef] [PubMed]
18. Rohrmann, S.; Shvetsov, Y.B.; Morimoto, Y.; Wilkens, L.R.; Monroe, K.R.; Le Marchand, L.; Franke, A.A.; Kolonel, L.N.; Maskarinec, G. Self-reported dietary flavonoid intake and serum markers of inflammation: The multiethnic cohort. *Cancer Causes Control* **2018**, *29*, 601–607. [CrossRef]
19. Irrera, N.; Pizzino, G.; D'anna, R.; Vaccaro, M.; Arcoraci, V.; Squadrito, F.; Altavilla, D.; Bitto, A. Dietary Management of Skin Health: The Role of Genistein. *Nutrients* **2017**, *9*, 622. [CrossRef]
20. Hwang, K.; Chung, R.S.; Schmitt, J.M.; Buck, D.; Winn, S.R.; Hollinger, J.O. The Effect of Topical Genistein on Soft Tissue Wound Healing in Rats. *J. Histotechnol.* **2001**, *24*, 95–99. [CrossRef]
21. Cooke, P.S.; Selvaraj, V.; Yellayi, S. Genistein, estrogen receptors, and the acquired immune response. *J. Nutr.* **2006**, *136*, 704–708. [CrossRef]
22. Sinkiewicz, I.; Śliwińska, A.; Staroszczyk, H.; Kołodziejska, I. Alternative Methods of Preparation of Soluble Keratin from Chicken Feathers. *Waste Biomass Valorization* **2017**, *8*, 1043–1048. [CrossRef]
23. Sadowska-Krowicka, H.; Mannick, E.E.; Oliver, P.D.; Sandoval, M.; Zhang, X.-J.; Eloby-Childess, S.; Clark, D.A.; Miller, M.J.S. Genistein and Gut Inflammation: Role of Nitric Oxide. *Proc. Soc. Exp. Boil. Med.* **1998**, *217*, 351–357. [CrossRef] [PubMed]
24. Kharwade, R.; Ali, N.; Gangane, P.; Pawar, K.; More, S.; Iqbal, M. DOE-Assisted Formulation, Optimization, and Characterization of Tioconazole-Loaded Transferosomal Hydrogel for the Effective Treatment of Atopic Dermatitis: In Vitro and In Vivo Evaluation. *Gels* **2023**, *9*, 303. [CrossRef] [PubMed]
25. Seibel, J.; Molzberger, A.F.; Hertrampf, T.; Laudenbach-Leschowski, U.; Diel, P. Oral treatment with genistein reduces the expression of molecular and biochemical markers of inflammation in a rat model of chronic TNBS-induced colitis. *Eur. J. Nutr.* **2009**, *48*, 213–220. [CrossRef] [PubMed]
26. Kharwade, R.S.; Mahajan, N.M. Formulation and Evaluation of Nanostructured Lipid Carriers Based Anti-Inflammatory Gel for Topical Drug Delivery System. *Asian J. Pharm. Clin. Res.* **2019**, *12*, 286–291. [CrossRef]
27. Tie, L.; An, Y.; Han, J.; Xiao, Y.; Xiaokaiti, Y.; Fan, S.; Liu, S.; Chen, A.F.; Li, X. Genistein accelerates refractory wound healing by suppressing superoxide and FoxO1/iNOS pathway in type 1 diabetes. *J. Nutr. Biochem.* **2013**, *24*, 88–96. [CrossRef]
28. Stipcevic, T.; Piljac, A.; Piljac, G. Enhanced healing of full-thickness burn wounds using di-rhamnolipid. *Burns* **2006**, *32*, 24–34. [CrossRef]
29. Woods, A. Syndecans: Transmembrane modulators of adhesion and matrix assembly. *J. Clin. Investig.* **2001**, *107*, 935–941. [CrossRef]
30. Greene, D.K.; Tumova, S.; Couchman, J.R.; Woods, A. Syndecan-4 Associates with α-Actinin. *J. Biol. Chem.* **2003**, *278*, 7617–7623. [CrossRef]
31. Diegelmann, R.F.; Cohen, I.K.; Kaplan, A.M. The Role of Macrophages in Wound Repair: A review. *Plast. Reconstr. Surg.* **1981**, *68*, 107–113. [CrossRef]

32. Gupta, A.; Perumal, R.; Yunus, R.B.M.; Kamarudin, N.B. Extraction of keratin protein from chicken feather. *J. Chem. Chem. Eng.* **2011**, *6*, 732.
33. Šafarič, R.; Zemljič, L.F.; Novak, M.; Dugonik, B.; Bratina, B.; Gubeljak, N.; Bolka, S.; Strnad, S. Preparation and Characterisation of Waste Poultry Feathers Composite Fibreboards. *Materials* **2020**, *13*, 4964. [CrossRef]
34. Wang, H.; Pampati, N.; McCormick, W.M.; Bhattacharyya, L. Protein Nitrogen Determination by Kjeldahl Digestion and Ion Chromatography. *J. Pharm. Sci.* **2016**, *105*, 1851–1857. [CrossRef] [PubMed]
35. Fekkes, D.; van Dalen, A.; Edelman, M.; Voskuilen, A. Validation of the determination of amino acids in plasma by high-performance liquid chromatography using automated pre-column derivatization with o-phthaldialdehyde. *J. Chromatogr. B Biomed. Sci. Appl.* **1995**, *669*, 177–186. [CrossRef] [PubMed]
36. Blackburn, S.; Lowther, A.G. The action of organic acids on some fibrous proteins: The oxidation of wool keratin. *Biochem. J.* **1951**, *49*, 554–559. [CrossRef] [PubMed]
37. Elhawary, Y.; Badria, F. Mode of Action of Centella Asiatica in Wound Healing Using Immunohistochemical Studies. *Egypt. Dent. J.* **2011**, *57*, 1.
38. Ullah, N.; Amin, A.; Farid, A.; Selim, S.; Rashid, S.A.; Aziz, M.I.; Kamran, S.H.; Khan, M.A.; Khan, N.R.; Mashal, S.; et al. Development and Evaluation of Essential Oil-Based Nanoemulgel Formulation for the Treatment of Oral Bacterial Infections. *Gels* **2023**, *9*, 252. [CrossRef]
39. Jia, Z.; Babu, P.V.A.; Si, H.; Nallasamy, P.; Zhu, H.; Zhen, W.; Misra, H.P.; Li, Y.; Liu, D. Genistein inhibits TNF-α-induced endothelial inflammation through the protein kinase pathway A and improves vascular inflammation in C57BL/6 mice. *Int. J. Cardiol.* **2013**, *168*, 2637–2645. [CrossRef]
40. Fule, R.; Kaleem, M.; Asar, T.O.; Rashid, A.; Shaik, R.A.; Eid, B.G.; Nasrullah, M.Z.; Ahmad, A.; Kazmi, I. Formulation, Optimization and Evaluation of Cytarabine-Loaded Iron Oxide Nanoparticles: From In Vitro to In Vivo Evaluation of Anticancer Activity. *Nanomaterials* **2023**, *13*, 175. [CrossRef]
41. Omer, A.B.; Dalhat, M.H.; Khan, M.K.; Afzal, O.; Altamimi, A.S.A.; Alzarea, S.I.; Almalki, W.H.; Kazmi, I. Butin Mitigates Memory Impairment in Streptozotocin-Induced Diabetic Rats by Inhibiting Oxidative Stress and Inflammatory Responses. *Metabolites* **2022**, *12*, 1050. [CrossRef]
42. Bahramsoltani, R.; Farzaei, M.H.; Rahimi, R. Medicinal plants and their natural components as future drugs for the treatment of burn wounds: An integrative review. *Arch. Dermatol. Res.* **2014**, *306*, 601–617. [CrossRef] [PubMed]
43. Midwood, K.S.; Williams, L.V.; Schwarzbauer, J.E. Tissue repair and the dynamics of the extracellular matrix. *Int. J. Biochem. Cell Biol.* **2004**, *36*, 1031–1037. [CrossRef]
44. Singh, A.K.; Bhadauria, A.S.; Kumar, U.; Raj, V.; Rai, A.; Kumar, P.; Keshari, A.K.; Kumar, D.; Maity, B.; Nath, S.; et al. Novel Indole-fused benzo-oxazepines (IFBOs) inhibit invasion of hepatocellular carcinoma by targeting IL-6 mediated JAK2/STAT3 oncogenic signals. *Sci. Rep.* **2018**, *8*, 5932. [CrossRef] [PubMed]
45. Park, E.; Lee, S.M.; Jung, I.-K.; Lim, Y.; Kim, J.-H. Effects of genistein on early-stage cutaneous wound healing. *Biochem. Biophys. Res. Commun.* **2011**, *410*, 514–519. [CrossRef]
46. Alghamdi, R.M.; Hassan, M.A.; Kaleem, M.; Kayali, A.; Halwani, M.A.; Zamzami, M.A.; Choudhry, H.; Alhosin, M. Targeting Itch/p73 pathway by thymoquinone as a novel therapeutic strategy for cancers with p53 mutation. *Eur. J. Cell Sci.* **2020**, *2*, 20–26. [CrossRef]
47. Al Khalaf, A.K.; Abdulrahman, A.O.; Kaleem, M.; Nur, S.M.; Asseri, A.H.; Choudhry, H.; Khan, M.I. Comparative Analysis of the Impact of Urolithins on the Composition of the Gut Microbiota in Normal-Diet Fed Rats. *Nutrients* **2021**, *13*, 3885. [CrossRef] [PubMed]
48. Kaleem, M.; Haque, S.E. Evaluation of Cardioprotective Role of Vinpocetine in Isoproterenolinduced Myocardial Infarction in Rats. *J. Pharm. Res.* **2015**, *9*, 408–414. Available online: https://www.researchgate.net/publication/342480421_Evaluation_of_Cardioprotective_role_of_Vinpocetine_in_Isoproterenol-induced_Myocardial_Infarction_in_Rats (accessed on 29 November 2021).

gels

MDPI

Article

Formulation of Intranasal Mucoadhesive Thermotriggered In Situ Gel Containing Mirtazapine as an Antidepressant Drug

Mohammed Ghazwani [1], Rajalakshimi Vasudevan [2], Geetha Kandasamy [3], Naredla. Manusri [4], Praveen Devanandan [5], Ranadheer Chowdary Puvvada [5], Vinoth Prabhu Veeramani [6], Premalatha Paulsamy [7], Krishnaraju Venkatesan [2], Kumarappan Chidmabaram [2] and Rajeshri Dhurke [4,*]

1 Department of Pharmaceutics, College of Pharmacy, King Khalid University, Abha 62529, Saudi Arabia
2 Department of Pharmacology and Toxicology, College of Pharmacy, King Khalid University, Abha 62529, Saudi Arabia
3 Department of Clinical Pharmacy, College of Pharmacy, King Khalid University, Abha 62529, Saudi Arabia
4 Department of Pharmaceutics, St. Peter's Institute of Pharmaceutical Sciences, Hanamkonda 506001, Telangana, India
5 Department of Pharmacy Practice, St. Peter's Institute of Pharmaceutical Sciences, Hanamkonda 506001, Telangana, India
6 Department of Pharmacy Practice, Faculty of Pharmacy, University of Tabuk, Tabuk 71491, Saudi Arabia
7 College of Nursing, King Khalid University, Abha 62529, Saudi Arabia
* Correspondence: rajeshri.dhurke@gmail.com; Tel.: +91-7893190191

Abstract: The purpose of the present work was to develop nanoemulsion-based formulations of mirtazapine for intranasal delivery using a spray actuator to target the brain for treating depression. Research on the solubility of medications in different oils, surfactants, co-surfactants, and solvents has been done. Using pseudo-ternary phase diagrams, the various ratios of the surfactant and co-surfactant mix were computed. Thermotriggered nanoemulsion was formulated using different concentrations of poloxamer 407 (i.e., 15%, 15.5%, 16%, 16.5% up to 22%). Similarly, mucoadhesive nanoemulsion using 0.1% Carbopol and water-based plain nanoemulsions were also prepared for comparative assessment. The developed nanoemulsions were analyzed for physicochemical properties, i.e., physical appearance, pH, viscosity, and drug content. Drug-excipient incompatibility was determined by Fourier transform infrared spectral (FTIR) analysis and differential scanning calorimetry (DSC). In vitro drug diffusion studies were conducted for optimized formulations. Among the three formulations, RD1 showed the highest percentage of drug release. Ex vivo drug diffusion studies were conducted on freshly excised sheep nasal mucosa with Franz diffusion cell simulated nasal fluid (SNF) for all three formulations up to 6 h, and the thermotriggered nanoemulsion (RD1) showed 71.42% drug release with 42.64 nm particle size and a poly dispersity index of 0.354. The zeta potential was found to be −6.58. Based on the above data, it was concluded that thermotriggered nanoemulsion (RD1) has great potential to be used as an intranasal gel for treating depression in patients. It can offer great benefits by reducing dosing frequency and improving bioavailability of mirtazapine by direct nose-to-brain delivery.

Keywords: intranasal; thermosensitive gels; mirtazapine; nose to brain delivery; nanoemulsions; antidepressant

Citation: Ghazwani, M.; Vasudevan, R.; Kandasamy, G.; Manusri, N.; Devanandan, P.; Puvvada, R.C.; Veeramani, V.P.; Paulsamy, P.; Venkatesan, K.; Chidmabaram, K.; et al. Formulation of Intranasal Mucoadhesive Thermotriggered In Situ Gel Containing Mirtazapine as an Antidepressant Drug. *Gels* **2023**, *9*, 457. https://doi.org/10.3390/gels9060457

Academic Editor: Mario Grassi

Received: 6 April 2023
Revised: 25 May 2023
Accepted: 26 May 2023
Published: 2 June 2023

1. Introduction

Intranasal delivery as a non-invasive drug delivery method has gained popularity in recent years. A wide range of medicinal chemicals may be delivered intranasally for topical, systemic, and central nervous system action since the nasal mucosa provides many advantages as a target tissue for drug delivery. It is currently acknowledged that intranasal drug delivery is a beneficial and trustworthy option compared to oral and parenteral routes. Without a doubt, intranasal medication administration has been utilized extensively for a long time for the symptomatic alleviation, prevention, and treatment of topical nasal diseases [1,2] The nasal mucosa has, however, recently made a significant comeback as a therapeutically effective route for systemic drug administration. In general, pharmacologically active substances with limited stability in gastrointestinal fluids, poor intestinal absorption, and/or significant hepatic first-pass elimination, such as peptides, proteins, and polar medicines, are among the main targets for intranasal delivery. The intranasal delivery method appears to be an effective strategy to get over the blood-brain barrier (BBB) barriers, enabling direct medication delivery of central nervous system (CNS)-active substances during their biophase [3,4].

A nanostructured drug delivery system known as a nanoemulsion (NE) is characterized by its small droplet size and kinetic stability, which prevents flocculation even when the substance is stored for extended periods of time. It is suitable for drug delivery because it has a number of benefits, such as ease of synthesis and optimization, high drug loading capacity, higher drug permeability across mucosal membranes, and increased bioavailability of drug molecules [5–8].

Antidepressant mirtazapine is used to treat moderate to severe depression. The Food and Drug Administration has approved it as the sole tetracyclic antidepressant to treat depression and anxiety. Being a powerful antagonist at postsynaptic 5-HT2 and 5-HT3 (serotonergic) and central noradrenergic receptors, it is also used to treat anxiety by boosting central noradrenergic and serotonergic (5-HT1) neurotransmission [9,10]. Although mirtazapine is quickly absorbed after oral administration, its absolute bioavailability is only 50% due to strong first-pass metabolism. These factors highlight the necessity for an alternate drug delivery technology that can target mirtazapine to the brain with precision. Intranasal administration methods may be predicted to decrease the extensive dispersion of medications to non-targeted areas such systemic/peripheral circulation since pharmaceuticals are preferentially transported to the brain. To ensure that the drug is quickly transported through the nasal mucosa and via the olfactory bulb, the delivery mechanism must be carefully constructed [11,12]. The preparation and characterization of a thermotriggered nanoemulsion loaded with mirtazapine for intranasal administration was the goal of this work. The research was done with the intention of delivering drugs to the brain for a faster onset of action than oral administration, minimizing adverse effects, maximizing therapeutic index, and lowering dosage and dosing frequency.

2. Result and Discussion

2.1. Saturation Solubility Studies

The saturation solubility studies of mirtazapine were carried out in different solvents. Oleic acid was selected as the oil phase from the solubility studies (Figure 1) based on the drug's solubility, and Tween 80 and ethanol were selected as the surfactant and co-surfactant, respectively. Tween 80 not only effectively dissolves drugs but also effectively forms emulsions.

2.2. Construction of Pseudo-Ternary Phase Diagrams

Microemulsions were prepared using oleic acid as the oil phase, Tween 80 as the surfactant, and ethanol as the co-surfactant. Figure 2 represents the pseudo-ternary phase diagrams of oleic acid with various ratios of Tween 80 and ethanol. Based on the pseudo-ternary phase diagrams, a higher ratio of Smix than oil, i.e., 9:1 ratio of Smix:oil was selected. The optimized ratio mixture of surfactant to co-surfactant was 1:3.

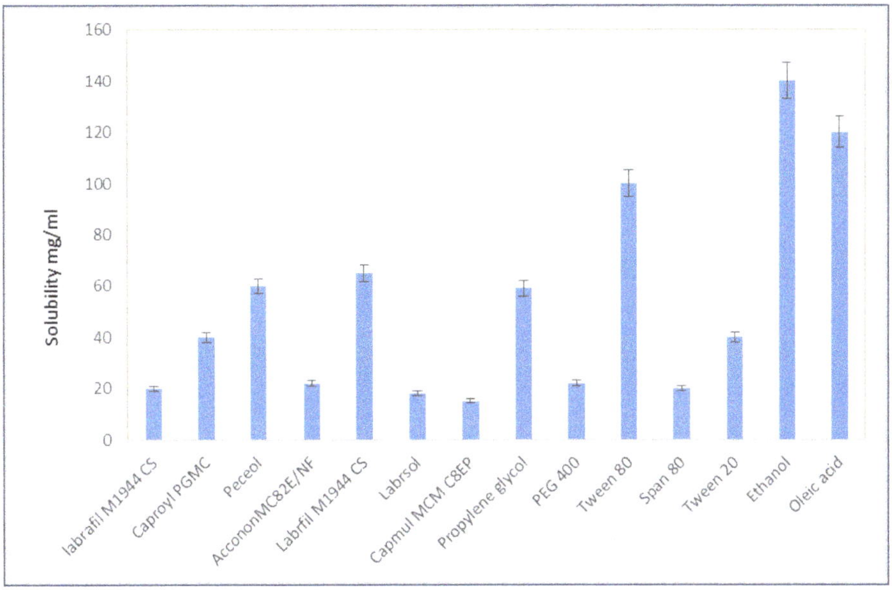

Figure 1. Saturation solubility studies of mirtazapine in different solvents and oils.

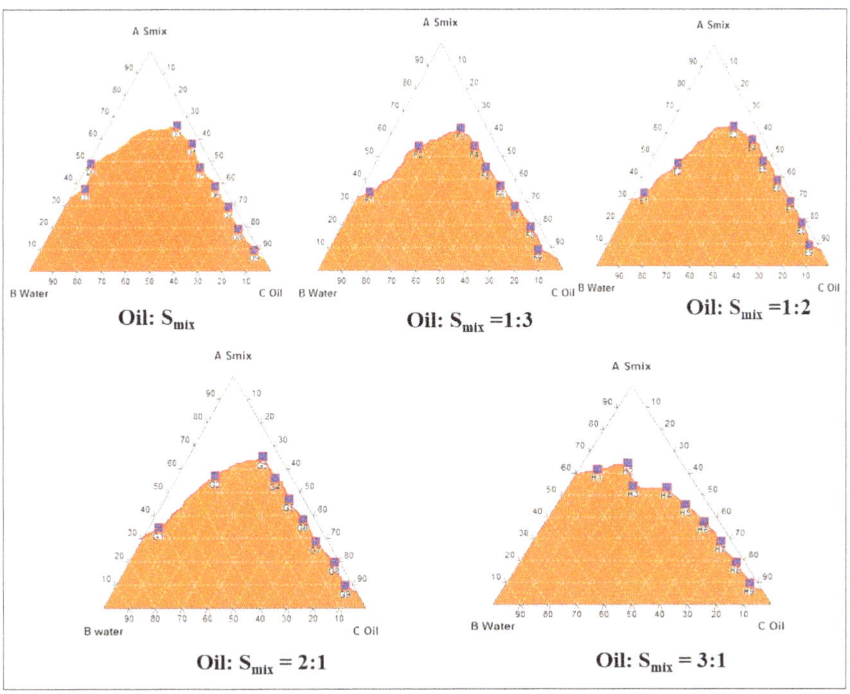

Figure 2. Pseudoternary phase diagrams.

2.3. Development of Microemulsion Formulations

Formulations were developed based on the highest microemulsion zone obtained from pseudo-ternary phase diagrams. A 1:3 ratio of Smix (Tween 80: ethanol) and a 9:1 ratio of oil and Smix was selected, and further various formulations like RD1, RD2, RD3 were formulated. Carbopol 934P was used as a mucoadhesive agent in the formulation, which helps the formulation to remain adhered to the mucosal membrane. A thermotriggered formulation was prepared by changing different concentrations of poloxamer 407, i.e.,15%, 16%, 16.5%, 17%, 17.5%, 18%, 18.5%, 19%, 19.5%, and 20%. Poloxamer 407 of 18.5% concentration was selected based on gelling time and gelling temperature. The optimized formulation (RD1) gave a gelling time of 4 sec and a gelling temperature of 32 °C.

2.4. Characterization of Nanoemulsion

2.4.1. Physicochemical Properties

All the developed nanoemulsion formulations were physically examined and were confirmed to be clear and transparent. The pH of all the formulations was measured using a pH meter (Remi Equipment Pvt. Ltd., Kolkata, India) and was found to be close to nasal pH, indicating an acceptable range, as evident from Table 1.

Table 1. Physicochemical properties of nanoemulsion formulations.

Sr.No	Formulation	Physical Appearance	pH Range	Drug Content (%)	Drug Entrapment Efficiency	Viscosity (cps)	Stability
1	RD1	Transparent	6.2 ± 0.2	99 ± 0.5	82.56 ± 2	96 ± 2	Stable
2	RD2	Transparent	6.0 ± 0.2	98 ± 0.4	72.56 ± 2	91 ± 2	Stable
3	RD3	Transparent	6.1 ± 0.1	99 ± 0.2	72.56 ± 2	87 ± 2	Stable

Thermotriggered nanoemulsion gel formation: The thermotriggered nanoemulsion formulation was added dropwise into distilled water by heating. A consistent thermal gel was formed at 32 °C and was stable.

2.4.2. In Vitro Drug Diffusion Studies

For the created formulations, in vitro release tests were carried out utilizing Franz diffusion cells and diffusion experiments. The study was carried out for up to 6 h for developed thermotriggered nanoemulsion (RD1), mucoadhesive nanoemulsion (RD2), and water-based nanoemulsion (RD3). The results are graphically represented in Figure 3. It can be observed that optimal formulation, i.e., thermotriggered nanoemulsion (RD1), has shown 71.42% drug release after 6 h, whereas mucoadhesive nanoemulsion (RD2) and water-based nanoemulsion (RD3) showed drug releases of 66.12% and 60%, respectively, after 6 h. This indicated that the thermotriggered nanoemulsion (RD1) was better among the three formulations. The high release may be because of the poloxamer 407, which is helping the formulation to form in situ gel as soon as it is sprayed in the nose, while the mucoadhesive properties of carbopol are keeping the nanoemulsion adhered to the mucosal lining. Formulation RD2 lacks poloxamer 407, hence it doesn't get sufficient time for contact with the mucosal area and drains. The least release with RD3 formulation is due to the absence of a mucoadhesive polymer as well as a thermosensitive polymer.

2.4.3. Ex Vivo Permeation Studies

Figure 4 illustrates how mirtazapine thermotriggered nanoemulsions (RD1), mucoadhesive nanoemulsions (RD2), and water-based nanoemulsions (RD3) permeate the nasal mucosa. From the figure, it is evident that pure drug dispersion could show only 25% drug difussion in 6 h, whereas the nanoemulsion formulation showed greatly improved drug diffusion compared with the pure drug. Formulation RD1 showed drug diffusion of more than 80%, whereas RD2 and RD3 only showed drug diffusion in the range of 70–72%. Due to a lack of mucoadhesion and thermosensitive characteristics, the drug in the water-based nanoemulsion (RD3) demonstrated the lowest permeability. As the contact time with the mucosal area was extended, drug penetration in mucoadhesive and thermotriggered nanoemulsions improved. The high transcellular uptake, high solubilization capacity, possibility for improved absorption by gel formation by mucoadhesion, and presence of a poloxamer may all contribute to the higher permeability. Carbopol 934 P may have also improved drug absorption; this may be because it opened tight junctions, which made it easier for pharmaceuticals to go through paracellular pathways [13].

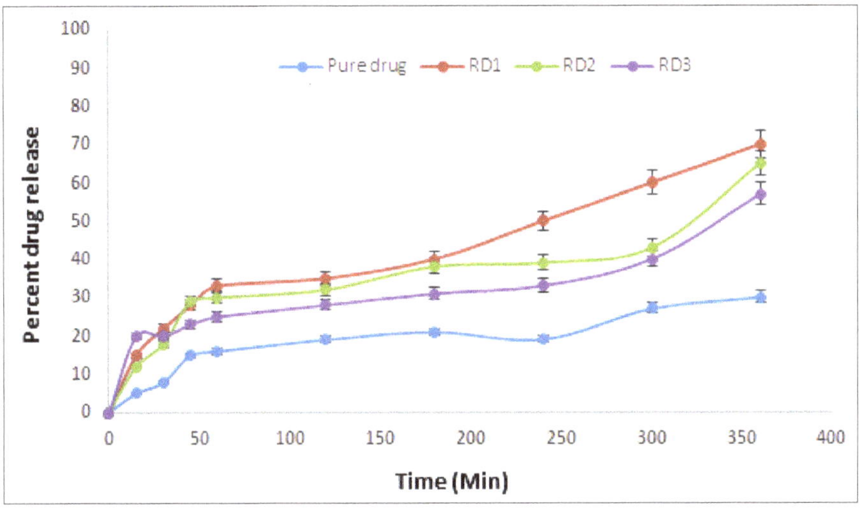

Figure 3. In vitro diffusion profiles of the formulations.

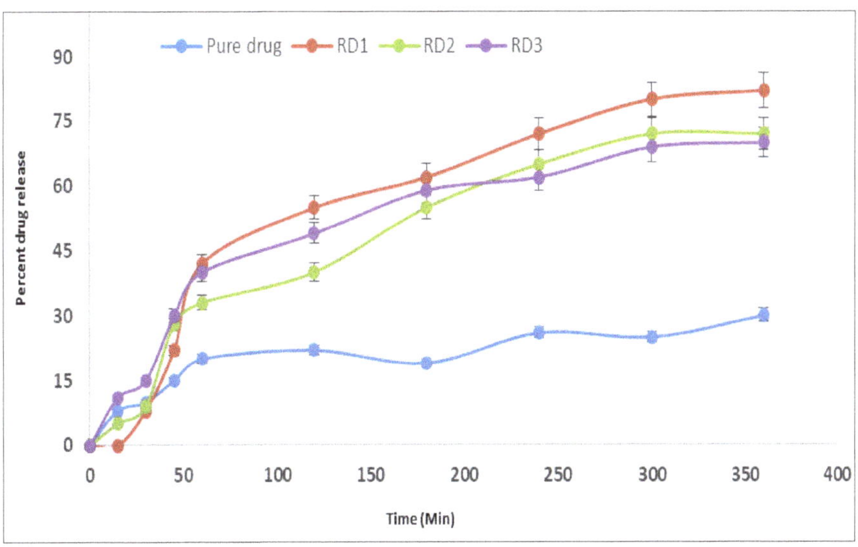

Figure 4. Ex vivo diffusion profiles of the formulations.

2.4.4. Fourier Transform Infrared Spectroscopy (FTIR)

FTIR studies were performed to determine the compatibility between pure drug and excipients. FTIR spectrum of pure mirtazapine, thermotriggered formulation, and poloxamer 407 characteristic peaks are given in Figure 5. From the results, it was concluded that there is no considerable change in the drug when mixed with the excipients. It can be concluded that there is no drug excipient incompatibility.

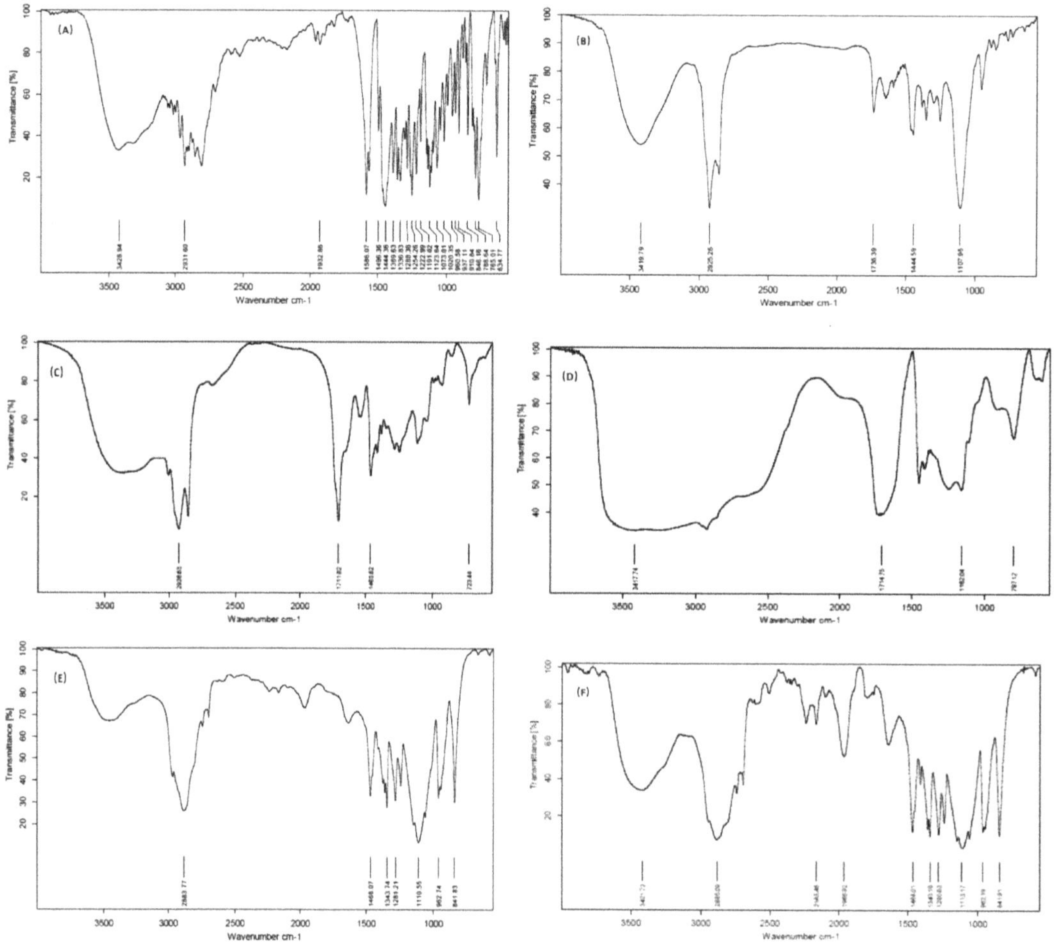

Figure 5. IR Spectra of (**A**) mirtazapine, (**B**) formulation, (**C**) oleic acid, (**D**) Carbapol934P, (**E**) poloxamer 407, and (**F**) PEG 6000.

2.4.5. Differential Scanning Calorimetry (DSC)

DSC studies were performed for pure mirtazapine, thermotriggered nanoemulsion, poloxamer 407, and other excipients used in the formulation to know the thermal behavior and the physical state of the drug in the formulation (crystalline/amorphous) through characteristic melting points.

The DSC thermogram of mirtazapine, thermotriggered nanoemulsion, and other excipients are shown in Figure 6. The thermogram of pure mirtazapine showed melting endotherm at 116 °C. The thermogram of thermotriggered nanoemulsion showed an endotherm peak at 162.85 °C, and poloxamer 407 showed an endotherm peak at 60.53 °C, which indicates that there is no significant change in the melting endotherm of the pure drug. The slightly broad peak in the formulation indicates the change in the state of the drug to an amorphous state.

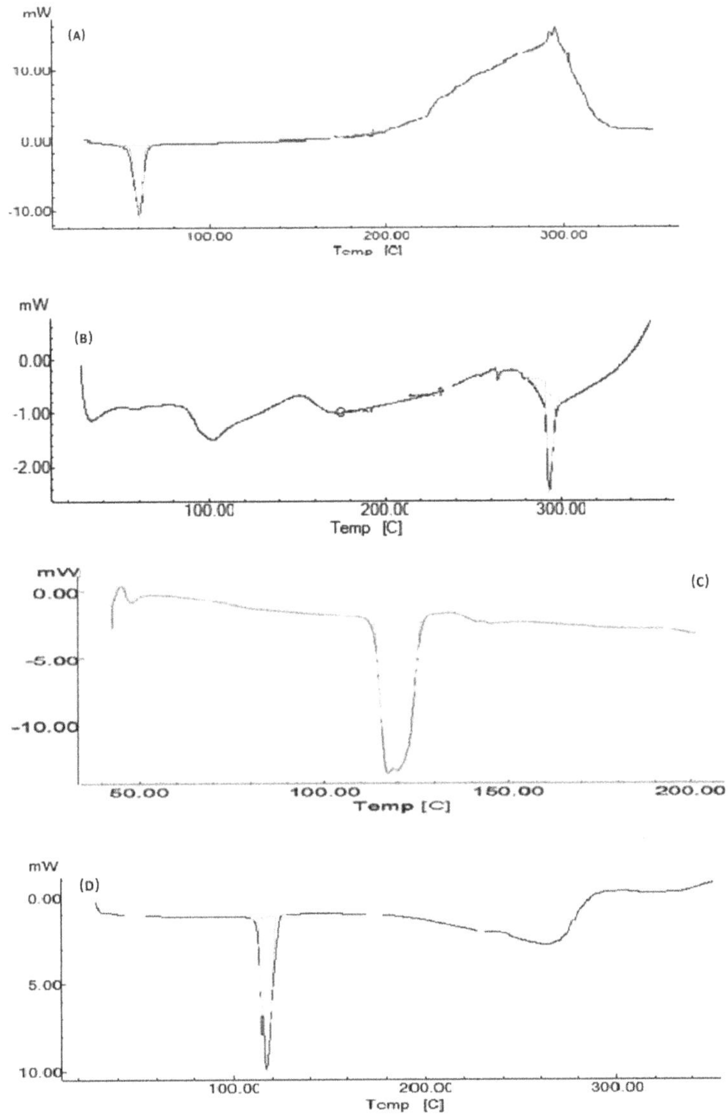

Figure 6. Reports of DSC thermograms of pure drug, formulation, and excipients. (**A**) Poloxamer 407, (**B**) Oleic acid, (**C**) Formulation, (**D**) Mirtazapine.

2.4.6. Measurement of Particle Size and Zeta Potential

The average sizes of the globules of RD1, RD2, and RD3 were found to be 42.6 nm, 135.2 nm, and 52.44 nm, respectively, as seen in Table 2 and depictated in Figure 7. The percentage of the intensity of distribution of the 42.64 nm-sized globules of thermotriggered nanoemulsion was high, i.e., 90.2%, whereas the average globule size and distribution percentage of RD2 and RD3 were found to be 50.8 and 60.5, respectively.

Table 2. Particle size and zeta potential for developed formulations.

Formulation	Particle Size (nm) (Mean ± SD)	Zeta Potential	Poly Dispersity Index
RD1	42.60 ± 1.24	−6.58	0.345
RD2	135.20 ± 3.29	−2.25	0.435
RD3	52.44 ± 1.41	−0.193	0.299

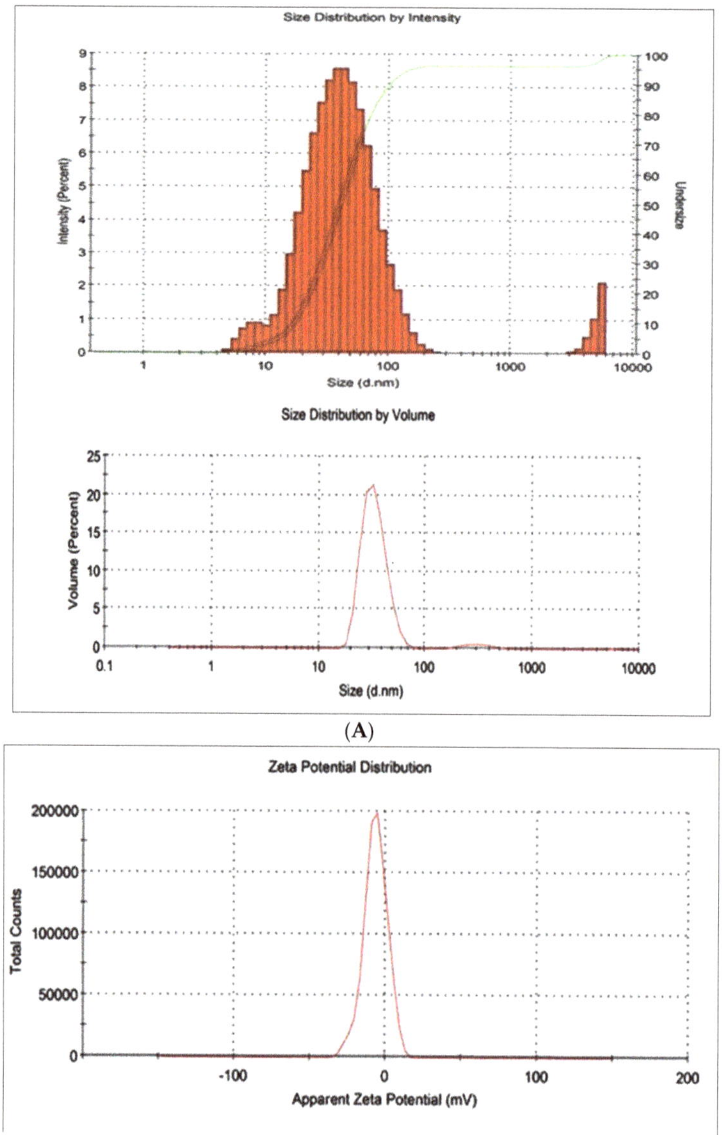

(B)

Figure 7. Particle size data of thermotriggered nanoemulsion (RD1), (**A**) size distribution and (**B**) zeta potential.

2.4.7. Transmission Electron Microscopy (TEM)

From the images obtained by TEM analysis, it is evident that thermotriggered nanoemulsion (RD1) showed highly uniform and spherical shapes globules. White, nearly spherical globules were seen against a dark background created by negative staining with phosphotungstic acid, as evident from Figure 8.

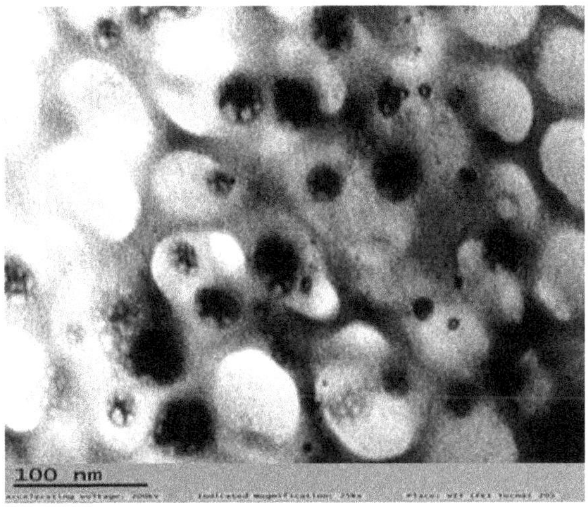

Figure 8. TEM Photographs of Thermotriggered Nanoemulsion (RD1).

3. Stability Studies

Nanoemulsions are considered to be thermodynamically stable systems that are formed with a particular concentration of oil, surfactant, and water, with no phase separation, creaming, or cracking. Selected formulations from the phase diagram were subjected to different stress stability tests, such as a heating-cooling cycle, centrifugation, and a freeze-thaw cycle. All nanoemulsion formulation cycles were found to be stable.

4. Statistical Analysis

The results of different formulations were analyzed by one-way ANOVA. The values were considered to be statistically significant when the p value was less than 0.05. It was observed that the p value of all the responses was found to be below 0.05, hence considered as significant.

5. Conclusions

Thermotriggered intranasal gel for the antidepressant mirtazapine was successfully developed to treat depression in patients. Three different nanoemulsions, including thermotriggered nanoemulsion (RD1), mucoadhesive nanoemulsion (RD2), and plain nanoemulsion, were developed, and comparative studies were done. After comparison of these three formulations for various physicochemical and analytical parameters, it was concluded that the thermotriggered nanoemulsion (RD1) consisting of carbopol and PEG 6000 was best suitable for intranasal delivery in terms of properties like excellent globule size, shape, zeta potential, polydispersity index, and percentage of drug release. From the results obtained in ex vivo studies conducted on freshly excised sheep nasal mucosa, it can be hypothesized that the thermotriggered nanoemulsion may have still better absorption capacity through the olfactory and trigeminal nerves, as it shows good paracellular and transcellular transport and can be efficiently used to treat major depressive disorder. To support the hypothesis, in vivo animal studies must be done; these studies would give a clear idea on brain-targeting efficiency of the formulation.

6. Materials

Mirtazapine (CAS: 85650-52-8) was procured as a kindly gifted sample from Wockhardt Pvt. Ltd., Aurangabad, Maharashtra, India. Poloxamer 407 (CAS: 691397-13-4) was obtained as a gifted sample from BASF India, Ltd., Mumbai, India. Labrafil M1944 CS (CAS: 69071-70-1), Capryol PGMC (CAS: 31565-12-5), Peceol, and Labrosol, were gift samples from Gattefosse India, Pvt. Ltd. (Mumbai, India), while Campul MCM C8 EP (CAS: 26402-26-6) and Acconon MC82E/NF (CAS: 85536-01-8) were a generous gifts from Abitech Ltd. Finar Chemicals, Hyderabad, India supplied the Tween 80, Tween 20, Span 80, propylene glycol, polyethylene glycol 400, and oleic acid. The remaining reagents were all of analytical grade.

7. Methods

7.1. Saturation Solubility Studies

The solubility studies of mirtazapine were carried out in different solvents and buffers, like double-distilled water, ethanol, dichloromethane, Labrafil M1944 CS, Caproyl PGMC, Peceol, Acconon MC82E/NF, Labrosol, Campul MCMC8EP, propylene glycol, PEG 400, Tween 80, Tween 20, Span 80, and oleic acid. In order to create saturated solutions of mirtazapine, excess amounts of the drug were added to 5 mL of each chosen vehicle and stirred on a rotary shaker for 48 h at 25 °C. After equilibrium was reached, samples were taken and centrifuged for 15 min at 10,000 rpm. Following the collection and proper dilution of an additional 100 L of supernatant with dichloromethane, mirtazapine samples were analyzed using UV-visible spectrophotometry at 228 nm [14].

7.2. Construction of Pseudo-Ternary Phase Diagram

Oleic acid was chosen as the oil phase, Tween 80 as the surfactant, and ethanol as the co-surfactant based on the solubility experiments. Tween 80 and ethanol were measured in the weight ratios (1:1, 1:2, 1:3, 2:1, and 3:1) of surfactant to co-surfactant (Smix). The pseudo-ternary phase diagrams were produced at room temperature by the water titration method. For each pseudo-ternary phase diagram, oil and Smix mixtures were created with weight ratios (*w*/*w*) of 1:9, 2:8, 3:7, 4:6, 5:5, 6:4, 7:3, 8:2, and 9:1. When the combination reached a particular point and began to break down into a macroemulsion, double-distilled water was gradually added to each ratio of Smix to oil while the mixture was being stirred magnetically. In order to complete the pseudo-ternary phase diagrams, the component concentrations were noted. The oil, surfactant, co-surfactant, and water contents were then chosen at the proper weight ratios based on the stability and transparency of the resulting microemulsions [14,15].

7.3. Development of Nanoemulsion Formulation

The criteria for selection of the microemulsion zone were carried out by observing the stability of the formulation from the pseudo-ternary phase diagrams. The microemulsion formulations were composed of S_{mix} (1:3) and oil. In accordance with the pseudo-ternary phase diagrams, a S_{mix} ratio of 1:9 was used. Oleic acid was used as oil in the microemulsion [14,15].

7.3.1. Formulation of Thermotriggered Nanogel

Thermotriggered nanogel was prepared by a cold method, different concentrations of polymers were used, such as poloxamer 407 (10.5–20.5% *w*/*v*) Carbopol 934 P (0.1–0.5% *w*/*w*), and PEG 6000 (0.1–0.3% *w*/*v*) to determine the optimum concentration of poloxamer 407 and other polymers required for thermotriggered gelling. After a series of combinations, it was found that poloxamer 407 (18.5%), Carbopol (0.1%), and PEG 6000 (0.3%) were found to have optimum concentration, which formed thermotriggered gel at a temperature of 32 °C.

Thermotriggered nanogel was prepared by dissolving 0.1% Carbopol 934 P and 0.3% PEG 6000 in the aqueous phase by continuously stirring on a magnetic stirrer at 400 rpm. To the aqueous phase, 18.5% of poloxamer 407 was slowly added under continuous stirring in cold conditions [16]. This aqueous phase was then added dropwise to the nanoemulsions. After complete solubilization of poloxamer 407, the solution was stored at 4 °C overnight to get the thermotriggered nanogel.

Carbopol 934P was used as the mucoadhesive agent for its mucoadhesive properties to make the formulation adhere to the mucosal system; poloxamer 407 was used as the thermosensitive agent to form in situ gel after spraying into the nostrils. The active drug was added in the formulations depending upon the amount of formulation being delivered by the spray pump. Table 3 shows the composition of the nanoemulsions.

Table 3. Composition of nanoemulsion formulations.

Sr.No:	Formulation	Oil Phase % *w/w*	S_{mix} % *w/w*	Water % *w/w*	Aqueous Phase % *w/w*		
					Poloxamer 407	Carbopol 934 P	PEG 6000
1	Thermotriggered nano emulsion (RD1)	3.846%	34.615%	61.53%	18.5%	0.1%	0.3%
2	Muco-adhesive nano emulsion (RD2)	3.846%	34.615%	61.53%	-	0.3%	-
3	Water based nano emulsion (RD3)	3.846%	34.615%	61.53%	-	-	-

7.3.2. Determination of Reproducibility of Dosage

The developed formulations were loaded into the 5 mL clean sterile glass container and were fitted with a spray actuator and crimpled.

The spray actuator was analyzed for determination of the reproducibility of the dosage. The formulation was sprayed in a test tube and weighed. The empty weight of the test tube was subtracted from the total weight. This was repeated a number of times, and the average weight of the dose was taken [17].

7.4. Characterization of Nanoemulsion

7.4.1. Measurement of Droplet Size and Zeta Potential

Using the zetasizer, a dynamic light scattering or photon correlation spectroscopy technique was used to measure the mean droplet size and zeta potential (Malvern Instruments version 7.01). Using filtered, twice-distilled water, each was diluted to the proper concentration. Analysis of globule size was carried out at 25 °C and with a detection angle of 90 °C. Nanoemulsions' size and polydispersity index were directly measured by the equipment [16–18].

7.4.2. pH Measurement

A pH meter was used to determine the pH values of the nanoemulsion samples (Remi Equipment Pvt. Ltd., Kolkata, India). Before each usage, the pH meter was calibrated using buffer solutions having pH values of 4.0, 7.0, and 9.0. The formulation's pH was measured three times, and the means of the measurements were computed.

7.4.3. Measurement of Viscosity

A Brookfield viscometer with spindle no. DV-II+ PRO was used to measure the viscosities of nanoemulsions. The spindle code for LV1 is 61. The spindle was dipped in the preparation and revolved for 5 min at 100 rpm at room temperature.

7.5. Determination of Entrapment Efficiency

7.5.1. Drug Content

Drug content was estimated by placing one dose (7.5 mg) equivalent microemulsion formulation in a 100 mL volumetric flask and adding a small volume of methanol, which solubilizes the microemulsion. Finally, the volume was made up to 100 mL, with a pH 6.4 phosphate buffer. It was shaken well for about 1 h in a shaker followed by centrifugation then analyzed. The resultant solution was filtered through Whatman filter paper, and the absorbance was measured at 228 nm using a UV-visible spectrophotometer [18]. To calculate the concentration of mirtazapine in the samples, the straight line equation obtained from regression anlaysis was used. The equation was obtained from the standard calibration plots, which were conducted in simulated nasal fluid, was well as a pH 6.4 phosphate buffer. The calibration curve of mirtazapine in the pH 6.4 phosphate buffer was found to obey Beer-Lambert's law within the linearity range of 10–60 µg/mL, with a regression coefficient value of 0.992 at 228 nm, while in simulated nasal fluid, it was within the linearity range of 2–20 µg/mL, with a regression coefficient value of 0.991 at 228 nm.

7.5.2. Drug Entrapment Efficiency

The drug entrapment efficiency of the mirtazapine nanoemulsion formulation was measured using the UV visible spectroscopic method. Each 1 mL sample was cooling centrifuged at 3500 rpm for 30 min. After centrifuge, the supernatant transparent layer was taken and diluted with 10 mL distilled water. The samples were measured at 228 nm using the UV-VIS spectroscopic method. Results were taken in triplicate, and the average was taken into consideration.

7.5.3. In Vitro Drug Diffusion Studies

A Franz diffusion cell and a dialysis membrane were used in an in vitro drug diffusion study (Hi Media, molecular weight 5000 Daltons). Simulated nasal fluid (SNF) was used to saturate the dialysis membrane overnight. The dialysis membrane was then attached to the Franz diffusion cell's upper (donor) and lower (receptor) compartments. On the donor compartment side of the dialysis membrane, the formulation corresponding to 7.5 mg of the dose was administered. An amount of simulated nasal fluid (SNF) equal to 18 mL was placed in the receptor compartment of the Franz diffusion cell. The diffusion cells (Remi, India) were maintained at 37 ± 0.5 °C throughout the experiment, with stirring happening at 600 rpm. At fixed time intervals, 1 mL of aliquots was withdrawn every 15 min, 30 min, 45 min, 60 min, 60 min, 120 min, 180 min, 240 min, 300 min, 360 min, and 420 min from the receptor compartment through a side tube, and an equal volume of SNF was replaced to maintain the sink condition [17–19] The samples withdrawn were analyzed through UV-visible spectrophotometry at 280 nm.

By dividing the slope of the steady state section of the line in the plot of drug amount penetrated per unit area of dialysis membrane verses time, the drug flux (g/hr/cm^2) at the steady state was estimated.

7.5.4. Ex Vivo Drug Diffusion Studies

Sheep nasal mucosa that had just been excised was used in a Franz diffusion cell for the ex vivo drug diffusion investigations of the mirtazapine nanoemulsion.

Sheep's nasal mucosa was the subject of experiments on drug diffusion that were done ex vivo. The sheep nasal mucosa was taken from a freshly removed sheep's nose at the butcher shop. Simulated nasal fluid (SNF) was used to soak this sheep's nasal mucosa before it was glued to a Franz diffusion cell with a 3.14 cm^2 diffusion area and then clamped. In the donor compartment, a formulation corresponding to a 7.5 mg dose was applied to the nasal mucosa, and simulated nasal fluid was placed in the receptor compartment. The diffusion cell was kept at body temperature throughout the experiment, and the stirring speed was kept at 600 rpm (Remi, India). At predetermined intervals, such as 15 min, 30 min, 45 min, 60 min, 120 min, 180 min, 240 min, 300 min, 360 min, and 420 min, the 1 mL aliquots were removed from the receptor compartment.

To keep the sink condition, the amounts of samples that were extracted were promptly replaced with equivalent amounts of SNF. At 228 nm, these samples were examined using a UV-visible spectrophotometer [19–21].

8. Fourier Transform Infrared Spectroscopy (FTIR) Studies

To evaluate any potential interactions that might have arisen during formulation between mirtazapine and other excipients, FTIR analysis was carried out. The infrared spectra of mirtazapine and thermotriggered nanoemulsion were acquired using a Bruker Alpha E, FTIR spectrometer (Bruker Alpha E, Opus-7.0.122), equipped with an ATR. They are typically evaluated using the single-reflection ATR accessories (attenuated total reflectance). The spectra were scanned in transmission mode spanning the 4000–650 cm^{-1} wavenumber range at room temperature.

8.1. *Differential Scanning Calorimeter*

To determine the physical condition of the medication in the nanoemulsion formulation, DSC experiments were carried out. A differential scanning calorimeter was used for the DSC measurements (Shimadzu, Kyoto, Japan, Thermal Analyzer DSC 60). Heat of fusion (enthalpy) (Hf) and peak transition temperature (Tm) were calculated and used in the analysis. For routine calibration, indium (Tm = 159.2 °C; Hf = 28.8 J/g) was utilized as the standard. Nitrogen (purity > 99.99%) was employed as the purge gas, and an empty aluminium pan served as the reference. Samples of 2.4–2.8 mg were weighed, put in open aluminium pans, and scanned over the temperature range of 30–350 °C at a rate of 10 °C/min.

8.2. *Transmission Electron Microscopy*

Transmission electron microscopy (TEM) (H-7500, Hitachi, Kyoto, Japan) was used to examine the morphology of the improved formulation in order to research the morphology of the resulting nanoemulsion. The nanoemulsion was placed on copper grids for observation after being dyed with 1% (w/v) phosphotungstic acid, and by examining the TEM images, the globule's shape was discovered [22].

8.3. *Stability Studies*

8.3.1. Thermodynamic Stability Tests

Selected formulations were subjected to different thermodynamic stability tests. Only those formulations that survived dispersion stability tests were selected for further study [21].

8.3.2. Heating Cooling Cycle

All samples were placed between refrigerator temperatures of 4 °C and 45 °C for six cycles with storage at each temperature for not less than 48 h and were then studied. Those formulations that were stable at these temperatures were subjected to centrifugation.

8.3.3. Centrifugation

Those formulations that passed were centrifuged at 3500 rpm for 30 min using centrifuge. The formulations that did not show any phase separation were taken for further tests.

8.3.4. Freeze-Thaw cycle

Those samples that passed the centrifugation test were placed between −21 °C and +25 °C for three freeze-thaw cycles with storage at each temperature for not less than 48 h.

8.3.5. Statistical Analysis

The statistical analysis was performed using one-way analysis of variance (ANOVA) and was used to compare the results of different formulations. A p value of 0.05 was considered to be statistically significant (GraphPad Prism version 6.03, Boston, MA, USA). The characterization data were expressed as the means of three experiments ± SD.

Author Contributions: Conceptualization, M.G., R.V. and R.D.; methodology, G.K.; software, N.M.; validation, P.D. and R.C.P.; formal analysis, V.P.V.; investigation, P.P.; resources, K.C.; data curation, K.V. and P.P.; writing—original draft preparation, M.G., R.V. and G.K.; writing—review and editing, R.D., N.M., P.D. and R.C.P.; visualization, M.G.; supervision, R.V.; project administration, G.K.; funding acquisition, G.K. All authors have read and agreed to the published version of the manuscript.

Funding: The authors extend their appreciation to the Deanship of Scientific Research at King Khalid University for funding this work through the Large Groups Project under grant number RGP2/150/44.

Institutional Review Board Statement: Not applicable.

Informed Consent Statement: Not applicable.

Data Availability Statement: Not applicable.

Acknowledgments: The authors extend their appreciation to the Deanship of Scientific research at King Khalid University, Saudi Arabia, for funding this work through the Large Groups Project under grant number RGP2/150/44.

Conflicts of Interest: The authors declare no conflict of interest.

References

1. Pires, V.A.; Fortuna, A.; Alves, G.; Falcão, A. Intranasal Drug Delivery: How, Why and What for? *J. Pharm. Pharm. Sci.* **2009**, *12*, 288–311. [CrossRef] [PubMed]
2. Xu, J.; Tao, J.; Wang, J. Design and Application in Delivery System of Intranasal Antidepressants. *Front. Bioeng. Biotechnol.* **2020**, *8*, 626882. [CrossRef] [PubMed]
3. Nguyen, T.T.; Maeng, H.J. Pharmacokinetics and Pharmacodynamics of Intranasal Solid Lipid Nanoparticles and Nanostructured Lipid Carriers for Nose-to-Brain Delivery. *Pharmaceutics* **2022**, *14*, 572. [CrossRef] [PubMed]
4. Devarajan, V.; Ravichandran, V. Nanoemulsions: As modified drug delivery tool. *Int. J. Compr. Pharm.* **2011**, *4*, 1–6.
5. Corazza, E.; di Cagno, M.P.; Brandl, A.B.; Abruzzo, A.; Bigucci, T.C.F.; Lupppi, B. Drug delivery to the brain: In situ gelling formulation enhances carbamazepine diffusion through nasal mucosa models with mucin. *Eur. J. Pharm. Sci.* **2022**, *179*, 1. [CrossRef]
6. Cirri, M.; Maestrelli, F.; Nerli, G.; Mennini, N.; D'Ambrosio, M.; Luceri, C.; Mura, P.A. Development of a Cyclodextrin-Based Mucoadhesive-Thermosensitive In Situ Gel for Clonazepam Intranasal Delivery. *Pharmaceutics* **2021**, *13*, 969. [CrossRef]
7. Hari Kumar, S.L.; Singh, V. Nanoemulsification—A novel targeted drug delivery tool. *J. Drug Del. Therap.* **2012**, *2*, 40–45. [CrossRef]
8. Mahajan, S.H.; Dinger, B.S. Design and in vitro evaluation of nanoemulsion for nasal delivery of artemether. *Indian J. Nov. Drug Del.* **2011**, *3*, 272–277.
9. Patel, P.R.; Joshi, J.R. An overview on nanoemulsion: A novel approach. *Int. J. Pharm. Sci. Res.* **2012**, *3*, 4640–4650.
10. Choudhary, R.; Goswami, L.; Kothiyal, P. Preparation of nanoparticles loaded nasal gel of Mirtazapine for treatment of depression. *J. Adv. Pharm. Sci.* **2013**, *3*, 2.
11. Benajeer, S.; Ramana, K.V.; Reddy, V.; Kumar, A. New Simple UV Spectrophotometric Method for Determination of Mirtazapine in Bulk and pharmaceutical dosage forms. *Int. J. Pharm. Sci. Res.* **2012**, *3*, 482–486.
12. Bahadur, S.; Pardhi, D.M.; Rautio, J.; Rosenholm, J.M.; Pathak, K. Intranasal Nanoemulsions for Direct Nose-to-Brain Delivery of Actives for CNS Disorders. *Pharmaceutics* **2020**, *12*, 1230. [CrossRef] [PubMed]
13. Pathan, S.A.; Iqbal, Z.; Zaidi, S.; Talegaonkar, S.; Vohra, D.; Jain, G.K.; Azeem, A.; Jain, N.; Lalani, J.R.; Khar, R.K.; et al. CNS Drug Delivery Systems: Novel Approaches. *Recent Pat. Drug Deliv. Formul.* **2009**, *3*, 71–89. [CrossRef]
14. Majithiya, J.R.; Ghosh, P.K.; Umrethia, M.L.; Murthy, R.S.R. Thermoreversible-mucoadhesive Gel for Nasal Delivery of Sumatriptan. *AAPS PharmSciTech* **2006**, *67*, 7. [CrossRef] [PubMed]
15. Thakkar, H.; Patel, A.; Chauhan, N. Formulation and optimization of mucoadhesive microemulsion containing mirtazapine for intranasal delivery. *Chron. Young Sci.* **2014**, *5*, 25–32. [CrossRef]
16. Bernardi, D.S.; Pereira, T.A.; Maciel, N.R.; Bortoloto, J.; Viera, G.S.; Oliveira, G.C.; Filho, P.A.D.R. Formation and stability of oil-in-water nanoemulsions containing rice bran oil: In vitro and in vivo assessments. *J. Nanobiotechnol.* **2011**, *9*, 44. [CrossRef]
17. Datta, R.; Bandyopadhyay, A.K. Development of a new drug delivery system of diazepam with natural mucoadhesive agent from *Trigonella foenum-graecum* L. *J. Sci. Res.* **2005**, *64*, 973–977.
18. Kumar, M.; Misra, A.; Babbar, A.K.; Mishra, A.; Mishra, P.; Kamla, P. Intranasal nanoemulsion based brain targeting drug delivery system of risperidone. *Int. J. Pharm.* **2008**, *358*, 285–291. [CrossRef]
19. Shah, R.R.; Chandrakant, S.M.; Patil, S.S.; Nilofar, S.N. Preparation and Evaluation of Aceclofenac Topical Microemulsion. *Iranian J. Pharm. Res.* **2010**, *9*, 5–11.
20. Subedia, R.K.; Ryoob, J.P.; Moonb, C.; Choi, H.K. Influence of formulation variables in transdermal drug delivery system containing zolmitriptan. *Int. J. Pharm.* **2011**, *419*, 209–214. [CrossRef]

21. Kamble, M.S.; Sandeep, M.D.; Bhalerao, K.K.; Pravin, D.C.; Ashok, V.B. Evaluation of brain targeting of drugs after administered intranasally. *J. Biomed. Pharm. Res.* **2012**, *1*, 33–38.
22. Menaka, M.; Pandey, V.P.; Smith, A. Colloidal dispersions as a potential nasal drug delivery system for ondansetron hydrochloride—In vitro and in vivo properties. *Asian J. Pharm. Clin. Res.* **2014**, *7*, 72–75.

gels

MDPI

Article

Preparation and Characterization of Biocompatible Iron/Zirconium/Polydopamine/Carboxymethyl Chitosan Hydrogel with Fenton Catalytic Properties and Photothermal Efficacy

Xiaoyi Zheng [1,2,†], Hang Wu [2,†], Shige Wang [3], Jiulong Zhao [2,*] and Lianghao Hu [2,*]

1 Postgraduate Training Base in Shanghai Gongli Hospital, Ningxia Medical University, Pudong New Area, No. 219 Miao Pu Road, Shanghai 200135, China

2 Department of Gastroenterology, Changhai Hospital, Naval Military Medical University, No. 168 Changhai Road, Shanghai 200433, China

3 School of Materials and Chemistry, University of Shanghai for Science and Technology, No. 516 Jungong Road, Shanghai 200093, China; sgwang@usst.edu.cn

* Correspondence: jlzhao9@163.com (J.Z.); lianghao-hu@smmu.edu.cn (L.H.)

† These authors contributed equally to this work.

Abstract: In recent years, multifunctional hydrogel nanoplatforms for the synergistic treatment of tumors have received a great deal of attention. Here, we prepared an iron/zirconium/polydopamine/carboxymethyl chitosan hydrogel with Fenton and photothermal effects, promising for future use in the field of synergistic therapy and prevention of tumor recurrence. The iron (Fe)–zirconium (Zr)@ polydopamine (PDA) nanoparticles were synthesized by a simple one-pot hydrothermal method using iron (III) chloride hexahydrate ($FeCl_3•6H_2O$), zirconium tetrachloride ($ZrCl_4$), and dopamine, followed by activation of the carboxyl group of carboxymethyl chitosan (CMCS) using 1-(3-Dimethylaminopropyl)-3-ethylcarbodiimide hydrochloride (EDC)/N(4)-hydroxycytidine (NHS). Finally, the Fe–Zr@PDA nanoparticles and the activated CMCS were mixed to form a hydrogel. On the one side, Fe ions can use hydrogen peroxide (H_2O_2) which is rich in the tumor microenvironment (TME) to produce toxic hydroxyl radicals (•OH) and kill tumor cells, and Zr can also enhance the Fenton effect; on the other side, the excellent photothermal conversion efficiency of the incorporated PDA is used to kill tumor cells under the irradiation of near-infrared light. The ability of Fe–Zr@PDA@CMCS hydrogel to produce •OH and the ability of photothermal conversion were verified in vitro, and swelling and degradation experiments confirmed the effective release and good degradation of this hydrogel in an acidic environment. The multifunctional hydrogel is biologically safe at both cellular and animal levels. Therefore, this hydrogel has a wide range of applications in the synergistic treatment of tumors and the prevention of recurrence.

Keywords: hydrogel; photothermal therapy; chemodynamic therapy; Fenton reaction

Citation: Zheng, X.; Wu, H.; Wang, S.; Zhao, J.; Hu, L. Preparation and Characterization of Biocompatible Iron/Zirconium/Polydopamine /Carboxymethyl Chitosan Hydrogel with Fenton Catalytic Properties and Photothermal Efficacy. *Gels* **2023**, *9*, 452. https://doi.org/10.3390/gels9060452

Academic Editor: Avinash J. Patil

Received: 6 May 2023
Revised: 25 May 2023
Accepted: 29 May 2023
Published: 31 May 2023

1. Introduction

The effective treatment of tumors remains a challenge for the biomedical field today [1]. Traditional clinical treatments include surgery [2], chemotherapy [3], radiotherapy [4,5], immunotherapy [6,7], etc. However, all these methods have certain limitations such as the risk of metastasis or infection due to the surgical trauma [8], the lack of targeting of some chemotherapeutic drugs to the lesion, and the killing of normal tissue cells [9,10]. These shortcomings have limited the prospects for their use in oncology treatment. Therefore, there is a pressing want to develop more effective therapeutic strategies to tackle the difficulties in tumor therapy.

In recent years, studies have proven that various kinds of tumor cells show off multiplied ranges of reactive oxygen species (ROS) and altered redox status due to genetic, metabolic, and microenvironmental alterations [11–14]. Stimulated by high ROS levels, oncogenes can induce the activation of various downstream signaling pathways to adapt to the oxidative stress environment [15–17]. This further leads to cell immortalization and contributes to the necessary conditions for the cells to reach a malignant growth state [18,19]. As a result, ROS-based antitumor strategies have gained widespread attention. Chemodynamic therapies (CDT) is a novel strategy for cancer treatment that involves multiple transition metal ions, such as iron (Fe), silver (Ag), molybdenum (Mo), ruthenium (Ru), cerium (Ce), and zinc (Zn) [20,21]. These metal ions can react with the endogenous hydrogen peroxide (H_2O_2) in the tumor microenvironment (TME) and produce a large number of hydroxyl radicals (•OH) to kill tumor cells [22–25]. Inspired by this promising therapeutic paradigm, iron-based nanoplatforms have been frequently studied for the treatment of tumors. Specifically, due to the anaerobic conditions at the tumor site, anaerobic enzymolysis and proton pumping at the cell membrane result in an acidic environment in the TME, and some TMEs exhibit high H_2O_2 expression [26]. CDT has several advantages over other therapies in that it is tumor-targeted, has few side effects on normal tissues, and can selectively kill tumor cells without relying on exogenous stimuli [23].

Photothermal therapy (PTT) has promising applications in the treatment of tumors. PTT refers to the use of photothermal agents with photothermal conversion efficiency and converts the near-infrared laser into heat [27–29]. This warmness can enlarge tumor temperature, cause irreversible injury to tumor cell membranes, and set off protein denaturation, leading to irreversible cell harm and subsequent tumor regression [30,31]. Currently, the main conjugated polymers (CPs) available for tumor PTT include polydopamine (PDA), polypyrrole, and polyaniline (PAI) [32–35]. Among them, PDA is a novel polymer assembled from dopamine monomers through oxidative autopolymerization [36,37]. It has been broadly used in tumor-targeted drug delivery systems due to its advantages such as good biocompatibility, excellent photothermal conversion properties, and multiple drug release response possibilities [36,38,39].

Hydrogels are three-dimensional porous structures produced from hydrophilic polymers by physical or chemical cross-linking methods [40]. Hydrogels can effectively encapsulate molecules and control their concentration at the lesion site [41]. In addition, hydrogels can respond to physical, biological, chemical, or external stimuli [42,43]. In this study, Fe ions are used as the classical Fenton agents [44,45], which can react with H_2O_2 at the tumor site to produce highly reactive •OH for tumor cells killing. Moreover, Zr ions that can synergize with Fe ions were introduced [46]. As a ligand for metal ions, PDA was employed to form noncovalent bonds with the nanoparticle surfaces and modulate the photothermal properties [47]. Then, the EDC/NHS activated carboxymethyl chitosan (CMCS) is used to combine with Fe–Zr@PDA nanoparticles to form a drug reservoir. Together with the CMCS hydrogel drug reservoir, this nanoplatform can be locally implanted to the target site to react with the H_2O_2, transform the laser to heat, and release the drug. Specifically, we introduced PDA with Fe–zirconium (Zr) to impart unique photothermal properties to the nanomedicine. In brief, Fe–Zr@PDA nanoparticles were mixed with carboxymethyl chitosan (CMCS) activated by 1-(3-Dimethylaminopropyl)-3-ethyl carbodiimide (EDC)/N-hydroxysuccinimide (NHS) to form a hydrogel.

In this work, an Fe-ions-mediated Fenton reaction was combined with the synergistic effect of Zr ions and the photothermal effect of PDA. Together with the CMCS hydrogel drug reservoir, this nanoplatform can be locally implanted to the target site to react with the H_2O_2, transform the laser to heat, and release the drug. Therefore, the composite hydrogel is expected to improve the efficacy of oncology treatments via preventing the postoperative tumor recurrence and increase patient compliance in future potential clinical applications. In summary, our aim is to combine the classical Fenton effect of Fe-based nanomaterials, the synergistic effect of Zr, the photothermal properties of PDA, and the bicompatible advantages of hydrogels to form a composite hydrogel. The composite system provides a new idea in broadening the application potential of hydrogels. While combining the Fenton effect and photothermal efficacy, the functionalized hydrogel can not only function as a tumor suppressor, but also plays an important role in preventing the recurrence of residual tumor tissue after local surgical excision (Scheme 1). The novelty of our work lies in the use of the classical Fenton reaction, combined with metal ion properties, to complement the hydrogel to form a composite multifunctional hydrogel, circumventing the disadvantages of applying nanoparticles alone and amplifying their advantages to provide a novel strategy in the field of tumor killing and prevention of local tumor recurrence.

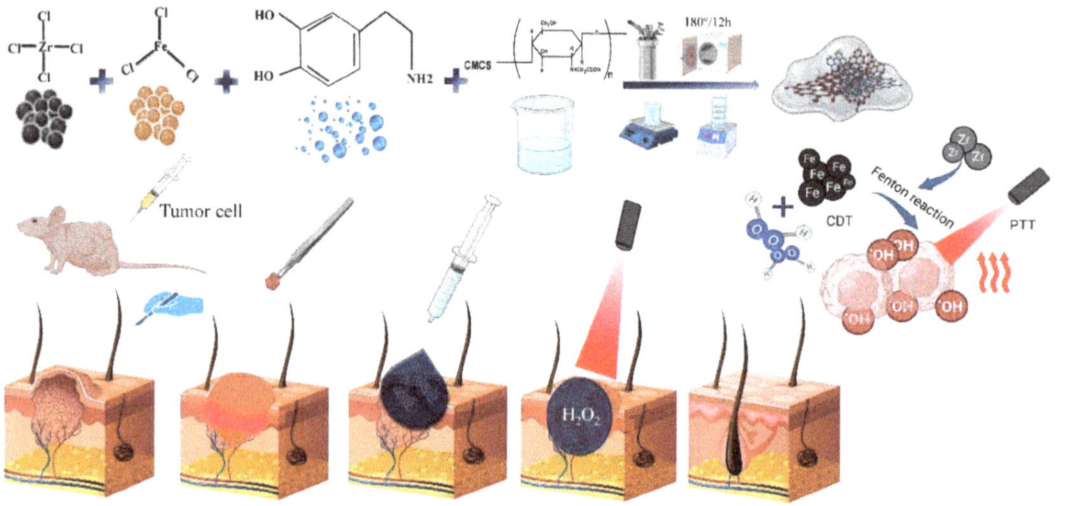

Scheme 1. A practical and working schematic of a multifunctional hydrogel that can be injected for future use in killing tumors and inhibiting their postoperative recurrence (CDT: chemodynamic therapies; PTT: photothermal therapy).

2. Results and Discussion

2.1. Preparation and Characterization of Fe–Zr@PDA and Fe–Zr@PDA@CMCS Hydrogel

To examine the microstructural morphology, the nanoparticle was investigated with a scanning electron microscope (SEM, Zeiss Sigma 300, Oberkochen, Germany). Fe–Zr@PDA was synthesized by a hydrothermal one-pot method. The Fe–Zr@PDA NPs exhibited a homogeneous spherical morphology with a particle size of approximately 42.4 ± 9.2 nm (Figure 1a). The hydrodynamic diameters of NPs dispersed in different media were measured using DLS. As shown in Figure 1b, the mean hydrodynamic diameters of Fe–Zr@PDA in water was 164 nm. The results of DLS also revealed that Fe–Zr@PDA tended to undergo certain aggregation. More importantly, the hydrodynamic diameters of Fe–Zr@PDA hardly changed within 24 h. To verify the existence of iron, Fe–Zr@PDA was mixed with the o-phenanthroline solution at room temperature. It was observed that the supernatant changed swiftly from colorless to orange–red and showed the characteristic absorption at 510 nm. However, there was no detectable color change when cultured without Fe–Zr@PDA (Figure 1c). These results suggest the presence of Fe in the composite. Fe–Zr@PDA@CMCS hydrogel was prepared by mixing the Fe–Zr@PDA with the EDC/NHS-activated CMCS. SEM analysis showed that the CMCS hydrogel showed a smooth surface (Figure 1d). Compared to the unactivated CMCS [48], a new peak at 1650 cm^{-1} assigned to the -C=N was observed in the FT-IR spectrum of CMCS hydrogel (Figure 1e). These results suggest that CMCS forms amide bonds through the activation of EDC/NHS, which facilitated the formation of hydrogels. A digital camera was used to record the changes of the black solution (Figure 1f, g), indicating the solution state and solid state of the hydrogel. The macroscopic and microscopic appearance of hydrogel was then recorded, which uncovered that the hydrogel is large and in macro scale. PDA is capable of generating thin film coatings on the surface of multiple materials. As shown in Figure 1d, h, CMCS hydrogel showed a smooth surface, while the Fe–Zr@PDA@CMCS hydrogel showed some loose porous pores.

This may be attributed to the formation of hydrogen bonds between PDA and CMCS. The element distribution diagram clearly showed that Zr were evenly distributed in the hydrogel, and there were no excess impurity elements in the hydrogel (Figure 1i).

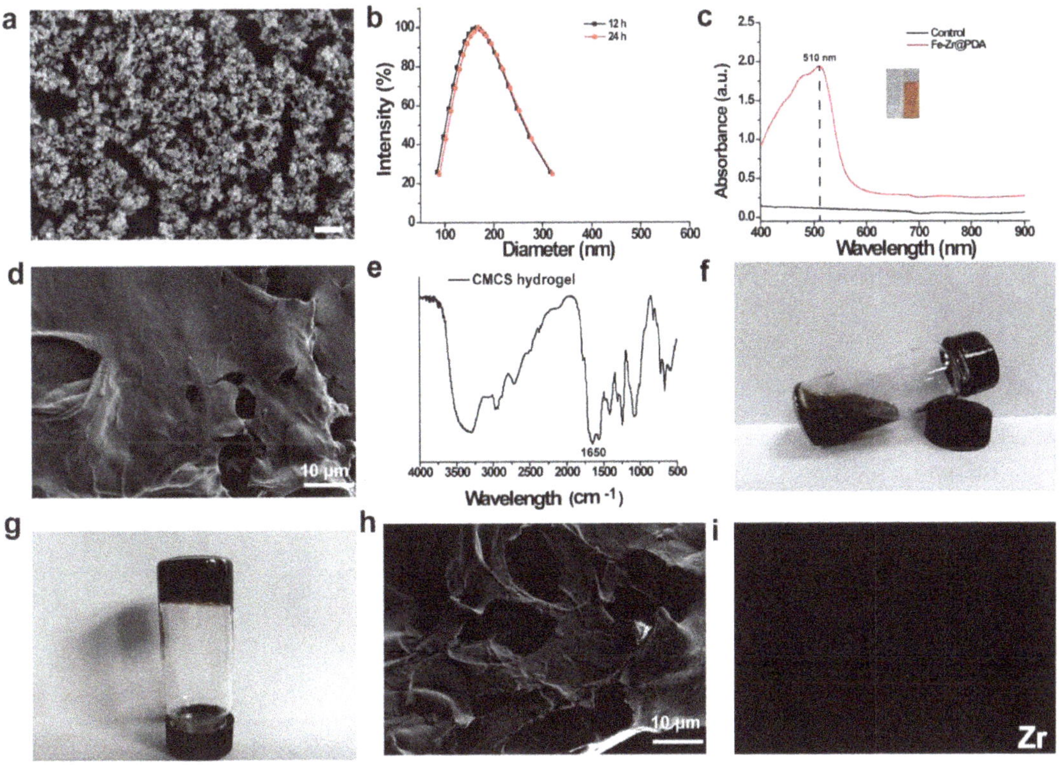

Figure 1. (**a**) SEM image of Fe–Zr@PDA nanoparticle; (**b**) the hydration kinetic diameter of Fe–Zr@PDA in H_2O; (**c**) the light absorption of phenanthroline solution with and without co-incubation with Fe–Zr@PDA; (**d**) SEM image of CMCS hydrogel; (**e**) FTIR spectra of CMCS hydrogel; (**f**) flowing solution state of Fe–Zr@PDA@CMCS hydrogel; (**g**) solidified state of Fe–Zr@PDA@CMCS hydrogel; (**h**) SEM image of Fe–Zr@PDA@CMCS hydrogel; (**i**) elemental distribution of Zr in Fe–Zr@PDA@CMCS hydrogel.

2.2. Photothermal Conversion Evaluation of Fe–Zr@PDA@CMCS Hydrogel

In recent years, great efforts have been made in developing photothermal agents for the therapy of various types of tumors [28]. PDA has good photothermal properties, can enhance the biocompatibility of nanomaterials, improve hydrophilicity, reduce cytotoxicity, and its many other advantages are widely used in the field of nanomaterials research [49,50]. The doping of PDA gives Fe–Zr@PDA@CMCS hydrogel the photothermal conversion properties to convert the absorbed NIR laser mild into heat for action. In this study, the temperatures of the hydrogel were recorded for the different groups of treatments to determine the overall photothermal performance of the hydrogels (Figure 2a–d). As shown in Figure 2a, as the concentration of Fe–Zr@PDA in the hydrogel increased (from 2 mg/mL to 8 mg/mL), and the ΔT gradually increased, with ΔT of 24.5 °C, 30 °C, and 42.1 °C, respectively, while the temperature of the control group only increased by 4.2 °C. The corresponding thermographs, likewise, illustrated that the concentration of PDA is closely related to the photothermal effect (Figure 2b). Further, the hydrogel was irradiated with an 808 nm NIR laser of exceptional energy densities (the specific power densities of the laser are 0.5 W/cm^2, 0.8 W/cm^2, and 1.0 W/cm^2) for 5 min and the ΔT of the hydrogel was measured. The results revealed that the higher the power density of the NIR laser was, the higher the ΔT was (Figure 2c). This suggests that the power density of the NIR laser is another factor affecting its photothermal performance. The thermographic images provided further evidence of the power-density-related photothermal performance (Figure 2d). In addition, Fe–Zr@PDA@CMCS hydrogel has good photothermal stability. After being irradiated by the laser for up to 120 min (six NIR irradiated cycles), the hydrogel showed a stable temperature rise trend, and the maximum ΔT was still greater than 50 °C (Figure 2e). Based on linear regression analysis, we calculated that the τs of Fe–Zr@PDA@CMCS hydrogel was 209.22 s (Figure 2f), and the η was 35.68% (Figure 2g), which was higher than the Au@Bi_2 Se_3 core–shell nanoparticle [42].

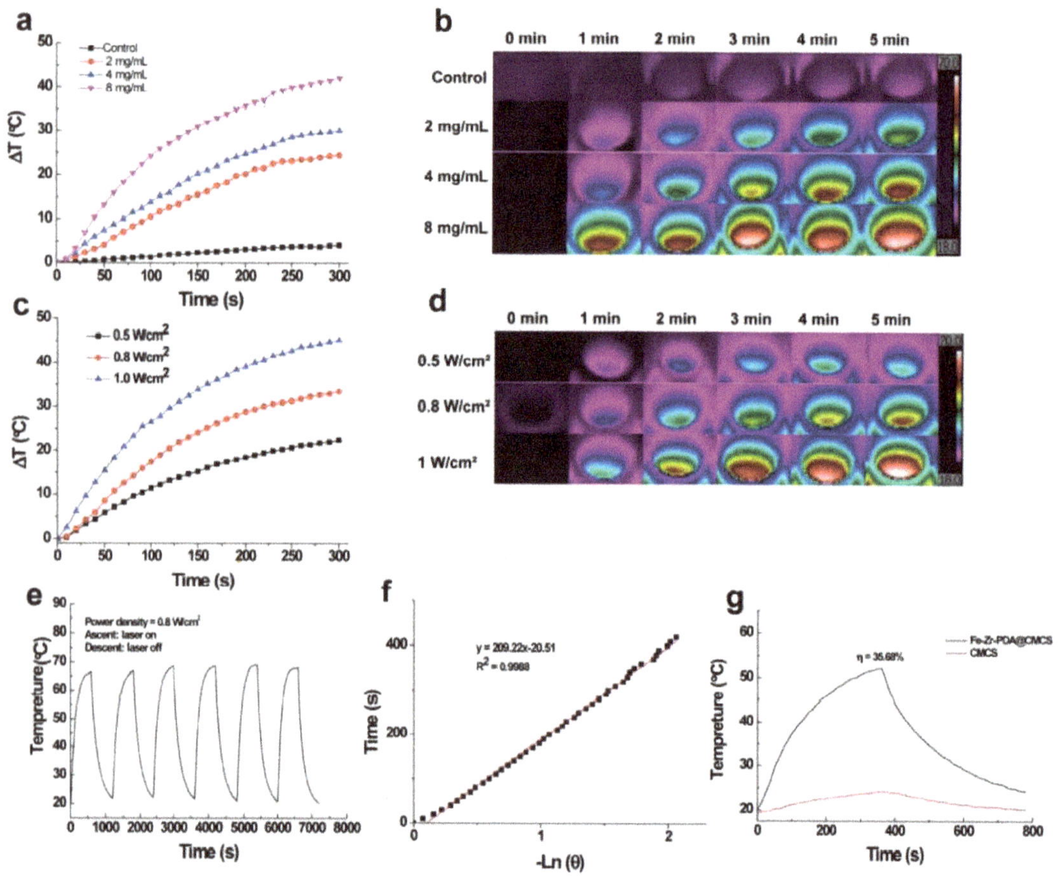

Figure 2. (**a**) Temperature variation of Fe–Zr@PDA@CMCS hydrogel with different Fe–Zr@PDA doping levels under 808 nm laser irradiation; (**b**) infrared thermographs correspond to (**a**), colors indicate different temperatures; (**c**) the temperature profiles of Fe–Zr@PDA@CMCS hydrogel irradiated with NIR laser with different power densities of 0.5, 0.8, and 1.0 W/cm^2; (**d**) infrared thermographs correspond to (**c**), colors indicate different temperatures; (**e**) results from photothermal cycling tests results of Fe–Zr@PDA@CMCS hydrogel after 8 irradiation and cooling cycles; (**f**) time constant curves for Fe–Zr@PDA@CMCS hydrogel (808 nm, 1.0 W/cm^2); (**g**) in vitro η values for Fe–Zr@PDA@CMCS hydrogel.

2.3. •OH Generating Capacity of Fe–Zr@PDA@CMCS Hydrogel

Therapeutic modalities based on reactive oxygen radicals and their derivatives have been demonstrated as emerging therapeutic strategies for tumors [43,51,52]. Among these, Fenton-based and Fenton-like responses are considered to be potential tumor cell therapy modalities and there are now several studies demonstrating the use of the Fenton effect to cause tumor-killing effects [44–47,53]. We first verified the ability of Fe–Zr@PDA to produce •OH; specifically, •OH can rapidly oxidize colorless TMB to a blue–green oxidized TMB compound (oxTMB), and further scanned the absorbance of the reacted solution at 652 nm by UV spectrophotometry. The magnitude of the absorbance gave a side view of the ability of the Fenton reaction to produce •OH. As shown in Figure 3a, the concentration of Fe–Zr@PDA was increased from 0 μg/mL to 100 μg/mL, and the results of the absorbance of the final solution after reaction with TMB were gradually increased, indicating a positive correlation between its ability to produce •OH and the concentration of Fe–Zr@PDA. The reaction of Fe–Zr@PDA with H_2O_2 shows that the colorless TMB is oxidized to the blue oxTMB, which further demonstrates the ability of Fe–Zr@PDA to generate •OH (Figure 3a). Further, we tested the •OH generation ability of Fe–Zr@PDA@CMCS hydrogel. The control hydrogel without Fe–Zr@PDA@CMCS had an absorbance value of almost zero at 652 nm and the mixture remained transparent (Figure 3b). In contrast, the absorbance of Fe–Zr@PDA@CMCS hydrogel increased with the concentration of Fe–Zr@PDA and the color of the solution deepened, which further indicated that the catalytic ability of Fe–Zr@PDA was well-preserved after doping in the hydrogel (Figure 3b).

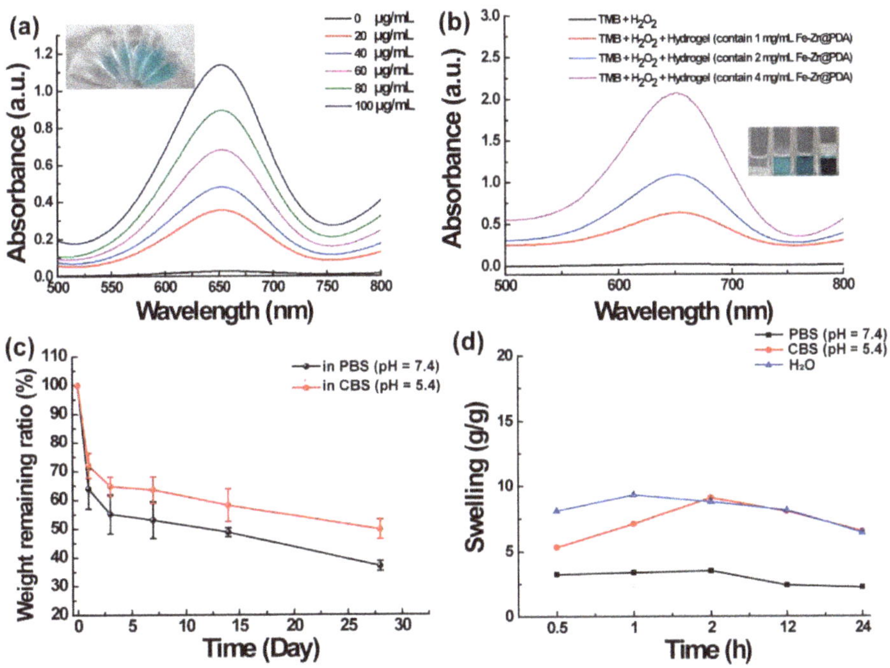

Figure 3. (**a**) Experimental results and corresponding digital camera photos of the reaction of different concentrations of Fe–Zr@PDA nanoenzymes with TMB and H_2O_2 to produce •OH; (**b**) absorbance values of different concentrations of Fe–Zr@PDA@CMCS after co-incubation with TMB and H_2O_2, and digital photographs of the supernatant photographic images; (**c**) degradation rates of Fe–Zr@PDA@CMCS hydrogel in CBS and PBS solutions; (**d**) swelling rate of Fe–Zr@PDA@CMCS hydrogel in ultrapure water, CBS solution, and PBS solution.

2.4. Evaluation of Hydrogel Degradation and Swelling Properties

The ability to swell and degrade swelling properties are important physical parameters of hydrogels, which affect the rate of release of loaded compounds and the absorption of tissue exudate, among other things [54–56]. In this section, we first investigated the degradation properties of Fe–Zr@PDA@CMCS hydrogel under different pH conditions. As shown in Figure 3c, Fe–Zr @PDA@CMCS hydrogel showed a fast weight loss in both PBS and CBS at the beginning and then degraded by more than 45% in the first three days. The degradation rate increased gradually with time, but the degradation rate was slower in the CBS than in the PBS. It is worth mentioning that the TME is acidic due to the altered metabolism of the tumor cells; therefore, the Fe–Zr@PDA@CMCS hydrogel can be retained in the acidic TME for a longer period to achieve prolonged drug release and tumor therapy.

Then, we evaluated the swelling properties of Fe–Zr@PDA@CMCS hydrogel by immersing them in ultrapure water, CBS, or PBS for 24 h. As shown in Figure 3d, the freeze-dried hydrogel immersed in PBS buffer and ultrapure water swelled hastily inside a brief length when they come upon the liquid. However, the swelling state of the hydrogel immersed in the PBS buffer solution also increased slowly with time. At the same time, the swelling of the hydrogel in pure water remained almost constant, and in the final experimental phase, the freeze-dried hydrogel reached swelling equilibrium. Interestingly, compared with the hydrogel soaked in PBS and pure water, Fe–Zr@PDA@CMCS hydrogel soaked in CBS showed little change in the swelling degrees. This will facilitate the presence of Fe–Zr@PDA@CMCS hydrogel in a slightly acidic environment.

2.5. In Vitro Cytocompatibility Results of Fe–Zr@PDA@CMCS Hydrogel

Biocompatibility refers to the ability of a material to elicit appropriate host and material responses in a given application environment [57,58]. The ideal composite hydrogel requires good biocompatibility both in vitro and in vivo [59,60]. We evaluated the cell safety of Fe–Zr@PDA@CMCS hydrogel through a series of in vitro cell experiments using the mouse fibroepithelial L929 cell line as a model. Adherently grown L929 cells were co-cultured with 1 mg/mL, 2.5 mg/mL, and 5 mg/mL of hydrogel extracts for 24 h and 48 h, and the results are shown in Figure 4a. Even at concentrations up to 5 mg/mL, the viability of L929 cells at 24 h and 48 h was higher than 90%, indicating that Fe–Zr@PDA@CMCS hydrogel has good cytocompatibility (Figure 4a). Following the CCK-8 cell viability assay, we further stained cells in each experimental and control group using AM-PI stain and observed the distribution of live and dead cells using inverted phase contrast fluorescence microscopy. As can be seen in Figure 4b, in the 24 h and 48 h images, both in the control and experimental groups, the l929 cells, which represent viable cells, show green fluorescence, while the red fluorescence, which represents dead cells, is negligible in the images. It is worth noting that there is no significant difference in the observation of cell morphology between the experimental group and the control group. The above staining results were consistent with the results of CCK-8 experiments, which further demonstrated that Fe–Zr@PDA@CMCS hydrogel has no significant effect on the cell morphology of L929 cells and has good cell safety.

2.6. In Vitro Hemocompatibility Results of Fe–Zr@PDA@CMCS Hydrogel

The validation of in vitro biocompatibility experiments is crucial to ensure the biomedical applications of the Fe–Zr@PDA@CMCS hydrogel. The hemocompatibility test was used to assess whether Fe–Zr@PDA@CMCS hydrogel would cause hemolysis. Briefly, Fe–Zr@PDA@CMCS hydrogel was co-cultured with rat red blood cells to investigate their hemocompatibility. As shown in Figure 4c, the hemolysis rate of erythrocytes was 2.56 ± 3.31%, 3.35 ± 0.41%, 3.98 ± 0.43%, and 4.61 ± 0.18% after being cultured with Fe–Zr@PDA@CMCS hydrogel with concentrations of 10, 20, 50, and 100 mg/mL respectively, which were all less than 5%. This result indicated that Fe–Zr@PDA@CMCS hydrogel has good hemocompatibility. In addition, we recorded digital photographs of the erythrocyte solution after co-culture and, in agreement with the previous results, the positive control group of erythrocytes co-cultured with ultrapure water appeared bright red. This demonstrated that the erythrocytes have all ruptured, whereas the supernatant of the experimental group was clearer and more transparent, indicating that most of the erythrocyte structure was still intact (Figure 4d).

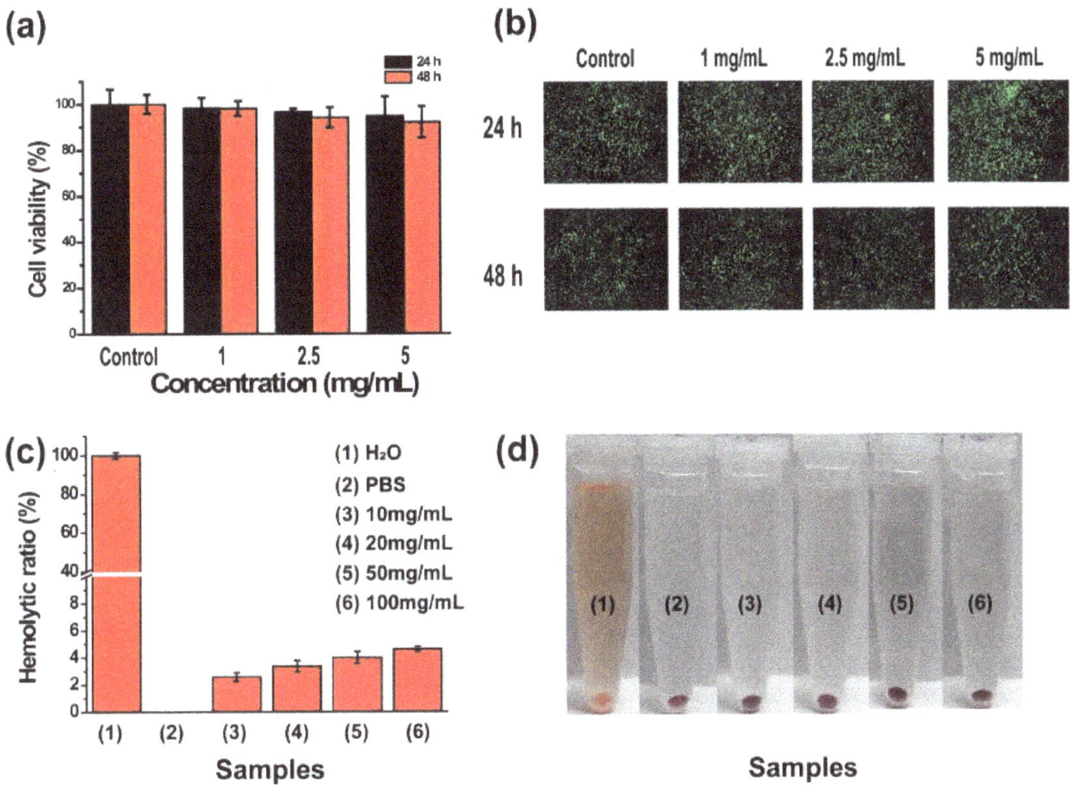

Figure 4. (**a**) Cell viability of Fe–Zr@PDA@CMCS hydrogel extracts at 1 mg/mL, 2.5 mg/mL, and 5 mg/mL co-cultured with mouse fibroblasts (L929 cell line) for 24 h and 48 h; (**b**) AM-PI stained images of L929 cell lines co-cultured with Fe–Zr@PDA@CMCS hydrogel extracts at 1 mg/mL, 2.5 mg/mL, and 5 mg/mL for 24 h and 48 h; (**c**) hemolysis ratio of rat erythrocytes after co-culture with different concentrations of Fe–Zr@PDA@CMCS hydrogel; (**d**) photos of rat erythrocytes after centrifugation.

2.7. In Vitro Biocompatibility Assessment Results of Fe–Zr@PDA@CMCS Hydrogel

After establishing the excellent in vitro cytocompatibility of Fe–Zr@PDA@CMCS hydrogel, the safety of the hydrogel was further assessed in vivo by subcutaneously embedding Fe–Zr@PDA@CMCS hydrogel in KM mice and recording the changes in body weight of the mice. The body weight of mice in both the experimental and control groups increased normally throughout the feeding cycle, and the body weight of mice in the experimental group fluctuated slightly for 21 days after encapsulation in the hydrogel; however, there was no statistical difference compared to the control group ($p > 0.05$) (Figure 5a). Further, we measured the blood biochemical parameters of the experimental and control mice on day 7 and 14 of the hydrogel implantation to assess whether the liver and kidney functions were affected. The serum biochemical parameters measured included TB (total bilirubin), ALT (alanine aminotransferase), AST (aspartate aminotransferase), UREA (urea), and CREA (creatinine). As shown in Figure 5b, the experimental and control groups of the hydrogel did not show a statistically significant difference in liver and kidney function ($p > 0.05$), which could further confirm the safety of the hydrogel. Next, we evaluated H&E staining sections of the major organs after 7 and 14 days of feeding, which showed no significant cellular hypertrophy, atrophy, or necrosis in any of the organ tissue sections. These results further demonstrated that the encapsulated Fe–Zr@PDA@CMCS hydrogel had no significant side effects on the normal physiological function of the major organs and had a good in vivo safety profile (Figure 5c). Finally, we examined the routine blood parameters of each group of mice, and the results are shown in Figure 6a–i. There was no statistical difference ($p > 0.05$) between the experimental and control groups of mice at 7 and 14 days.

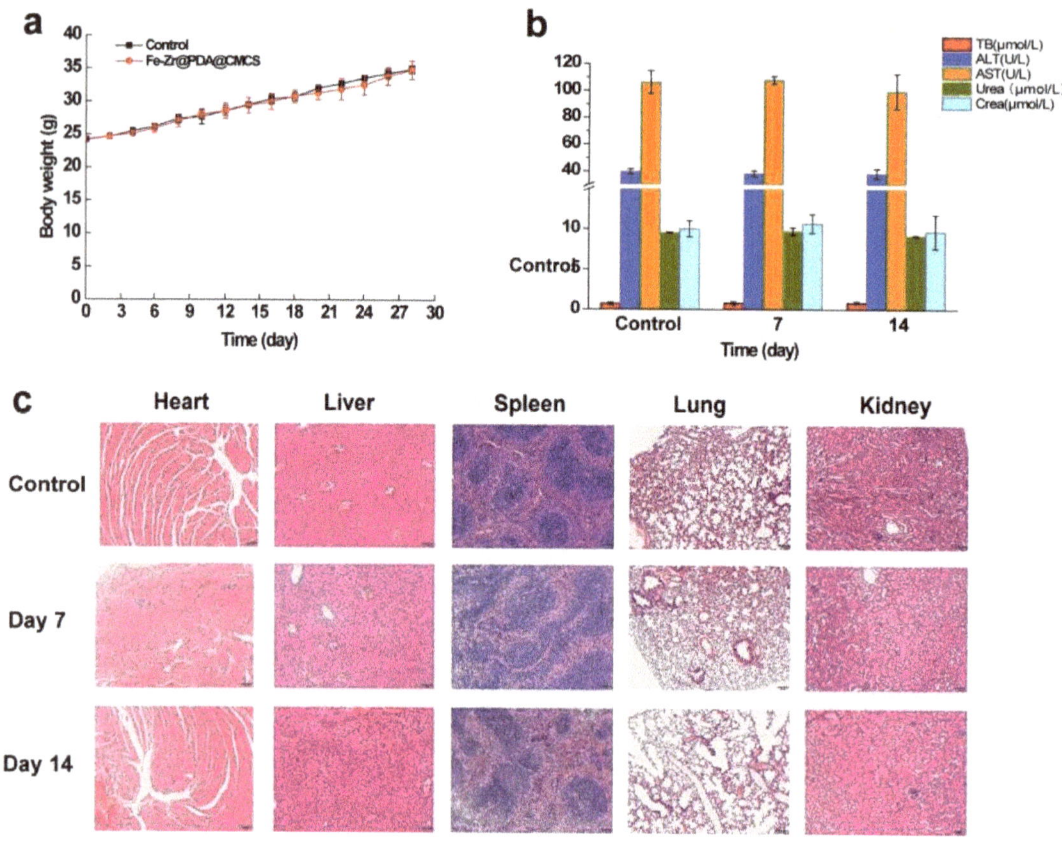

Figure 5. (**a**) Changes in body weight of Fe–Zr@PDA@CMCS hydrogel treated mice and normal mice (control) at day 28; (**b**) results of serum biochemical parameters of Fe–Zr@PDA@CMCS hydrogel treated mice and normal mice (control); (**c**) H&E staining of the main organs of the Fe–Zr@PDA@CMCS hydrogel treated mice and normal mice (control). Bar = 200 μm.

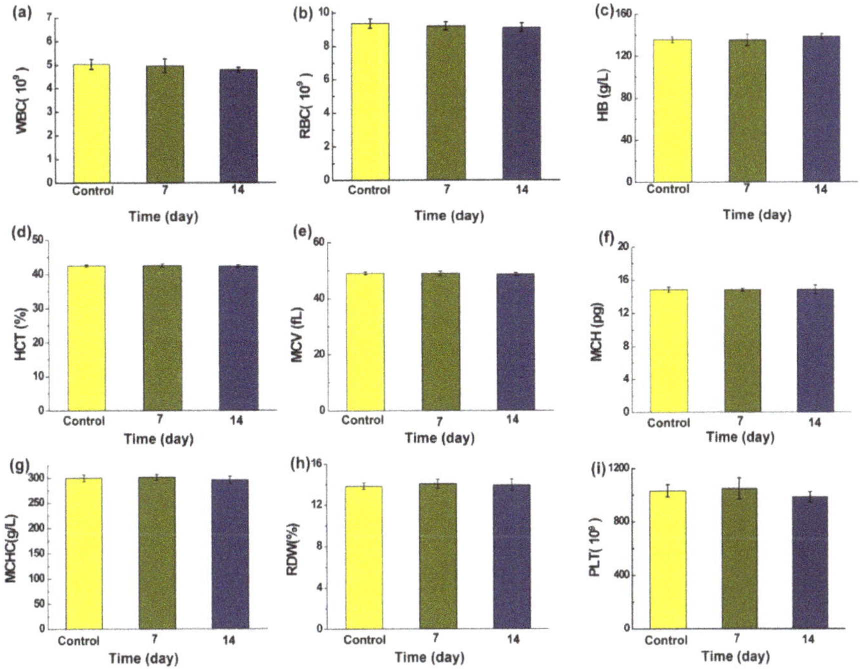

Figure 6. Routine blood test data of Fe–Zr@PDA@CMCS hydrogel-treated mice and normal mice (control): (**a**) white blood cell (WBC), (**b**) red blood cell (RBC), (**c**) hemoglobin (HB), (**d**) red blood cell-specific volume (HCT), (**e**) mean corpuscular volume (MCV), (**f**) mean corpuscular hemoglobin (MCH), (**g**) mean corpuscular hemoglobin concentration (MCHC), (**h**) red cell volume distribution width (RDW), and (i) PLATELET (PLT).

2.8. In Vitro Anticancer Potential of Fe–Zr@PDA@CMCS Hydrogel

To investigate their potential effectiveness in antitumor therapy, we selected SW1990 cells for in vitro evaluation of tumor therapy. The CCK-8 results (Figure 7a) showed that there was a decrease in the survival rate of SW1990 cells in the Fe–Zr@PDA nanoparticles group, proving that Fe–Zr@PDA caused some damage to the normal growth of cancer cells. The Fe–Zr@PDA@CMCS hydrogel extract group also showed a decrease in cell survival rate, suggesting that CMCS-doped functionalized hydrogels also have some tumor-cell-killing effect. In contrast, the survival rate of SW1990 cells receiving photothermal treatment alone was significantly lower, with almost half of the cells being inactive. Most importantly, the Fe–Zr@PDA@CMCS hydrogel extract combined with photothermal treatment showed an even more pronounced reduction in cell survival, indicating that the synergistic Fe–Zr@PDA@CMCS hydrogel and photothermal treatment can kill even more tumor cells. Afterwards, we further validated the in vitro anticancer effects by live–dead cell staining. The red fluorescence of Fe–Zr@PDA nanoparticles, Fe–Zr@PDA@CMCS hydrogel extracts treated group and photothermal treatment group all showed a significant increase compared to the control group (Figure 7b–e). Notably, there was a significant increase in red fluorescence and a significant decrease in green fluorescence after combining Fe–Zr@PDA@CMCS hydrogel with photothermal treatment, indicating that most of the SW1990 tumor cells were killed (Figure 7f). In conclusion, the combination of multifunctional hydrogel and photothermal therapy can effectively kill tumor cells, which is also expected to further enable tumor treatment at the in vivo level again.

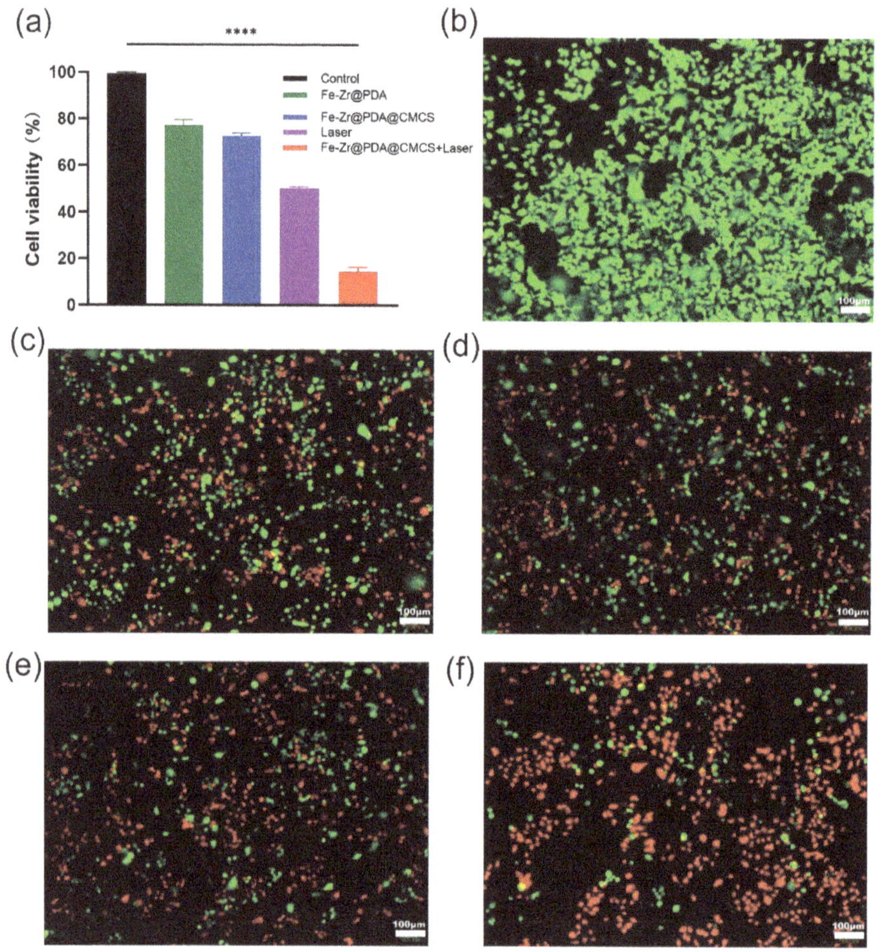

Figure 7. (**a**) CCK-8 results of cells after different treatments; (**b–f**) results of live and dead cell staining of SW1990 cells, morphological staining images of cells corresponding to CCK-8 results (**b**): L929 cells co-incubated with fresh medium; (**c**): L929 cells co-incubated with 100 μg/mL Fe–Zr@PDA nanoparticles; (**d**): L929 cells co-incubated with 100 μg/mL Fe–Zr@PDA@CMCS hydrogel extract; (**e**): L929 cells irradiated by 808 nm NIR, 1 W/cm^2, 5 min pure photothermal irradiation; (**f**): fluorescent staining images of L929 cells after treatment with 100 μg/ mL Fe–Zr@PDA@CMCS hydrogel extract and 808 nm NIR, 1 W/cm^2, by 5 min of light. Bar = 100 μm, **** $p < 0.0001$.

3. Conclusions

In summary, we successfully designed and prepared multifunctional Fe–Zr@PDA@CMCS hydrogel with the Fenton effect and photothermal conversion properties using Fe, Zr, PDA, and CMCS. Fe–Zr@PDA@CMCS hydrogel combined the advantages of hydrogel and nanoparticles. The photothermal conversion, degradation, and swelling capabilities of the hydrogel and Fenton's catalytic ability under different conditions were investigated. The results showed that the composite hydrogel retains its photothermal and Fentonian catalytic properties while protecting Fe–Zr@PDA nanoparticles from degradation. In addition, the good biocompatibility of the hydrogel was demonstrated at the cellular and animal levels. Further results demonstrated good therapeutic effects at the cellular level, and will be validated at the animal level in the future. Unfortunately, there are limitations to our study and we need further validation at the animal level and clinical level. In summary, multifunctional composite hydrogels can be used as carriers of drugs for multimodal tumor therapy and tissue regeneration for biomedical applications. Based on the good degradability and therapeutic effects of functionalized hydrogels, multifunctional composite hydrogels could provide novel ideas and show great promise for future clinical oncology treatments. The study is also expected to provide a reference for future clinical studies of novel hydrogels against tumor recurrence.

4. Materials and Methods

4.1. Materials

All chemicals were used without further purification. Zirconium tetrachloride ($ZrCl_4$), ferric chloride hexahydrate ($FeCl_3$-$6H_2O$), CMCS, PDA, ethylene glycol (EG), EDC, and NHS were purchased from Shanghai Aladdin Reagent Co., Ltd. (Shanghai, China). Mouse fibroblast cells (L929) and human pancreatic cancer cell lines (SW1990) were bought from the Institute of Biochemistry and Cell Biology, Chinese Academy of Sciences (Shanghai, China). Dulbecco's Modified Eagle Medium (DMEM), citrate buffer solution (CBS), and phosphate buffer saline (PBS) were procured from Gibco Co., Ltd. (Shanghai, China). The Cell counting kit-8 (CCK-8) was purchased from Vazyme Biotechnology Co., Ltd. (Nanjing, China). Calcein-AM/PI Live/Dead kit was purchased from Solarbio Technology (Beijing, China) Co., Ltd. Kunming (KM) mice (female, 4–6 weeks, 20–25 g) and Sprague-Dawley (SD) rats were ordered from Shanghai Slac Laboratory Animal Center (Shanghai, China). All experimental mice were fed strictly following the rules and regulations of the Ministry of Health of the People's Republic of China.

4.2. Preparation of Fe–Zr@PDA

The Fe–Zr@PDA nanoparticles were synthesized by a simple one-pot hydrothermal method. Specifically, 0.2 g of $FeCl_3$–$6H_2O$ and 0.075 g of $ZrCl_4$ were first weighed in scales and dissolved in 15 mL of ethylene glycol with stirring at room temperature, and the dissolved solution was named solution A. Next, 0.492 g of sodium acetate and 0.2 g of dopamine were dissolved in 10 mL of ethylene glycol (to form solution B). Solution A and solution B were then mixed well and added to a 100 mL autoclave lined with polytetrafluoroethylene, which was placed in an oven set at 180 °C for 12 h. Finally, after the reaction had cooled naturally, the product was washed 3 times with deionized water and anhydrous ethanol, centrifuged (9000 rpm, 8 min) to obtain the desired product, freeze-dried in a lyophilizer, and stored for subsequent experiments.

4.3. Synthesis of CMCS Hydrogel and Fe–Zr@PDA@CMCS Hydrogel

To synthesize the CMCS hydrogel, as a first step, 0.1 g of CMCS was introduced to 2 mL of ultrapure water, then dissolved while stirring in a water bath set to 50 °C. Next, 0.1 g of the activator EDC (0.05 g)/NHS (0.05 g) was added to 1 mL of pure water and the solution containing the activator was added to the CMCS solution and further stirred to form a gel. To synthesize the Fe–Zr@PDA@CMCS hydrogel, in brief, 0.05 g of EDC and 0.05 g of NHS were delivered to a solution of Fe–Zr@PDA nanomaterials (1 mL with a concentration of 0.01 g/mL) and finally mixed with the CMCS solution and stirred to form the hydrogel.

4.4. Characterizations of Fe–Zr@PDA@CMCS and Fe–Zr@PDA

To observe the microscopic morphology and elemental distribution of the CMCS and Fe–Zr@PDA@CMCS hydrogel, firstly, the conductive glue was carefully pasted on the test sample table, and the freeze-dried CMCS and Fe–Zr@PDA@CMCS hydrogel were cut into thin pieces and placed on the conductive glue. Then, an infrared baking lamp was used to dry the surface of the hydrogel. After that, its morphology was examined using a scanning electron microscope (SEM, Zeiss Sigma 300). Similarly, the morphology and particle size of Fe–Zr@PDA were observed using SEM, and the elemental composition of Fe–Zr@PDA nanoparticles was further analyzed using an X-ray energy spectrometer.

4.5. Characteristics of Photothermal Conversion

To explore the influence of different power densities on the temperature change, Fe–Zr@PDA@CMCS hydrogel was irradiated with the near-infrared laser electricity with strength densities of 0.5 W/cm^2, 0.8 W/cm^2, or 1 W/cm^2. Then, we recorded the temperature change using an $FLIR^{E60}$ camera. Finally, the photothermal stability of the hydrogel was explored. Specifically, the near-infrared laser with a power density of 0.8 W/cm^2 was used to irradiate the hydrogel for 6 cycles, and the cycle was set at 20 min each time. During the experiment, the thermal images and temperature changes of the experimental each cycle were recorded using a near-infrared thermal imager. The photothermal conversion efficiency (η) of Fe–Zr@PDA@CMCS was calculated by the modified Korgel formula (1):

$$\eta(\%) = \frac{hS(T_{max} - T_{surr}) - Q_{in.surr}}{I\left(1 - 10^{(-A(\lambda))}\right)} \tag{1}$$

In the above formula, η is the photothermal conversion efficiency, h is the warmness switch coefficient, I replace the power of the laser, T_{max} is the highest temperature of Fe–Zr@PDA@CMCS solution under the laser irradiation for 6 min, T_{surr} represents the ambient temperature throughout the test period, hS is the heat transfer coefficient, $A(\lambda)$ is the absorbance of Fe–Zr@PDA@CMCS at wavelength 808 nm, S is the effective area of laser radiation, and $Q_{in.surr}$ is the heat loss of Fe–Zr@PDA@CMCS in the heating process.

4.6. •OH Generating Capacity

Firstly, we explored the ability of different concentrations of Fe–Zr@PDA to produce •OH. Specifically, a 3,3′,5,5′-Tetramethylbenzidine (TMB) solution with a concentration of 3.2 mM was prepared using ultrapure water and 300 μL was mixed with different concentrations of Fe–Zr@PDA and H_2O_2. The total volume of the mixed solution was 1.5 mL, with the final concentration of H_2O_2 set at 8 mM and the final concentrations of Fe–Zr@PDA set at 0 μg/mL, 20 μg/mL, 40 μg/mL, 60 μg/mL, 80 μg/mL, and 100 μg/mL. The reaction lasted for 15 min at room temperature accompanied by ultra-high-speed centrifugation (15,000 rpm, 10 min). After the reaction, the supernatant was collected and photographed with a digital camera. At the same time, the absorbance of each sample supernatant at the wavelength of 652 nm was measured by an ultraviolet spectrometer.

The •OH formation ability of Fe–Zr@PDA@CMCS hydrogel was also determined by the color development properties of TMB after oxidation. Fe–Zr@PDA@CMCS hydrogel with different concentrations (1, 2, or 4 mg/mL) was mixed with TMB (mM) and H_2O_2 (0.3%). The group of TMB and H_2O_2 without Fe–Zr@PDA@CMCS hydrogel was set as the control. After being co-cultured at room temperature for 1 h, the supernatant was accumulated and photographed with a digital camera. The absorption value of the supernatant at λ = 652 nm was measured by UV–Vis–NIR spectroscopy (Lambda 25, Perkin Elmer, Waltham, MA, USA).

4.7. Degradation Analysis of Fe–Zr@PDA@CMCS Hydrogel

To explore the in vitro degradation ability, lyophilized Fe–Zr@PDA@CMCS hydrogel (100 mg) was immersed in 5 mL of PBS (pH = 7.4) or CBS (pH = 5.4). Then, the Fe–Zr@PDA@CMCS hydrogel samples were removed at predetermined time points (1, 3, 7, 14, and 28 days), washed with ultrapure water, and finally lyophilized and weighed in a lyophilizer. The degradation percentage of Fe–Zr@PDA@CMCS hydrogel was once calculated by way of the following formula:

$$\text{Weight remaining ratio}(\%) = \frac{R_t}{R_0} \times 100\% \tag{2}$$

where R_t represents the actual gram weight of Fe–Zr@PDA@CMCS hydrogel and R_0 represents the original weight of Fe–Zr@PDA@CMCS hydrogel.

4.8. Swelling Analysis of Hydrogel

The swelling rate (SR) and stability of Fe–Zr@PDA@CMCS hydrogel in different solutions (H_2O, PBS, and CBS) were determined by the swelling test. In short, 10 mg of Fe–Zr@PDA@CMCS hydrogel was placed into a sealed centrifuge tube (n = 3) and added to 20 mL of H_2O, PBS, and CBS respectively. The tubes were placed at 37 °C. The hydrogel was removed from the solution at predetermined time points (0.5, 1, 2, 12, and 24 h), and the surface water of the hydrogel was absorbed by filter paper and weighed again. The swelling kinetics curve of Fe–Zr@PDA@CMCS hydrogel was drawn, and the SR of the hydrogel reached swelling equilibrium was calculated according to the following formula:

$$\text{Swelling ratio } (\%) = \frac{m_t - m_0}{m_0} \times 100\% \tag{3}$$

where m_t is the mass of hydrogel after swelling at different points in time and m_0 represents the initial condition of this hydrogel.

4.9. In Vitro Cytocompatibility of Fe–Zr@PDA@CMCS Hydrogel

The cell safety of Fe–Zr@PDA@CMCS was evaluated using mouse fibroblasts (L929 cells). First, the L929 cells were inoculated on 96-well clear cell culture plates (5000 cells per well) and cultured for 24 h in a constant temperature incubator with 37 °C and 5% CO_2. At the same time, Fe–Zr@PDA@CMCS hydrogel was positioned in DMEM cell medium (1, 2.5, or 5 mg hydrogel, in 1 mL DMEM medium) and incubated at 37 °C in a single day to attain the extract with the attention of 5 mg/mL. On the second day, the old medium was replaced with 100 μL of the leaching solution, and the control group was added with a fresh DMEM medium. The extract concentration in the leaching solution was 1 mg/mL, 2.5 mg/mL, or 5 mg/mL. Finally, the medium was removed at 24 h or 48 h culture, and the cells were washed with PBS twice. Cell viability was determined by the CCK-8 kit. In addition, staining was performed using an AM-PI staining kit for both living and dead cells, and staining images were collected using an inverted phase contrast microscope (Leica DM IL, Weztlar, Germany).

4.10. In Vitro Blood Compatibility of Fe–Zr@PDA@CMCS Hydrogel

The feeding and testing of animals were conducted at Changhai Hospital of Naval Medical University in strict accordance with the program and policies of the Ministry of Health. To test whether Fe–Zr@PDA@CMCS can cause rupture of red blood cells and hemolysis, the blood compatibility of these cells was evaluated by using rat red blood cells. Firstly, blood was collected from the anesthetized heart of SD rats into anticoagulant collection vessels and centrifuged (4000 rpm, 5 min) to collect the red blood cells. Finally, the purified rat red blood cells were diluted to 2% with PBS and stored in the refrigerator at 4 °C for subsequent use. For the hemolysis experiment, we first mixed previously stored diluted erythrocyte suspensions with PBS buffer containing 20, 40, 100, and 200 mg of Fe–Zr@PDA@CMCS hydrogel, respectively. In addition, a negative control group (specifically, 1 mL of diluted PBS buffer mixed with 1 mL blood cells) and a positive control group (that is, 1 mL of diluted red cell suspension mixed with 1 mL ultrapure water) were set up, respectively. Finally, the above-mixed solutions were incubated in a water bath at 37 °C for 2 h and then centrifuged immediately to obtain the supernatant (3000 rpm, 5 min). The absorbance at 541 nm was measured with the UV–visible–near-infrared spectrometer (Lambda 25, Perkin Elmer, Waltham, MA, USA) of all the supernatants collected. Then, the hemolysis percentage (HP) was calculated according to the following formula, and the supernatant was photographed with a camera.

$$\text{Hemolytic ratio}(\%) = \frac{B_{sample} - B_{negative}}{B_{positive} - B_{negative}} \times 100\% \tag{4}$$

where B_{sample} is the absorbance at 541 nm of red blood cell suspension after treatment with hydrogel extract, $B_{negative}$ is the absorbance of erythrocyte suspension dealt with PBS buffer, and $B_{positive}$ is the absorbance of blood cell suspension after treatment with ultrapure water.

4.11. In Vivo Animal Tissue Safety of Fe–Zr@PDA@CMCS Hydrogel

To explore the safety of Fe–Zr@PDA@CMCS hydrogel in animals, KM mice were used as the animal model. Fe–Zr@PDA@CMCS hydrogel was embedded under the skin of mice to observe whether it would cause damage to mice. Specifically, KM mice (from the Laboratory in Shanghai Changhai Hospital, China) were randomly divided into control and experimental groups with three animals in each group ($n = 3$). Mice anesthetized with pentobarbital sodium were subcutaneously implanted with 200 mg of Fe–Zr@PDA@CMCS hydrogel in the experimental group. Normal healthful mice were used in the control group. The mice were weighed every 2 days after embedding to investigate the adjustments in body weight. The mice were sacrificed on the 7th, 14th, and 28th days, respectively. We collected blood by removing the eyes of mice, and major organs such as lungs, spleen, liver, heart, and kidney were collected and fixed in 4% paraformaldehyde for HE staining. Meanwhile, the collected blood was used to measure the values of the blood routine indexes of the two groups by the blood routine analyzer. An ELISA kit was used to detect the related indexes of kidney function and liver function in the serum. HE-stained sections of major organs were used to observe whether there were lesions such as inflammation and necrosis. All animal experimental operations were carried out in strict accordance with the protocols authorized by the hospital's Comprehensive Laboratory Animal Centre.

4.12. In Vitro Antitumor Effects of Fe–Zr@PDA@CMCS Hydrogels

We selected the human pancreatic cancer cell line SW1990 as a model to assess the in vitro therapeutic effect of Fe–Zr@PDA@CMCS. SW1990 cells were firstly inoculated into the 96-well culture plate at a density of 9000 cells/well. After 24 h of incubation, the original medium was replaced with 100 μL of fresh DMEM solution containing different substances and subjected to various treatments. The treatments were as follows: (a) control group (fresh DMEM), (b) Fe–Zr@PDA group (100 μg/mL Fe–Zr@PDA), (c) Fe–Zr@PDA@CMCS group (100 μg/mL Fe–Zr@PDA hydrogel extract), (d) laser group (808 nm NIR, 1 W/cm^2, 5 min), (e) Fe–Zr@PDA@CMCS + laser group (100 μg/mL Fe–Zr@PDA hydrogel extract, 808 nm NIR, 1 W/cm^2, 5 min). The groups were treated as described above and incubation was continued for 12 h, after which the survival of SW1990 cells was assessed by the CCK-8 method and the live–dead cell assay.

4.13. Statistical Analysis

All results are expressed as mean ± standard deviation and one-way ANOVA statistical analysis was used to assess the significance of the experimental data. The value of 0.05 was used as the significance level data, with probabilities less than 0.05 ($p < 0.05$), 0.01 ($p < 0.01$), and 0.001 ($p < 0.01$) indicated by (*), (**), and (***), respectively. The sample size was 3 unless stated ($n = 3$).

Author Contributions: Conceptualization, L.H., J.Z. and S.W.; methodology, X.Z. and H.W.; software, J.Z.; validation, X.Z., H.W. and L.H.; formal analysis, X.Z., J.Z., S.W. and L.H.; investigation, S.W.; resources, L.H. and J.Z.; data curation, H.W. and X.Z.; writing—original draft preparation, X.Z. and H.W.; visualization, J.Z. and S.W.; writing—review and editing, supervision, project administration, and funding acquisition, X.Z., H.W., S.W., J.Z. and L.H. All authors have read and agreed to the published version of the manuscript.

Funding: National Natural Science Foundation of China (82103696).

Institutional Review Board Statement: The animal study protocol was approved by the Institutional Review Board of The First Affiliated Hospital of Naval Medical University of the People's Liberation Army (SYXK (Shanghai) 2020-0033).

Informed Consent Statement: Not applicable.

Data Availability Statement: The data that support the findings of this study are available from the corresponding author upon reasonable request.

Conflicts of Interest: The authors declare no conflict of interest.

References

1. Dolgin, E. Cancer's new normal. *Nat. Cancer* **2021**, *2*, 1248–1250. [CrossRef] [PubMed]
2. Wyld, L.; Audisio, R.A.; Poston, G.J. The evolution of cancer surgery and future perspectives. *Nat. Rev. Clin. Oncol.* **2015**, *12*, 115–124. [CrossRef] [PubMed]
3. Frei, E., III. Curative cancer chemotherapy. *Cancer Res.* **1985**, *45*, 6523–6537.
4. Allen, C.; Her, S.; Jaffray, D.A. Radiotherapy for Cancer: Present and Future. *Adv. Drug Deliv. Rev.* **2017**, *109*, 1–2. [CrossRef] [PubMed]
5. Jarosz-Biej, M.; Smolarczyk, R.; Cichoń, T.; Kułach, N. Tumor Microenvironment as A "Game Changer" in Cancer Radiotherapy. *Int. J. Mol. Sci.* **2019**, *20*, 3212. [CrossRef]
6. Riley, R.S.; June, C.H.; Langer, R.; Mitchell, M.J. Delivery technologies for cancer immunotherapy. *Nat. Rev. Drug Discov.* **2019**, *18*, 175–196. [CrossRef]
7. Kennedy, L.B.; Salama, A.K.S. A review of cancer immunotherapy toxicity. *CA Cancer J. Clin.* **2020**, *70*, 86–104. [CrossRef]
8. Chen, Z.; Zhang, P.; Xu, Y.; Yan, J.; Liu, Z.; Lau, W.B.; Lau, B.; Li, Y.; Zhao, X.; Wei, Y.; et al. Surgical stress and cancer progression: The twisted tango. *Mol. Cancer* **2019**, *18*, 132. [CrossRef]
9. Sun, T.; Zhang, Y.S.; Pang, B.; Hyun, D.C.; Yang, M.; Xia, Y. Engineered nanoparticles for drug delivery in cancer therapy. *Angew. Chem. Int. Ed.* **2014**, *53*, 12320–12364. [CrossRef]
10. Mhaskar, R.; Clark, O.A.; Lyman, G.; Engel Ayer Botrel, T.; Morganti Paladini, L.; Djulbegovic, B. Colony-stimulating factors for chemotherapy-induced febrile neutropenia. *Cochrane Database Syst. Rev.* **2014**, *2014*, Cd003039. [CrossRef]
11. Sarmiento-Salinas, F.L.; Perez-Gonzalez, A.; Acosta-Casique, A.; Ix-Ballote, A.; Diaz, A.; Treviño, S.; Rosas-Murrieta, N.H.; Millán-Perez-Peña, L.; Maycotte, P. Reactive oxygen species: Role in carcinogenesis, cancer cell signaling and tumor progression. *Life Sci.* **2021**, *284*, 119942. [CrossRef] [PubMed]
12. Srinivas, U.S.; Tan, B.W.Q.; Vellayappan, B.A.; Jeyasekharan, A.D. ROS and the DNA damage response in cancer. *Redox Biol.* **2019**, *25*, 101084. [CrossRef] [PubMed]
13. Renaudin, X. Reactive oxygen species and DNA damage response in cancer. *Int. Rev. Cell Mol. Biol.* **2021**, *364*, 139–161. [CrossRef]
14. Moloney, J.N.; Cotter, T.G. ROS signalling in the biology of cancer. *Semin. Cell Dev. Biol.* **2018**, *80*, 50–64. [CrossRef]
15. Gorrini, C.; Harris, I.S.; Mak, T.W. Modulation of oxidative stress as an anticancer strategy. *Nat. Rev. Drug Discov.* **2013**, *12*, 931–947. [CrossRef] [PubMed]
16. Ray, P.D.; Huang, B.W.; Tsuji, Y. Reactive oxygen species (ROS) homeostasis and redox regulation in cellular signaling. *Cell Signal.* **2012**, *24*, 981–990. [CrossRef]
17. Zhang, Z.R.; Zhao, J.L.; Chen, Z.; Wu, H.; Wang, S.G. A molybdenum-based nanoplatform with multienzyme mimicking capacities for oxidative stress-induced acute liver injury treatment. *Inorg. Chem. Front.* **2023**, *10*, 1305–1314. [CrossRef]
18. Cheung, E.C.; Vousden, K.H. The role of ROS in tumour development and progression. *Nat. Rev. Cancer* **2022**, *22*, 280–297. [CrossRef]
19. Kirtonia, A.; Sethi, G.; Garg, M. The multifaceted role of reactive oxygen species in tumorigenesis. *Cell. Mol. Life Sci.* **2020**, *77*, 4459–4483. [CrossRef]
20. Li, M.; Zhang, W.; Xu, X.; Liu, G.; Dong, M.; Sun, K.; Zhang, P. Nanosystems for chemodynamic based combination therapy: Strategies and recent advances. *Front. Pharmacol.* **2022**, *13*, 1065438. [CrossRef]
21. Ouyang, Y.; Zhao, J.; Wang, S. Multifunctional hydrogels based on chitosan, hyaluronic acid and other biological macromolecules for the treatment of inflammatory bowel disease: A review. *Int. J. Biol. Macromol.* **2023**, *227*, 505–523. [CrossRef] [PubMed]
22. Zhang, L.; Li, C.X.; Wan, S.S.; Zhang, X.Z. Nanocatalyst-Mediated Chemodynamic Tumor Therapy. *Adv. Healthc. Mater.* **2022**, *11*, e2101971. [CrossRef] [PubMed]
23. Jia, C.; Guo, Y.; Wu, F.G. Chemodynamic Therapy via Fenton and Fenton-Like Nanomaterials: Strategies and Recent Advances. *Small* **2022**, *18*, e2103868. [CrossRef] [PubMed]
24. Galadari, S.; Rahman, A.; Pallichankandy, S.; Thayyullathil, F. Reactive oxygen species and cancer paradox: To promote or to suppress? *Free Radic. Biol. Med.* **2017**, *104*, 144–164. [CrossRef]
25. Xu, X.; Zeng, Y.; Chen, Z.; Yu, Y.; Wang, H.; Lu, X.; Zhao, J.; Wang, S. Chitosan-based multifunctional hydrogel for sequential wound inflammation elimination, infection inhibition, and wound healing. *Int. J. Biol. Macromol.* **2023**, *235*, 123847. [CrossRef]
26. Vilema-Enríquez, G.; Arroyo, A.; Grijalva, M.; Amador-Zafra, R.I.; Camacho, J. Molecular and Cellular Effects of Hydrogen Peroxide on Human Lung Cancer Cells: Potential Therapeutic Implications. *Oxid. Med. Cell. Longev.* **2016**, *2016*, 1908164. [CrossRef]
27. Chen, M.C.; Lin, Z.W.; Ling, M.H. Near-Infrared Light-Activatable Microneedle System for Treating Superficial Tumors by Combination of Chemotherapy and Photothermal Therapy. *ACS Nano* **2016**, *10*, 93–101. [CrossRef]

28. Li, X.; Lovell, J.F.; Yoon, J.; Chen, X. Clinical development and potential of photothermal and photodynamic therapies for cancer. *Nat. Rev. Clin. Oncol.* **2020**, *17*, 657–674. [CrossRef]
29. Chen, Z.; Zheng, X.; Zhao, J.; Tang, J.; Hu, L.; Wang, S. Glucose oxidase-loaded colloidal stable WS_2 nanobowls for combined starvation/photothermal therapy of colorectal tumors. *Int. J. Pharm.* **2023**, *636*, 122848. [CrossRef]
30. Zhao, L.; Zhang, X.; Wang, X.; Guan, X.; Zhang, W.; Ma, J. Recent advances in selective photothermal therapy of tumor. *J. Nanobiotechnol.* **2021**, *19*, 335. [CrossRef]
31. Chen, Z.; Yao, J.; Zhao, J.; Wang, S. Injectable wound dressing based on carboxymethyl chitosan triple-network hydrogel for effective wound antibacterial and hemostasis. *Int. J. Biol. Macromol.* **2023**, *225*, 1235–1245. [CrossRef]
32. Capáková, Z.; Radaszkiewicz, K.A.; Acharya, U.; Truong, T.H.; Pacherník, J.; Bober, P.; Kašpárková, V.; Stejskal, J.; Pfleger, J.; Lehocký, M.; et al. The biocompatibility of polyaniline and polypyrrole 2: Doping with organic phosphonates. *Mater. Sci. Eng. C Mater. Biol. Appl.* **2020**, *113*, 110986. [CrossRef] [PubMed]
33. Yang, J.; Wang, S. Polysaccharide-Based Multifunctional Hydrogel Bio-Adhesives for Wound Healing: A Review. *Gels* **2023**, *9*, 138. [CrossRef] [PubMed]
34. Fan, P.; Zeng, Y.; Zaldivar-Silva, D.; Agüero, L.; Wang, S. Chitosan-Based Hemostatic Hydrogels: The Concept, Mechanism, Application, and Prospects. *Molecules* **2023**, *28*, 1473. [CrossRef] [PubMed]
35. Musaie, K.; Abbaszadeh, S.; Nosrati-Siahmazgi, V.; Qahremani, M.; Wang, S.; Eskandari, M.R.; Niknezhad, S.V.; Haghi, F.; Li, Y.; Xiao, B.; et al. Metal-coordination synthesis of a natural injectable photoactive hydrogel with antibacterial and blood-aggregating functions for cancer thermotherapy and mild-heating wound repair. *Biomater. Sci.* **2023**, *11*, 2486–2503. [CrossRef]
36. Ambekar, R.S.; Kandasubramanian, B. A polydopamine-based platform for anti-cancer drug delivery. *Biomater. Sci.* **2019**, *7*, 1776–1793. [CrossRef]
37. Yang, X.; Wang, S.; Zhang, X.; Ye, C.; Wang, S.; An, X. Development of PVA-based microsphere as a potential embolization agent. *Mater. Sci. Eng. C Mater. Biol. Appl.* **2022**, *135*, 112677. [CrossRef]
38. An, P.; Fan, F.; Gu, D.; Gao, Z.; Hossain, A.M.S.; Sun, B. Photothermal-reinforced and glutathione-triggered in Situ cascaded nanocatalytic therapy. *J. Control. Release* **2020**, *321*, 734–743. [CrossRef]
39. Wang, T.; Yi, W.; Zhang, Y.; Wu, H.; Fan, H.; Zhao, J.; Wang, S. Sodium alginate hydrogel containing platelet-rich plasma for wound healing. *Colloids Surf. B Biointerfaces* **2023**, *222*, 113096. [CrossRef]
40. Cai, J.; Guo, J.; Wang, S. Application of Polymer Hydrogels in the Prevention of Postoperative Adhesion: A Review. *Gels* **2023**, *9*, 98. [CrossRef]
41. Ouyang, Y.; Zhao, Y.; Zheng, X.; Zhang, Y.; Zhao, J.; Wang, S.; Gu, Y. Rapidly degrading and mussel-inspired multifunctional carboxymethyl chitosan/montmorillonite hydrogel for wound hemostasis. *Int. J. Biol. Macromol.* **2023**, *242*, 124960. [CrossRef]
42. Mohammadniaei, M.; Lee, T.; Bharate, B.G.; Yoon, J.; Choi, H.K.; Park, S.J.; Kim, J.; Kim, J.; Choi, J.W. Bifunctional Au@Bi_2Se_3 Core-Shell Nanoparticle for Synergetic Therapy by SERS-Traceable AntagomiR Delivery and Photothermal Treatment. *Small* **2018**, *14*, e1802934. [CrossRef] [PubMed]
43. Liu, P.; Peng, Y.; Ding, J.; Zhou, W. Fenton metal nanomedicines for imaging-guided combinatorial chemodynamic therapy against cancer. *Asian J. Pharm. Sci.* **2022**, *17*, 177–192. [CrossRef] [PubMed]
44. Shen, Z.; Liu, T.; Li, Y.; Lau, J.; Yang, Z.; Fan, W.; Zhou, Z.; Shi, C.; Ke, C.; Bregadze, V.I.; et al. Fenton-Reaction-Acceleratable Magnetic Nanoparticles for Ferroptosis Therapy of Orthotopic Brain Tumors. *ACS Nano* **2018**, *12*, 11355–11365. [CrossRef]
45. Gao, H.; Cao, Z.; Liu, H.; Chen, L.; Bai, Y.; Wu, Q.; Yu, X.; Wei, W.; Wang, M. Multifunctional nanomedicines-enabled chemodynamic-synergized multimodal tumor therapy via Fenton and Fenton-like reactions. *Theranostics* **2023**, *13*, 1974–2014. [CrossRef]
46. Zhou, Y.; Fan, S.; Feng, L.; Huang, X.; Chen, X. Manipulating Intratumoral Fenton Chemistry for Enhanced Chemodynamic and Chemodynamic-Synergized Multimodal Therapy. *Adv. Mater.* **2021**, *33*, e2104223. [CrossRef]
47. Mohammed, D.F.; Madlool, H.A.; Faris, M.; Shalan, B.H.; Hasan, H.H.; Azeez, N.F.; Abbas, F.H. Harnessing inorganic nanomaterials for chemodynamic cancer therapy. *Nanomedicine* **2022**, *17*, 1891–1906. [CrossRef]
48. Fang, Y.; Zhang, L.; Chen, Y.; Wu, S.; Weng, Y.; Liu, H. Polysaccharides based rapid self-crosslinking and wet tissue adhesive hemostatic powders for effective hemostasis. *Carbohydr. Polym.* **2023**, *312*, 120819. [CrossRef] [PubMed]
49. Yang, C.; Mi, X.; Su, H.; Yang, J.; Gu, Y.; Zhang, L.; Sun, W.; Liang, X.; Zhang, C. GE11-PDA-Pt@USPIOs nano-formulation for relief of tumor hypoxia and MRI/PAI-guided tumor radio-chemotherapy. *Biomater. Sci.* **2019**, *7*, 2076–2090. [CrossRef]
50. Li, Y.; Jiang, C.; Zhang, D.; Wang, Y.; Ren, X.; Ai, K.; Chen, X.; Lu, L. Targeted polydopamine nanoparticles enable photoacoustic imaging guided chemo-photothermal synergistic therapy of tumor. *Acta Biomater.* **2017**, *47*, 124–134. [CrossRef]
51. Cui, Q.; Wang, J.Q.; Assaraf, Y.G.; Ren, L.; Gupta, P.; Wei, L.; Ashby, C.R., Jr.; Yang, D.H.; Chen, Z.S. Modulating ROS to overcome multidrug resistance in cancer. *Drug Resist. Updates* **2018**, *41*, 1–25. [CrossRef] [PubMed]
52. Su, X.; Shen, Z.; Yang, Q.; Sui, F.; Pu, J.; Ma, J.; Ma, S.; Yao, D.; Ji, M.; Hou, P. Vitamin C kills thyroid cancer cells through ROS-dependent inhibition of MAPK/ERK and PI3K/AKT pathways via distinct mechanisms. *Theranostics* **2019**, *9*, 4461–4473. [CrossRef] [PubMed]
53. Manivasagan, P.; Joe, A.; Han, H.W.; Thambi, T.; Selvaraj, M.; Chidambaram, K.; Kim, J.; Jang, E.S. Recent advances in multifunctional nanomaterials for photothermal-enhanced Fenton-based chemodynamic tumor therapy. *Mater. Today Bio* **2022**, *13*, 100197. [CrossRef] [PubMed]

54. Webber, M.J.; Pashuck, E.T. (Macro)molecular self-assembly for hydrogel drug delivery. *Adv. Drug Deliv. Rev.* **2021**, *172*, 275–295. [CrossRef] [PubMed]
55. McKenzie, M.; Betts, D.; Suh, A.; Bui, K.; Kim, L.D.; Cho, H. Hydrogel-Based Drug Delivery Systems for Poorly Water-Soluble Drugs. *Molecules* **2015**, *20*, 20397–20408. [CrossRef]
56. Flegeau, K.; Toquet, C.; Rethore, G.; d'Arros, C.; Messager, L.; Halgand, B.; Dupont, D.; Autrusseau, F.; Lesoeur, J.; Veziers, J.; et al. In Situ Forming, Silanized Hyaluronic Acid Hydrogels with Fine Control Over Mechanical Properties and In Vivo Degradation for Tissue Engineering Applications. *Adv. Healthc. Mater.* **2020**, *9*, e2000981. [CrossRef]
57. Zor, F.; Selek, F.N.; Orlando, G.; Williams, D.F. Biocompatibility in regenerative nanomedicine. *Nanomedicine* **2019**, *14*, 2763–2775. [CrossRef]
58. Crawford, L.; Wyatt, M.; Bryers, J.; Ratner, B. Biocompatibility Evolves: Phenomenology to Toxicology to Regeneration. *Adv. Healthc. Mater.* **2021**, *10*, e2002153. [CrossRef]
59. Riehemann, K.; Schneider, S.W.; Luger, T.A.; Godin, B.; Ferrari, M.; Fuchs, H. Nanomedicine-challenge and perspectives. *Angew. Chem. Int. Ed.* **2009**, *48*, 872–897. [CrossRef]
60. Adabi, M.; Naghibzadeh, M.; Adabi, M.; Zarrinfard, M.A.; Esnaashari, S.S.; Seifalian, A.M.; Faridi-Majidi, R.; Tanimowo Aiyelabegan, H.; Ghanbari, H. Biocompatibility and nanostructured materials: Applications in nanomedicine. *Artif. Cells Nanomed. Biotechnol.* **2017**, *45*, 833–842. [CrossRef]

gels MDPI

Article

Quality-by-Design-Assisted Optimization of Carvacrol Oil-Loaded Niosomal Gel for Anti-Inflammatory Efficacy by Topical Route

Mohammed Ghazwani [1,*], Umme Hani [1], Aftab Alam [2] and Mohammed H. Alqarni [2]

[1] Department of Pharmaceutics, College of Pharmacy, King Khalid University, Abha 61441, Saudi Arabia; uahmed@kku.edu.sa
[2] Department of Pharmacognosy, College of Pharmacy, Prince Sattam Bin Abdulaziz University, Al Kharj 11942, Saudi Arabia; a.alam@psau.edu.sa (A.A.); m.alqarni@psau.edu.sa (M.H.A.)
* Correspondence: myghazwani@kku.edu.sa

Abstract: Niosomes are multilamellar vesicles that effectively transfer active ingredients into the skin's layers. To improve the active substance's penetration across the skin, these carriers are frequently utilized as topical drug delivery systems. Essential oils (EOs) have garnered significant interest in the field of research and development owing to their various pharmacological activities, cost-effectiveness, and simple manufacturing techniques. However, these ingredients undergo degradation and oxidation over time, leading to a loss of functionality. Niosome formulations have been developed to deal with these challenges. The main goal of this work was to create a niosomal gel of carvacrol oil (CVC) to improve its penetration into the skin for anti-inflammatory actions and stability. By changing the ratio of drug, cholesterol and surfactant, various formulations of CVC niosomes were formulated using Box Behnken Design (BBD). A thin-film hydration technique using a rotary evaporator was employed for the development of niosomes. Following optimization, the CVC-loaded niosomes had shown: 180.23 nm, 0.265, −31.70 mV, and 90.61% of vesicle size, PDI, zeta potential, and EE%. An in vitro study on drug release discovered the rates of drug release for CVC-Ns and CVC suspension, which were found to be 70.24 ± 1.21 and 32.87 ± 1.03, respectively. The release of CVC from niosomes best fit the Higuchi model, and the Korsmeyer–Peppas model suggests that the release of the drug followed the non-Fickian diffusion. In a dermatokinetic investigation, niosome gel significantly increased CVC transport in the skin layers when compared to CVC–conventional formulation gel (CVC-CFG). Confocal laser scanning microscopy (CLSM) of rat skin exposed to the rhodamine B-loaded niosome formulation showed a deeper penetration of 25.0 µm compared to the hydroalcoholic rhodamine B solution (5.0 µm). Additionally, the CVC-N gel antioxidant activity was higher than that of free CVC. The formulation coded F4 was selected as the optimized formulation and then gelled with carbopol to improve its topical application. Niosomal gel underwent tests for pH determination, spreadability, texture analysis, and CLSM. Our findings imply that the niosomal gel formulations could represent a potential strategy for the topical delivery of CVC in the treatment of inflammatory disease.

Keywords: carvacrol; essential oil; niosomes; confocal laser scanning microscopy; anti-inflammatory; dermatokinetics; in vitro drug release studies

Citation: Ghazwani, M.; Hani, U.; Alam, A.; Alqarni, M.H. Quality-by-Design-Assisted Optimization of Carvacrol Oil-Loaded Niosomal Gel for Anti-Inflammatory Efficacy by Topical Route. *Gels* **2023**, *9*, 401. https://doi.org/10.3390/gels9050401

Academic Editor: Shige Wang

Received: 10 April 2023
Revised: 25 April 2023
Accepted: 6 May 2023
Published: 10 May 2023

1. Introduction

Due to their low cost and simple manufacturing processes, the utilization of EO constituents for biomedical applications has attracted global interest [1]. Their numerous qualities, such as viricidal, bactericidal, fungicidal, local anaesthetic, anti-inflammatory, and analgesic properties, have made them widely utilized for centuries. Because they are relatively safer for the skin than synthetically made traditional drugs, EO are often chosen as priority [2]. The components derived from plants still have the potential to cause allergies, but they are used in much lower concentrations than directly applying the oils to the skin, making them a safer option [3]. EOs are mostly made up of volatile substances such as aliphatic, aromatic, terpenoids and phenolic compounds. However, these chemicals could slowly deteriorate and oxidize, which would reduce their functionality [4]. Additionally, the bioavailability, flavour, and colour of the other ingredients may change in response to one another [5]. EOs are substances that, from a physiological standpoint, rapidly evaporate and/or decay even while the temperature is milder than the room temperature. This is because their boiling temperatures are extremely low, and they break down when directly exposed to light or oxygen, in general. Additionally, EO formulations frequently result in skin sensitivity and irritation [5]. The plant species *Thymus vulgaris, Origanum vulgare, Ocimum gratissimum, Carum copticum, Oliveria decumbens, Trachyspermum ammi, Monarda didyma, Nigella sativa, Origanum onites, Origanum syriacum, Plectranthus amboinicus, Lavandula multifida,* and *Satureja thymbra* are the primary sources of thymol (2-isopropyl-5-methylphenol; $C_{10}H_{14}O$), an isomeric form of carvacrol (5-isopropyl-2-methylphenol; $C_{10}H_{14}O$) [6,7]. Carvacrol and thymol have similar chemical structures and biological functions, which may indicate that they have related action mechanisms and therapeutic benefits. The CVC possesses a wide range of therapeutic activity, including those of an immunomodulator, an antioxidant, an antiseptic, an anti-diabetic, a cancer prophylactic agent, an anti-inflammatory compound, an anti-spasmodic molecule, and a growth promoter [8–15]. CVC is a promising drug for the treatment of local skin damage [16,17]. However, its pharmaceutical use is severely constrained by its lower bioavailability as a result of its poor water solubility, high volatility, and instability [10]. CVC exhibited anti-inflammatory activity by reducing cyclooxygenase-2 (COX-2) and inhibiting interleukin-1β (IL-1β), and potentially induced interleukin-10 (IL-10) release [18]. The release of IL-10 would result in an anti-inflammatory impact because IL-10 is an anti-inflammatory cytokine that inhibits several cell surface molecules, such as the MHC class II proteins and co-stimulatory molecules [19,20]. Many strategies have been put up to overcome this solubility and bioavailability. One of the flexible approaches involves encasing the EOs in other nanoparticles, such as liposomes [21], niosomes [22], graphene oxide [23], and others. Niosomes can extend drug retention duration, increase solubility, and improve targeted delivery [24].

Niosome is a biocompatible bilayer system based on non-ionic surfactants and cholesterol (as a lipid) [25]. The amphiphilic non-ionic surfactant self-assembles to produce niosomes in aqueous fluids. Niosomes also have several benefits, such as the ability to encapsulate both hydrophilic and hydrophobic components, improved therapeutic performance, long storage times, easy surface modification, biodegradability, and non-immunogenicity [26]. Entrapping EOs in the hydrophobic areas of niosomes can increase their permeability and stability qualities, lowering their potential for skin irritation [27]. These nonionic surfactant vesicles enhance the potential of the entrapped drug when applied topically by modifying the properties of the stratum corneum, facilitating the transdermal transport of trapped substances and decreasing trans-epidermal water loss [28]. Niosomal formulations have been created for a variety of targeted drug delivery purposes throughout the past decade [29]. To produce small-sized niosomes with maximal encapsulation efficiency for any given drug, however, the size and drug encapsulation efficiency of the niosomes must be optimized by altering the formulation (surfactants and lipids) [30,31]. Niosomal formulations of compounds containing CVC oil have been developed for anti-inflammatory and other studies, but a CVC-loaded niosomal formulation has not yet been developed [32–36]. The primary objective of this study was to investigate the potential of CVC for the development of niosomes. Niosome efficacy can be affected by many factors related to formulation and processing. Investigation into these aspects of niosome preparation would thus significantly improve the formulation. The quality by design (QbD) strategy involves the design and development of a product with manufacturing procedures that meet predetermined product criteria [37]. To ensure that selection criteria for desired responses are established for a range of vesicle sizes (<200 nm), polydispersity index (PDI) (<0.3), and entrapment efficiency (>70%) for formulation optimization, response surface methodology (RSM) was employed [38]. For the preparation of niosomes in drug delivery, a variety of nonionic surfactants, including polysorbates, alkyl esters, and alkyl ethers, are usually used. An important factor influencing the physicochemical properties of niosomes, such as bilayer rigidity, vesicle size, stability, and drug release rate, is the type of surfactant used [39]. Tween 80 and Span 80 have undergone testing as drug delivery systems, and specifically, they can be used to create highly stable niosomes [40]. In this formulation of niosomes, tween 80 was used as a surfactant. The stabilization of niosomes that have both hydrophilic and lipophilic domains was accomplished using the surfactant Tween 80. Its longer aliphatic tail causes adsorption on the drug's surface and allows its hydrophilic portion to potentially reach the aqueous phase. This would completely cover the surface and lessen the interfacial friction between hydrophobic drug vesicles [41]. The larger unsaturated side chain of the tween 80's oleate component may increase the poorly water-soluble drug CVC's ability to be enclosed more densely in niosome bilayers [42]. The developed niosomal formulation was incorporated into the gel for better retention time. The vesicle morphology, in vitro drug release, CLSM, dermatokinetic and optimized gel formulation texture analysis was also assessed. It can be anticipated that, the use of quality-by-design approach to optimise the CVC-loaded niosomal gel formulation is a novel approach that can help to ensure the formulation's quality, safety, and efficacy. Furthermore, using a topical niosomal formulation can improve carvacrol skin penetration and absorption, making it more beneficial for inflammatory skin disorders that require targeted therapy without systemic absorption. This approach may lead to the development of a more efficient and effective treatment for inflammatory skin disorders.

2. Results and Discussion

2.1. *Optimization of Carvacrol-Loaded Niosomes (CVC-Ns)*

The formulation optimization of the CVC-Ns was carried out utilizing a Box–Behnken experimental design with three parameters and three levels. Different independent variables' individual and combined effects on the responses were assessed. Table 1 displays the minimum (−1), average (0), and maximum (+1) levels of the three factors (drug as A, surfactant as B, and cholesterol quantity as C). Table 1 displays the data of the evaluation of the 17 different compositions of CVC-loaded niosomes from the BBD design.

Table 1. Responses observed in BBD for development and optimization of CVC-Niosome formulations with a summary of parameters for responses (Y_1, Y_2, Y_3) by Design Expert Software.

Formulations	Independent Variables			Dependent Variables		
	A	B	C	Y_1	Y_2	Y_3
1	5	120	7.5	205.05	0.291	78.49
2	10	120	5	180.98	0.259	88.97
3	10	150	7.5	186.98	0.285	77.95
4	10	120	5	180.23	0.265	90.41
5	10	120	5	181.07	0.261	89.48
6	10	150	2.5	185.89	0.281	78.62
7	15	120	7.5	208.95	0.294	76.38
8	5	120	2.5	204.62	0.286	81.64
9	15	90	5	206.99	0.279	82.84
10	15	120	2.5	208.95	0.283	80.51
11	15	120	5	181.28	0.262	89.55
12	5	150	5	200.01	0.272	84.18
13	10	90	2.5	195.54	0.279	79.98
14	5	90	5	199.94	0.279	81.55
15	10	120	5	181.08	0.26	90.01
16	10	90	7.5	191.08	0.288	76.08
17	15	150	5	205.54	0.271	83.17

A, drug (mg); B, surfactant (mg); C, cholesterol (mg); Y_1, vesicle size (nm); Y_2, PDI; Y_3, EE (%).

As indicated in Table 2 and the above-mentioned equation, the CVC-loaded noisome vesicle size ranged from 180.23 to 206.99 nm. Figures 1A and 2 show 3D and contour plots that illustrate the effect of factors on size. The sizes of the niosomes slightly increase when the drug concentration (A) rises. The rise in the concentration of the drug (A) being trapped in the surfactant (B) and cholesterol (C) may be the cause of the size increase [43]. Our research found that the outcomes are consistent with previously published findings: the size increases as the drug concentration in the formulation increases [44]. As per the above equation, surfactant concentration had a negative impact on the size of the vesicles, as increasing the concentration of the surfactant reduces the size of the droplet until the concentration of the surfactant reaches a certain threshold, after which the size of the vesicle increases due to the formation of aggregates by the excess surfactant [45]. Cholesterol, on the other hand, is an important factor responsible for the formation of vesicles: as cholesterol levels rise, vesicle size decreases. This may be because cholesterol above a certain concentration can disrupt the bi-layered structure of vesicular membranes, resulting in drug loss from vesicles [40].

Table 2. Results of regression analysis of Responses Y_1, Y_2 and Y_3.

Quadratic Model	R^2	Adjusted R^2	Predicted R^2	SD	%CV	*p*-Value
Response (Y_1)	0.9882	0.9731	0.8162	1.85	0.9395	<0.0001
Response (Y_2)	0.9777	0.9490	0.7883	0.0026	0.7818	<0.0001
Response (Y_3)	0.9890	0.9749	0.8673	0.7946	0.9582	<0.0001

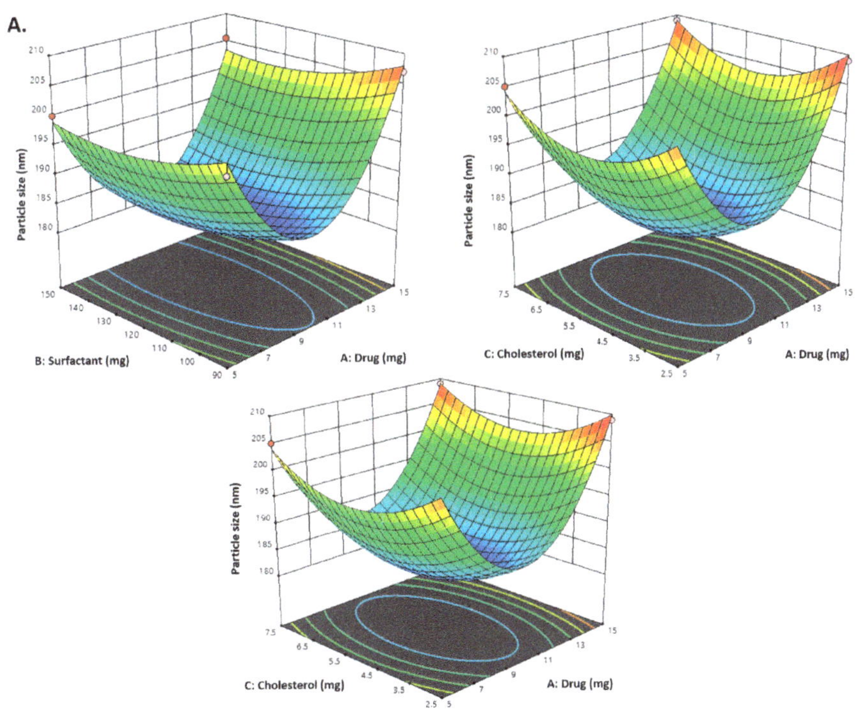

Figure 1. *Cont.*

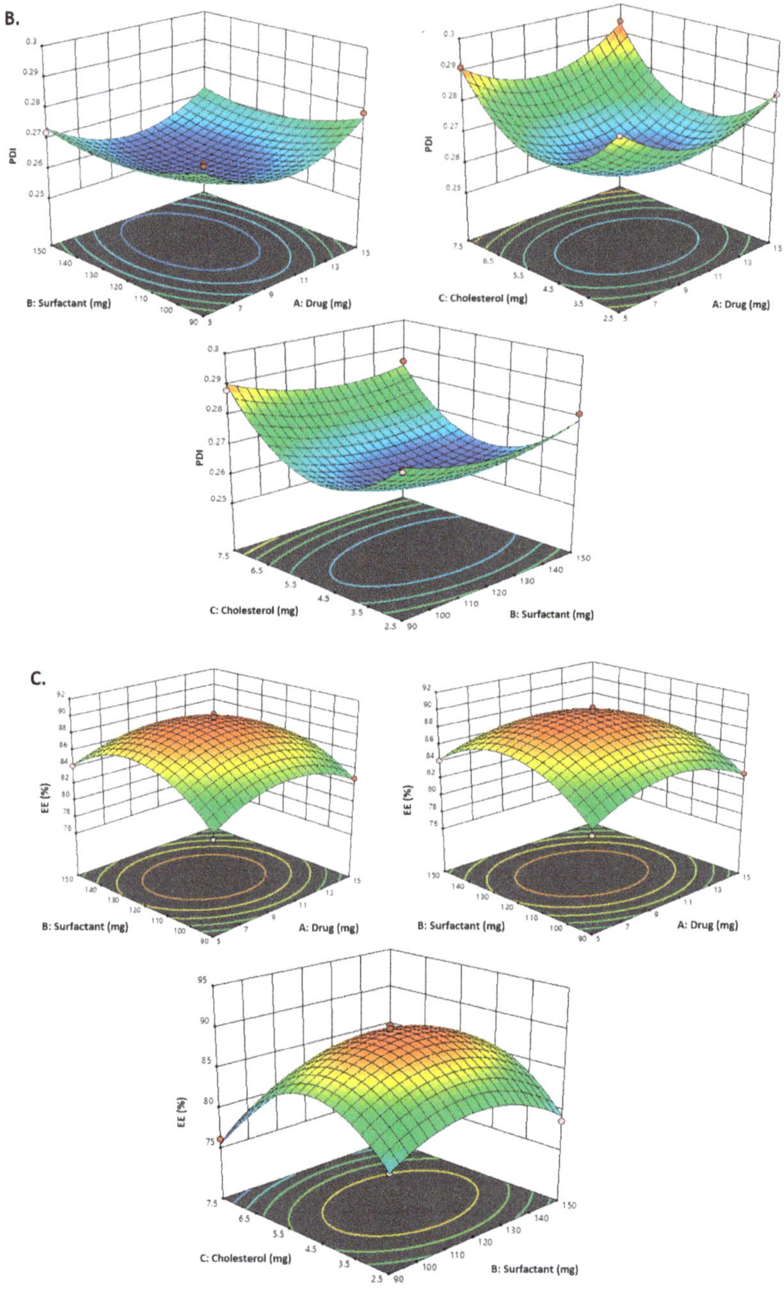

Figure 1. (**A**) Representation of 3D plots of independent variables impacting the dependent variable. Effect of drug conc. (A), surfactant conc. (B) and cholesterol (C) on the PDI (Y_3). (**B**) Representation of 3D plots of independent variables impacting the dependent variable (PDI). Effect of drug conc. (A), surfactant conc. (B) and cholesterol (C) on entrapment efficiency (%) (Y_2). (**C**) Representation of 3D plots of independent variables impacting the dependent variable (EE%). Effect of drug conc. (A), surfactant conc. (B) and cholesterol (C) on the vesicle size (Y_1).

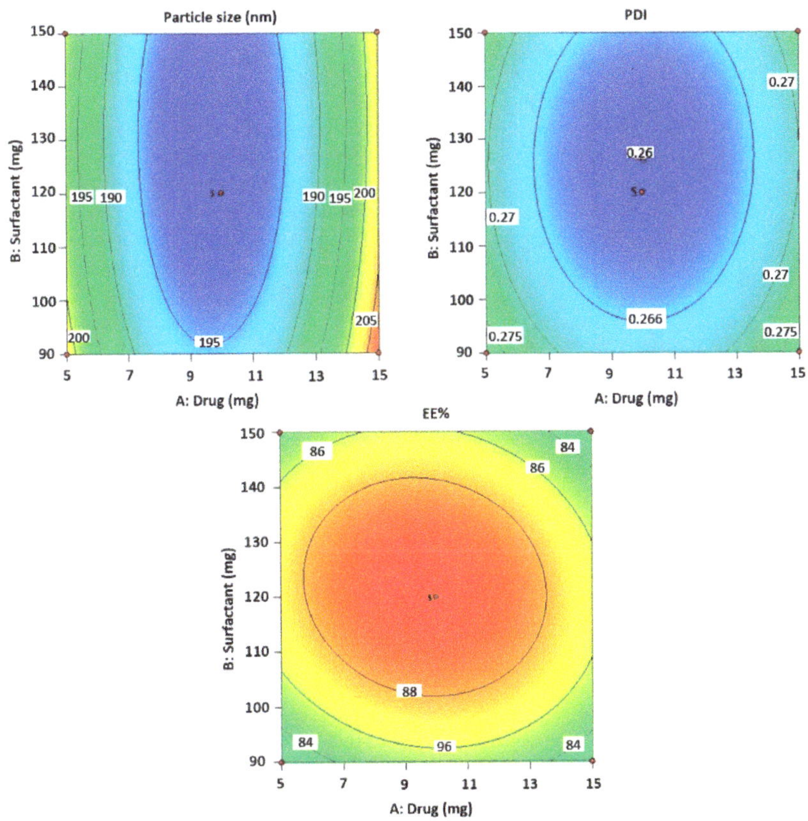

Figure 2. The counterplots for responses for vesicle size, PDI and entrapment efficiency of CVC-Ns.

According to the equation below, the size's interaction terms is displayed by the following equation:

$$\text{Vesicles size} = +166.43 + 2.60\text{A} - 1.89\text{ B} - 0.367\text{ C} - 0.380\text{ AB} - 1.0723.40\text{ AC} + 1.39\text{ BC} + 19.61\text{ A}^2 + 2.59\text{ B}^2 + 6.36\text{ C}^2.$$

All seventeen trials' PDI values ranged between 0.259 and 0.294. Figures 1B and 2 show 3D and contour plots that illustrate the effect of factors on PDI. The experimental data indicated that the remainder of the independent variable's surfactant and cholesterol had a negative effect on PDI, indicating that PDI decreased with increasing surfactant concentration until a certain level, as shown in the equation below.

$$\text{PDI} = +0.2614 + 0.001\text{ A} - 0.0020\text{ B} - 0.0036\text{ C} - 0.0003\text{ AB} + 0.001\text{ AC} - 0.001\text{ BC} + 0.0101\text{ A}^2 + 0.004\text{ B}^2 + 0.018\text{ C}^2$$

Cholesterol is an essential factor responsible for vesicle formation: as cholesterol levels rise, vesicle size decreases, resulting in a decrease in PDI. This may be because cholesterol above a certain concentration can disrupt the bi-layered structure of vesicular membranes, resulting in drug loss from vesicles [45].

The following equation displays the EE interaction terms:

$$\text{EE} = +29.68 - 0.3700\text{ A} + 0.4338\text{ B} - 1.48\text{ C} - 0.5750\text{ AB} - 0.2540\text{ AC} + 0.8075\text{ BC} - 2.83\text{ A}^2 - 3.92\text{ B}^2 - 7.60\text{ C}^2$$

According to the above-mentioned equation, the developed CVC-niosomes had an entrapment efficiency that ranged from 76.08 to 90.41% (Table 2). As per the above equation, the drug concentration increases, and the entrapment efficiency decreases. Figures 1C and 2 show 3D and contour plots that illustrate how factor variables influence the efficiency of encapsulation. The encapsulation effectiveness of the independent factors varied significantly. The effectiveness of the niosome's encapsulation is favourably impacted by the factor surfactant (B). The encapsulation efficiency increases with a rise in surfactant concentration due to the formation of multiple layers of surfactant molecules around the vesicles and resulting in extra space for incorporating the drug [46]. Moreover, the use of surfactant tween 80 also helped in enhancing the solubilization of the drug molecule by incorporating them [47]. To obtain stable niosomes, cholesterol is the most common additive added to the formulation. It stabilizes bilayers, prevents leakage, and slows the permeation of solutes contained within the aqueous interior of these vesicles. In the present study, the above equation indicates that cholesterol has had a negative effect on entrapment efficiency. The entrapment efficiency of the drug in niosomes was decreased by increasing the cholesterol concentration from 2.5 to 7.5 mg. This may be because cholesterol above a certain concentration can disrupt the bi-layered structure of the vesicular membranes, resulting in drug loss from the vesicles [45].

2.2. Design Validation

To develop the ideal formulation with the lowest globule size, PDI, and highest entrapment efficiency, BBD was used. Table 3 lists the predicted values for all responses and variables derived from the design. Therefore, CVC-N formulations based on the runs, process variables, and responses as shown in Table 2 were created, and the experimental findings attained using the formulations were compared to the expected responses. The fact that the experimental and predicted numbers were nearly closed supports the validity of the optimization process [48].

Table 3. In vitro drug release kinetics of CVC-loaded niosomes.

Kinetic Models	*X*-Axis	*Y*-Axis	CVC-Ns (R^2)
Zero-order	Fraction of drug released	Time	0.9268
First order	Log % drug remaining	Time	0.9878
Higuchi matrix	Fraction of drug release	Square root time	0.9933
Korsmeyer–Peppas	Log fraction of drug released	Log time	0.9927

2.3. Optimized Composition

The optimized formula obtained from BBD for formulating CVC-loaded niosomes, drug (9.76 mg), surfactant (131.02 mg), and cholesterol (4.62 mg). Optimized formulation was further characterized for various parameters. The chosen formulation for the optimized niosomes was formulation code F4, which contains the drug (mg) (A) 10, surfactant (mg) (B) 120, and cholesterol (mg) (C) 5.

2.4. Characterization of CVC-Ns

For measurements of the vesicle size, Polydispersity Index (PDI), and zeta potential (ZP), the optimal formulation of CVC-Ns was carefully developed based on the characteristics of obtaining the ideal vesicle size and PDI with zeta potential. For the optimized niosomal formulation, the vesicle size distribution (PDI) was 0.265 ± 0.11, and the mean vesicle size was 180.23 ± 1.21 nm. Figure 3A,B, respectively, show the size and zeta potential curve of the optimized formulation. The optimized zeta potential was found to be −31.70 ± 1.11 mV; the zeta potential magnitude gives stability potential of the formulation. All particles in suspension with a large negative or positive zeta potential will resist each other and thus have a low tendency to aggregate [49].

Entrapment Efficiency and Drug Loading

The % entrapment efficiency and % drug loading of optimized CVC-Ns was found to be 90.61 ± 2.14% and 60.40 ± 0.99%.

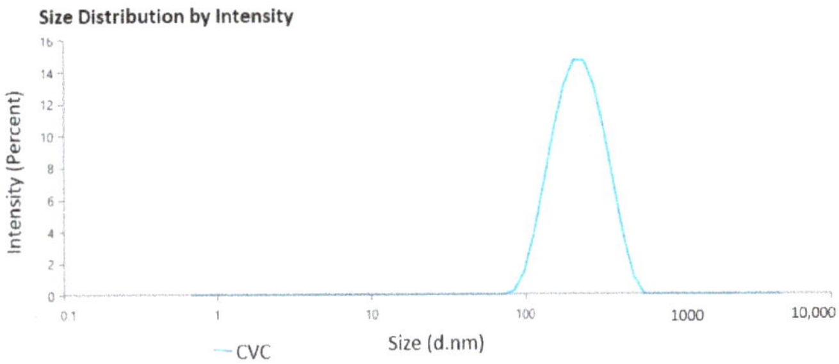

Statistics Table

Name	Mean	S. Deviation	RSD	Minimum	Maximum
Z-Average (nm)	180.23	—	—	180.23	180.23
Polydispersity Index (PI)	0.265	—	—	0.265	0.265
Intercept	0.977	—	—	0.977	0.977

(**A**)

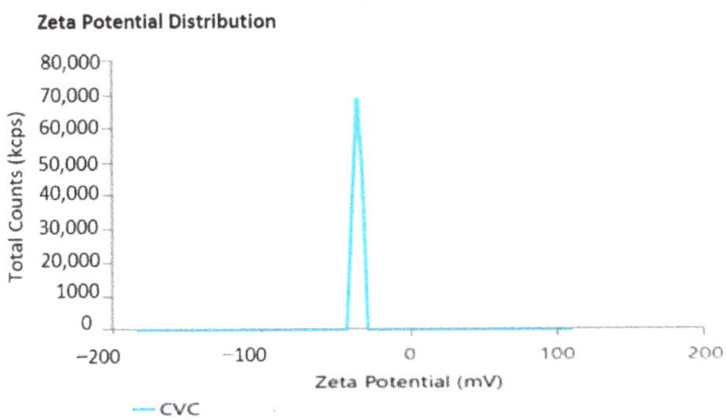

Statistics Table

Name	Mean	Standard Deviation	RSD
Zeta Potential (mV)	−31.70	—	—
Zeta Peak I Mean (mV)	−31.70	—	—
Conductivity (Ms/cm)	0.01807	—	—

(**B**)

Figure 3. (**A**). Vesicle size and distribution of CVC-loaded niosomes. (**B**). Zeta potential of CVC-loaded niosomes.

2.5. Morphology of CVC-Ns

The optimized CVC-Ns formulation's TEM picture displays the produced vesicles to be well-defined sealed structures with uniform size distribution and spherical morphologies (Figure 4). The spherical vesicles indicate that CVC is entrapped by niosomes. These vesicles are spherical in shape and homogeneous. The other spots represent the small amount of CVC entrapment in the niosomes [50].

2.6. In Vitro Release Study

The CVC-loaded optimized niosomal formulation represents a release of CVC through the dialysis bag of 70.24 ± 1.21%, compared to the in vitro CVC release of CVC-suspension, which was found to be 32.87 ± 1.03% (Figure 5A). More than 10–15% of the drug was released in the early two hours, followed by a sustained release for the next 24 h. Initially, the fast release was caused by the rapid release of CVC from the niosome surface, but later, a slower release phase was discovered. The delayed phase in drug release was controlled by diffusion through swollen niosomal bilayers [51]. The results of the in vitro drug release experiment were analysed by applying several mathematical kinetic models to the data. It was discovered that the value of the correlation coefficient (R^2) for the release of CVC from optimized CVC-loaded niosomes for the Higuchi matrix model had the greatest value (R^2 = 0.9933) (Figure 5B). First-order (R^2 = 0.9878) and zero-order (R^2 = 0.9268) models were found to have the lowest R^2 values, as shown in Table 3 and Figure 5C,D. Consequently, after obtaining the highest value of the correlation coefficient, the optimized CVC-Ns indicated Higuchi's model as the best-fit model. The Korsmeyer–Peppas model was used to fit data to analyses of the release mechanism of CVC from optimized CVC-Ns (Figure 5E). The R^2 value was found to be 0.9927, and the n value was found to be $0.45 < n < 0.89$, indicating that the release mechanism of CVC from optimized CVC-N follows non-Fickian diffusion [52].

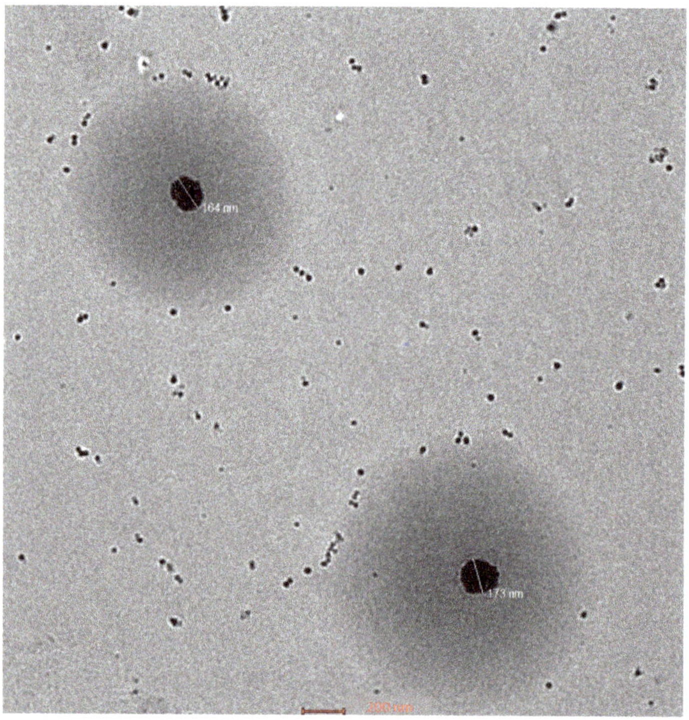

Figure 4. Surface morphology of optimized niosomes.

2.7. Analysis of the Gel and Texture of the Optimized CVC-N Gel

The pH for the CVC-N gel formulation was 6.01, which is quite close to the skin's pH [52]. According to data on texture analysis, as shown in Figure 6, the CVC-N gel has the following properties: firmness of 239.09 g, consistency of 1587.00 gm sec, cohesiveness of −89.17 gm, and work of cohesion of −645.47 gm sec, as shown in Table 4 [53]. Peak or maximum force is used to measure firmness; the higher the number, the thicker the sample's consistency. The sample's stickiness or cohesiveness is determined by measuring the greatest negative force. The stiffer the sample is, the more negative the number [54].

2.8. Confocal Laser Scanning Microscopy

The stratum corneum thickness of rat or human skin is reported to be 18 and 17 μm, respectively, and the active site for most of the therapeutic complications associated with the skin lies below the stratum corneum; thus, a drug moiety must penetrate at least

20–200 μm across the skin to exert therapeutic effect [55]. The results showed that niosomal gel loaded with rhodamine B penetrated a deeper layer to a depth of 25.0 μm (Figure 7A). Rhodamine B hydroethanolic solution only exhibited infiltration up to 5.0 μm, suggesting that the dye was only present in the top layers of the rat's skin (Figure 7B). In contrast, because the fluorescence intensity is greater in the centre of the skin, it is probable that the formulation was maintained in the lower epidermal area of the rat skin. Many skin conditions that affect this lower epidermal skin region require the formulation loaded with rhodamine B to remain in the skin. Thus, it could be said that the created niosomal gel effectively carried rhodamine B dye deeper into the rat's skin layers.

Table 4. Physiochemical characterization of CVC-based niosomal gel (CVCNG).

Homogeneity	**Appearance**	**Washability**	**Separation of Phase**	**Odour**
Homogeneous	Translucent	Yes	No	odourless
Colour	**Content of Drug (%)**	**pH**	**Spreadability (gm. cm/sec)**	
off-white	90.11 ± 0.98	6.01 ± 1.01	18.27 ± 1.24	
Cohesiveness (gm)	**Consistency (gm.Sec)**	**Firmness (gm)**	**Work of cohesion g, sec**	
−89.17	1587.00	239.09	−645.47	

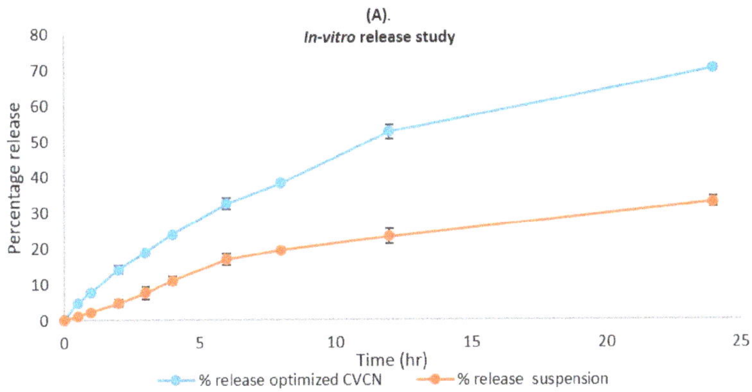

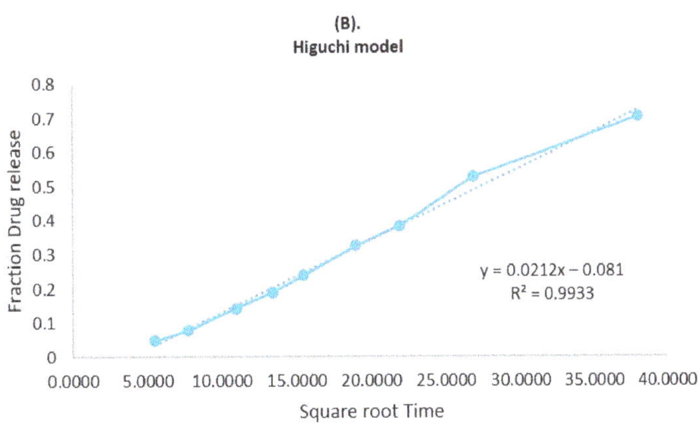

Figure 5. *Cont.*

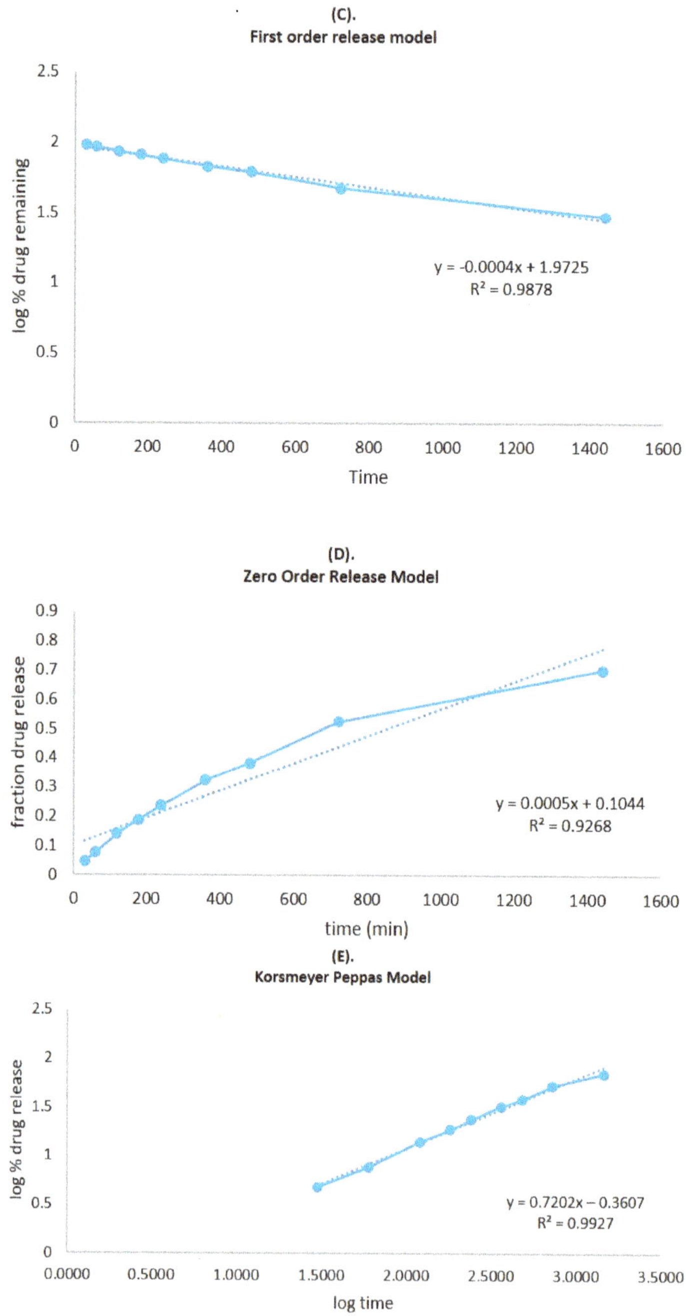

Figure 5. (**A**) Comparison of the optimized CVC-Ns formulation's in vitro release profile with that of CVC suspension. The investigation was performed in triplicate, and the data are depicted as mean ± SD. (**B**) Higuchi release kinetics of CVC-niosome formulation. (**C**) Release kinetics for a first order and (**D**) release kinetics for zero order of CVC-loaded niosomes. (**E**) Korsmeyer–Peppas model for predicting the release of the drug.

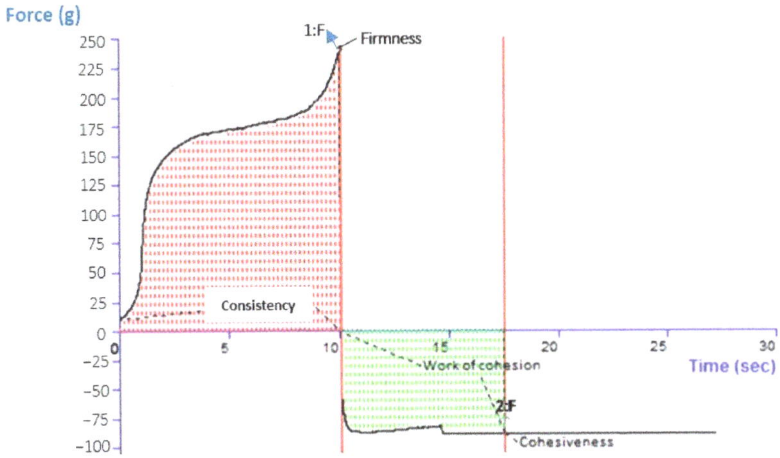

Figure 6. Texture analysis of optimized CVC-N gel.

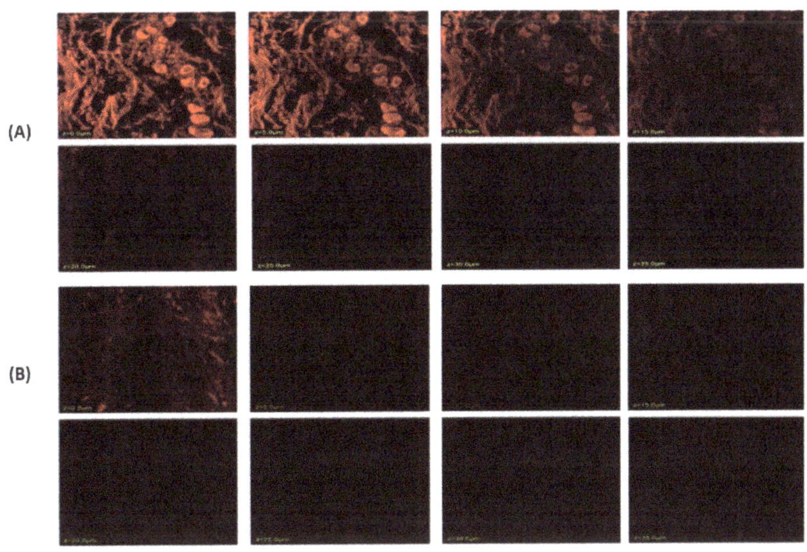

Figure 7. Confocal laser scanning microscopy; (**A**) niosome-formulation-loaded rhodamine B dye; (**B**) hydroalcoholic solution of rhodamine B dye.

2.9. Dermatokinetic Studies

As a result of the application of CVC-CF gel and CVC-N gel at a predefined time interval, the CVC concentration in the dermis and epidermis of the rat's skin is shown in Figure 8. Table 5 displays the values of the dermatokinetic parameters. The rat skin that had been exposed to CVC-CF gel displayed $C_{Skin\ max}$ values of 179.04 ± 0.96 μg/cm^2 in the epidermis and 160.13 ± 0.64 μg/cm^2 in the dermis. A $C_{Skin\ max}$ value of 283.54 ± 1.01 μg/cm^2 was found in the epidermis, and 262.64 ± 1.12 μg/cm^2 was found in the dermis of a rat whose skin had been treated with CVCN gel. $AUC_{0\text{-}t}$ values of 677.47 ± 0.28 and 572.23 ± 0.31 μg/cm^2 h, respectively, were seen in the epidermis and dermis of a rat whose skin had been treated with CVC-CF gel. $AUC_{0\text{-}t}$ values in the epidermis and dermis of rat skin treated with CVC-N gel, in contrast, were 1135.5 ± 0.64 and 1158.7 ± 1.08 μg/cm^2 h, respectively. When rat skin was treated with CVC-N gel in contrast to the CVC-CF gel formulation, a greater percentage of CVC was maintained in both rat skin layers. The epidermis and dermis both showed greater $C_{Skin\ max}$ and $AUC_{0\text{-}t}$ values, which demonstrate that the CVC-N gel improved drug absorption in both the epidermal and dermis layers because of its nanosized vesicles' ease of penetration into the skin's lipid layers, enhancing its therapeutic efficacy in the treatment of inflammatory diseases [56].

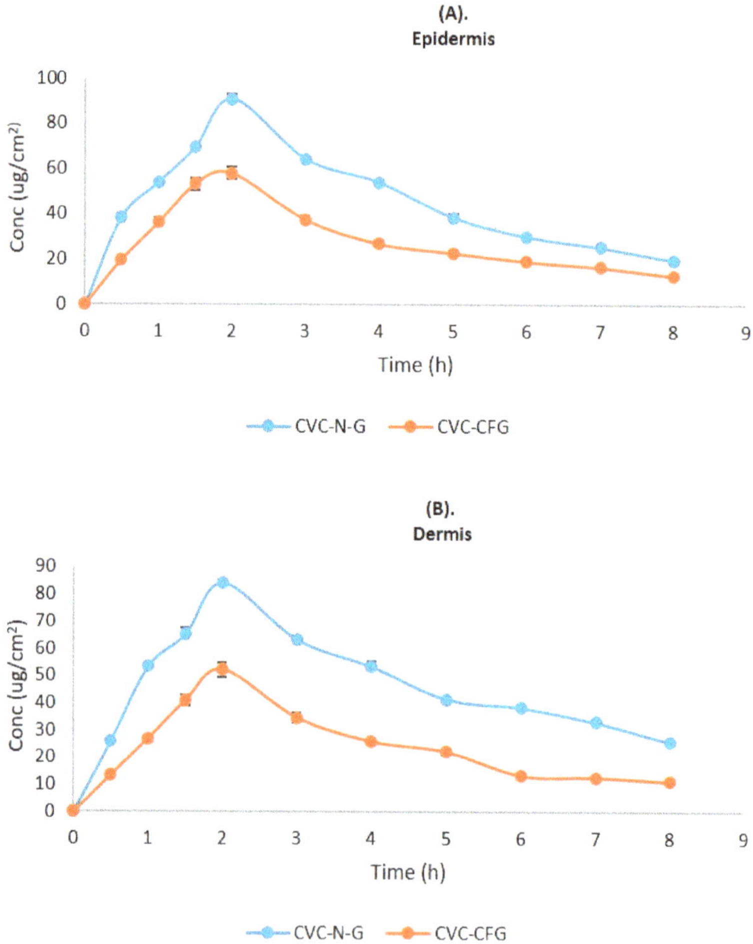

Figure 8. CVC concentration on the (**A**) epidermis and (**B**) dermis of excised rat skin following application to the skin of CVC-N gel and CVC-CF gel.

Table 5. Dermatokinetic parameters (mean ± SD) of CVC-N gel and CVC-CF gel.

Dermatokinetics Parameters	CVC-N Gel		CVC-CF Gel	
	Epidermis Mean ± SD	Dermis Mean ± SD	Epidermis Mean ± SD	Dermis Mean ± SD
$T_{skin\ max}$ (h)	2	2	2	2
$C_{skin\ max}$ ($\mu g/cm^2$)	283.54 ± 1.01	262.64 ± 1.12	179.04 ± 0.96	160.13 ± 0.64
$AUC_{0\text{-}8}$ ($\mu g/cm^2$ h)	1135.5 ± 0.64	1158 ± 1.08	677.47 ± 0.28	572.23 ± 0.31
Ke (h^{-1})	0.171 ± 1.11	0.119 ± 0.91	0.146 ± 0.47	0.119 ± 0.78

2.10. Ferric-Reducing Antioxidant Power (FRAP)

The analysis indicated that CVC has a significant amount of antioxidant capacity and that the agent is effective at scavenging free radicals, and preventing the oxidation of antioxidant activity was found to be 60.14 ± 1.11% in free CVC, 88.41 ± 2.32% in an ascorbic acid solution, and 71.24 ± 3.31% in a CVC-N-optimized formulation. As the results indicate, CVC-Ns have been shown to have a significant antioxidant effect.

2.11. Stability Studies

The data of experiments into short-term accelerated stability were evaluated for six months. Physical appearance, shape, vesicle size, PDI, zeta potential, colour appearance, phase separation, clarity, homogeneity, pH, and drug content were detected (Tables 6 and 7). This study thus validates the notion that niosomes are durable over long periods of storage.

Table 6. Evaluation of the CVC-loaded optimized niosome (CVC-N) formulation's short-term accelerated stability.

Evaluation Parameters	Initial	1 Month		3 Months		6 Months	
		4 ± 2 °C	25 ± 2 °C/ 60 ± 5% RH	4 ± 2 °C	25 ± 2 °C/ 60 ± 5% RH	4 ± 2 °C	25 ± 2 °C/ 60 ± 5% RH
Appearance	+++	+++	+++	++	++	++	++
Phase separation	NO PHASE SEPARATION						
Shape	Spherical in shape						
PDI	0.259	0.259	0.264	0.271	0.277	0.280	0.284
Vesicle size (nm)	180.23	180.23	180.99	181.11	182.14	183.09	183.21
Zeta potential (mV)	−31.70	−31.70	−32.01	−32.45	−33.41	−33.39	−33.34

++ good, +++ excellent.

Table 7. Short-term accelerated stability evaluation of CVC-loaded optimized niosomal gel formulation.

Evaluation Parameters	Initial	1 Month		3 Months		6 Months	
		4 ± 2 °C	25 ± 2 °C/ 60 ± 5% RH	4 ± 2 °C	25 ± 2 °C/ 60 ± 5% RH	4 ± 2 °C	25 ± 2 °C/ 60 ± 5% RH
Colour	Slightly off white						
Appearance	Translucent						
Phase Separation	NO PHASE SEPARATION						
Clarity	√	√	√	√	√	√	√
pH	6.01	6.01	6.29	6.31	6.39	6.55	6.67
Homogeneity	***	***	**	***	**	***	*
Washability	Washable						
Odour	NO						

* satisfactory, ** good, *** excellent.

3. Conclusions

Niosomes were effectively developed using thin film hydration. Using BBD, the niosomal formulation was optimized by taking independent and dependent factors. The optimized formulation achieved the highest amount of CVC entrapping, suggesting that it was the most ideal among all formulations, i.e., batch code F4. The optimized CVC-loaded niosomes had reduced PDI values and were in the colloidal size range, which indicated that the formulations were homogeneous. The TEM images demonstrated the spherical shapes of the vesicles, providing evidence of an entrapped CVC. According to in vitro drug release studies, the amount of CVC released from CVC-loaded niosomes was two times higher than it was from CVC suspension solution, demonstrating improved drug release from the niosome formulation due to nanosized vesicles. A CLSM study on rat skin has provided insight into the penetration of rhodamine-loaded B hydroalcoholic and rhodamine-loaded niosome formulations, and the results showed that rhodamine-B-loaded niosomal formulations penetrated the skin much more effectively than hydroalcoholic solutions of the rhodamine B dye, indicating that niosomes have improved in vivo prospects for anti-inflammatory treatment. Compared to CVC-CFG, dermatokinetic studies show that a higher concentration of CVC-NG reaches the epidermis and dermis (Cskin max and AUC0-t) due to the nanosized vesicles providing a means to cross the stratum corneum to have a profound effect. The antioxidant studies revealed that the niosome formulation has a greater potential for antioxidants than the pure drug, showing a capability to reduce the level of free radicals and reactive oxygen stress and hence to reduce inflammation. Moreover, the results of this experiment confirmed that the developed, optimized niosome formulation is an effective means of delivering CVC drugs topically and can therefore be used to treat anti-inflammatories more successfully. However, the actual skin penetration of carvacrol may vary depending on the formulation and the distinctive characteristics of the skin. While carvacrol is usually regarded as safe, the safety profile of carvacrol-containing niosomes for topical distribution has not been fully established, and additional research may be required to determine their possible toxicity and side effects and to validate the in vitro and skin permeation results by using appropriate animal models in preclinical studies.

4. Materials and Methods

4.1. Materials

Carvacrol oil, polyethylene glycol 400 (PEG), triethanolamine rhodamine-123, Ascorbic acid and potassium ferricyanide were procured from Sigma-Aldrich (St. Louis, MI, USA). Cholesterol and tween 80 were provided by S D Fine Chemicals Ltd. (Mumbai, India). Methanol and chloroform were provided by Merck Mumbai, India. B.S. Goodrich in Pleveland generously provided carbopol-934 as a gift sample. Other agents used in the experiment such as disodium hydrogen phosphate, potassium dihydrogen phosphate, and sodium chloride of analytical grade were provided by S D Fine Chemicals Limited, Mumbai, India.

4.2. Method

4.2.1. Preparation of CVC-Loaded Niosomes (CVC-Ns)

Niosomes loaded with CVC (CVC-Ns) were made using the thin-layer hydration method. Briefly, the drug (CVC), cholesterol, and surfactant were dissolved in a round-bottomed flask using a chloroform-to-methanol ratio of 2:1. The organic phase was then evaporated in a rotary evaporator (120 rpm, 60 °C, 1 h) to generate a thin layer. Phosphate-buffered saline solution (PBS pH—6.8, 10 mL) was then used to rehydrate the desiccated thin film at room temperature (30 °C) for one hour (200 rpm). CVC-loaded niosomes were then sonicated for 2 min by a probe sonicator. The samples were preserved in a refrigerator at 4 °C for further investigation [57].

4.2.2. CVC-Loaded Niosome Optimization through Experimental Design

The Box–Behnken design (BBD) of the Design-Expert software (Stat-Ease Inc.'s, Design Expert®, Version 13) was used to examine the relationships between the three selected input variables (drug, surfactant, cholesterol). The output variables studied were the vesicle size (Y_1), polydispersity index (PDI, Y_2) and entrapment efficiency (Y_3). The input variables were chosen, at their three levels, i.e., their minimum, average, and maximum values, which are shown in Table 8 to explore the final optimal formulation [58].

Table 8. Factors and responses employed in BBD to create CVC-Ns.

Factor	The Level Used, Actual Coded		
Factors	**Low (−1)**	**Medium (0)**	**High (+1)**
A = Drug (mg)	05	10	15
B = Surfactant (mg)	90	120	150
C = Cholesterol (mg)	2.5	05	7.5
Responses		**Aim**	
Y_1 = globule size (nm)		<200 nm	
Y_2 = PDI		<0.3	
Y_3 = Entrapment Efficiency (%)		>70%	

A, drug (mg); B, surfactant (mg); C, cholesterol (mg); Y_1, Vesicle size (nm); Y_2, PDI; Y_3, EE (%); BBD, Box–Behnken experimental design; Ns, Niosomes; CVC, carvacrol.

4.3. Characterization of CVC-Ns

4.3.1. Determination of Globule Size and Zeta Potential

To determine the zeta potential and vesicle size distribution, the dynamic light scattering (DLS) method was utilized. Before analysis, samples were 100-fold diluted in double-distilled water, and impurities were removed using 0.45 m membrane filters. The Malvern particle size analyser equipment (Malvern Instrument Ltd., Malvern, UK) was utilized to determine the zeta potential and to quantify the vesicle size [59].

4.3.2. Measurement of % Entrapment Efficiency and % Drug Loading

The drug loading and entrapment efficiency of CVC-loaded niosome vesicles were determined by ultra-centrifuging (Beckman Coulter India Pvt. Ltd., Mumbai, India) formulations for one hour at 15,000 r/min and 4 °C. The upper layer of liquid (supernatant) was withdrawn using a pipette, diluted, and filtered (0.25 µm), and the CVC concentration was measured employing UV spectroscopy (UV 1601, Shimadzu, Nagoya, Japan) at 280 nm [60]. The experiment was performed in triplicate. For evaluating the drug-loading capability of the niosomes, drug entrapment efficiency is a crucial factor. This parameter depends on the developmental methodology, the physicochemical features of the drug, and the formulation factors [61].

The equation was used to obtain the entrapment efficiency (EE%) and drug loading (%) [62]:

$$\text{Encapsulation efficiency } (\%) = \frac{(\text{total amount } - \text{ free amount})}{\text{total amount}} \times 100$$

$$\text{Drug loading } (\%) = \frac{\text{amount of drug entraped}}{\text{total amount of drug and lipid}} \times 100$$

4.4. Morphological Studies

Transmission electron microscopy (TEM) analysis was carried out on a JEOL-2000 Ex II TEM (Akishima, Japan) to morphologically characterise the niosomes. The excess sample was taken off the carbon-coated copper grid using filter paper after a drop of the niosomal formulation was applied to it. The carbon-coated copper grid was then applied with a drop of 2% (*w*/*v*) PTA (phosphotungstic acid solution) and kept undisturbed for 2 min. The excess of the staining agent was removed with the help of a filter paper, and the sample was air-dried before being examined on the transmission electron microscope [63].

4.5. Formulation of CVC Niosomal Gel (CVCNG)

Niosomal dispersions have a very low viscosity, which makes them unsuitable for topical use; therefore, gelling was performed. Carbopol 934 was spread out by slowly adding distilled water and letting it sit in the dark so that it could fully swell. The dispersion was further made neutral by adding triethanolamine drop by drop to achieve the intended result of a viscous transparent gel. Finally, the optimized CVC-N formulation was added slowly to the gel with constant agitation to create CVC-NG [64].

4.6. In Vitro Drug Release Study

The optimized CVC-N formulation and carvacrol suspension (drug (10 mg)) was dissolved in methanol and the volume was made up to 10 mL with adequate solvent. The drug release investigation was conducted using the cellulose dialysis membrane bag (molecular weight cut off–12,000 kDa). The dialysis bag was filled with both of the prepared formulations (2 mL of each containing 2 mg of the drug), and the ends were knotted. As part of the experiment, the dialysis bag was submerged in 100 mL of the phosphate buffer (pH 6.8; medium). The release media's temperature was maintained at 37 ± 2 °C throughout the investigation, which was conducted with constant stirring. The sample (1 mL) was taken at 0, 0.5, 1, 2, 4, 6, 8, 12 and 24 h, and precisely the right amount of fresh medium was added to keep the sink conditions. The in vitro drug release study was conducted for 24 h to determine the drug release pattern, the mechanism it follows and how long it remains in the target area. The obtained samples were filtered (0.22 µm) and further diluted to determine the content of the drug by utilizing a UV spectrophotometer at 280 nm. To evaluate the release mechanism, the drug release data were fitted to various release kinetic models, including zero-order, first-order, Higuchi, and Korsmeyer–Peppas models [65].

4.7. Characterization of Gel

Evaluation of pH and Texture of CVC-N Gel

A digital pH meter electrode (Mettler Toledo, Chiyoda, Japan) was utilized to directly measure the pH of the CVC-N gel. TA.XT Plus Texture analyser (Stable Micro Systems Ltd., Surrey, UK) was used to determine the gel's texture. The texture analysis curve produced at the end of the trials was used to identify parameters including hardness, consistency, cohesiveness, and gel viscosity index [66].

4.8. Spreadability Studies

The parallel plate method was used to determine the niosomal gel spreadability. In this procedure, 500 mg of the formulated gel was added to a circle diameter of 1 cm, atop a glass plate that had been pre-marked and another glass plate had been placed on top. Five minutes were given for a weight of 500 gm to lay on the top plate of the glass. It was seen that the gel spreading caused the diameter to expand [67].

4.9. Permeation Depth Study by Loading Rhodamine B Dye

Franz diffusion cell was utilized in the confocal studies to evaluate the infiltration in the skin by rhodamine-B-dye-loaded niosomal formulation and standard solution of rhodamine B dye. A 1 cm^2 sample of rat skin was cut from the abdomen, and the hairs were removed using purified water and a phosphate-buffered saline solution; the rat skin was meticulously washed and rinsed until it was completely clean. The stratum corneum was then facing up and the dermis was facing down when the skin was then put on the diffusion cell. The niosome formulation was mixed with rhodamine B, while the standard was hydroalcoholic rhodamine B solution. Both were put into the donor compartment and kept at a temperature of 32 ± 0.5 °C for six hours. Additionally, slides with the stratum corneum pointing upward were examined using a confocal microscope and an optical excitation of 488 nm. Above 560 nm, both the argon laser beam and the fluorescence emission could be seen. Confocal laser scanning microscopy (CLSM) was used to compare the penetration and distribution of improved gel formulation with traditional gel (TCS SP5II; Leica Micro System Ltd., Wetzlar, Germany) [68].

4.10. Dermatokinetics

The dermatokinetic analysis was performed to determine the quantity of drug in different layers of excised rat skin after applying the CVC-NG formulation mounted on a diffusion cell apparatus. The research was carried out following the protocol described in the CLSM study. In this case, the Franz diffusion cell skin samples were obtained for this study at 0, 1, 2, 4, and 8 h. The skin was then washed with normal saline and kept at 60 ± 0.5 °C for 2–3 min. The skin layers (epidermis and dermis) were separated using forceps and sliced into tiny fragments before being soaked in methanol for 24 h to extract the drug contained in the layers. The methanolic extracts of the drug were further filtered (0.22 μm), and the drug concentration was determined using a UV spectrophotometer. All the measurements were performed in triplicate [69].

4.11. Ferric-Reducing Antioxidant Power (FRAP)

In brief, 2.5 mL of 0.2 M sodium phosphate buffer (pH 6.6) and 2.5 mL of 1% (w/v) potassium ferricyanide solution were combined with 1 mL of sample solution (10 μg/mL) in methanol. Employing a method similar to Oyaizu's with a few minor alterations, the reduction power (antioxidant) of CVC in the niosomal formulation was determined. The mixture was then incubated for 20 min at 50 °C before being treated with 2.5 mL of 10% trichloroacetic acid. Then, 2.5 mL from the top layer of the mixture, 2.5 mL of distilled water, and 0.5 mL of $FeCl_3$ (0.1%) were added after the mixture had been centrifuged for 10 min at 3000 rpm. The amount of ascorbic acid that was utilized as the standard was 10 μg/mL, and a spectrophotometer was used to measure the absorbance at 700 nm. Each test was performed in triplicate [70].

4.12. Stability Studies

Optimized CVC-N and CVC-N gel formulations underwent short-term accelerated stability evaluation in line with the International Conference on Harmonization (ICH) Q1 A R2 guidelines. Lyophilized niosomes (CVC-Ns) were placed in an Eppendorf, sealed, and maintained in a stability chamber at 25 ± 2 °C and RH 60 ± 5% at 4 ± 2 °C. The formulation along with the gel was then assessed for changes in physical appearance, shape, vesicle size, PDI, zeta potential, clarity, homogeneity, pH and phase separation for six months [71].

Author Contributions: Conceptualization, A.A. and M.G.; methodology, M.G. and U.H.; software, A.A.; validation, M.H.A. and U.H.; formal analysis, U.H. and M.G.; investigation, U.H. and M.G.; resources, U.H.; data curation, M.H.A.; writing—original draft preparation, M.H.A. and A.A.; writing—review and editing, M.G. and U.H.; visualization, U.H.; supervision, M.G. and A.A.; project administration, M.H.A. and M.G.; funding acquisition, M.H.A. and M.G. All authors have read and agreed to the published version of the manuscript.

Funding: The authors extend their appreciation to the Deanship of Scientific Research at King Khalid University for funding this work through Small Groups Project under grant number (RGP1/399/44).

Institutional Review Board Statement: The animal study was duly approved by the Standing Committee of Bioethics Research (SCBR-020-2023), Prince Sattam bin Abdulaziz University Al-Kharj, Saudi Arabia.

Informed Consent Statement: Not applicable.

Data Availability Statement: Not applicable.

Conflicts of Interest: There is no conflict of interest.

References

1. Schrooyen, P.M.M.; van der Meer, R.; Kruif, C.G. De Microencapsulation: Its Application in Nutrition. *Proc. Nutr. Soc.* **2001**, *60*, 475–479. [CrossRef] [PubMed]
2. Souto, E.B.; Fernandes, A.R.; Martins-Gomes, C.; Coutinho, T.E.; Durazzo, A.; Lucarini, M.; Souto, S.B.; Silva, A.M.; Santini, A. Nanomaterials for Skin Delivery of Cosmeceuticals and Pharmaceuticals. *Appl. Sci.* **2020**, *10*, 1594. [CrossRef]

3. Sindle, A.; Martin, K. Art of prevention: Essential oils-natural products not necessarily safe. *Int. J. Women's Dermatol.* **2021**, *7*, 304–308. [CrossRef] [PubMed]
4. Khatibi, S.A.; Misaghi, A.; Moosavy, M.-H.; Amoabediny, G.; Basti, A.A. Effect of Preparation Methods on the Properties of Zataria Multiflora Boiss. Essential Oil Loaded Nanoliposomes: Characterization of Size, Encapsulation Efficiency and Stability. *Pharm. Sci.* **2015**, *20*, 141–148.
5. Hosseini, S.F.; Zandi, M.; Rezaei, M.; Farahmandghavi, F. Two-Step Method for Encapsulation of Oregano Essential Oil in Chitosan Nanoparticles: Preparation, Characterization and in Vitro Release Study. *Carbohydr. Polym.* **2013**, *95*, 50–56. [CrossRef] [PubMed]
6. Hassan Aubais-aljelehawy, Q.; Mohammadi, S.; Mohamadian, E.; Raji Mal Allah, O.; Mirzaei, A.; Ghahremanlou, M. Antimicrobial, anticancer, antidiabetic, antineurodegenerative, and antirheumatic activities of thymol: Clarification of mechanisms. *Micro Nano Bio Asp.* **2023**, *2*, 1–7.
7. Aljelehawy, Q.; Maroufi, Y.; Javid, H.; Mohammadi, M.R.; Raji Mal Allah, O.; Taheri, S.V.; Mohammadzade, H. Anticancer, antineurodegenerative, antimicrobial, and antidiabetic activities of carvacrol: Recent advances and limitations for effective formulations. *Nano Micro Biosyst.* **2023**, *2*, 1–10.
8. Guimarães, A.G.; Xavier, M.A.; De Santana, M.T.; Camargo, E.A.; Santos, C.A.; Brito, F.A.; Barreto, E.O.; Cavalcanti, S.C.H.; Antoniolli, Â.R.; Oliveira, R.C.M.; et al. Carvacrol Attenuates Mechanical Hypernociception and Inflammatory Response. *Naunyn. Schmiedebergs. Arch. Pharmacol.* **2012**, *385*, 253–263. [CrossRef]
9. Guimarães, A.G.; Silva, F.V.; Xavier, M.A.; Santos, M.R.V.; Oliveira, R.C.M.; Oliveira, M.G.B.; Oliveira, A.P.; De Souza, C.C.; Quintans-Júnior, L.J. Orofacial Analgesic-like Activity of Carvacrol in Rodents. *Z. Naturforsch. Sect. C J. Biosci.* **2012**, *67C*, 481–485. [CrossRef]
10. Lima, M.D.S.; Quintans-Júnior, L.J.; De Santana, W.A.; Martins Kaneto, C.; Pereira Soares, M.B.; Villarreal, C.F. Anti-Inflammatory Effects of Carvacrol: Evidence for a Key Role of Interleukin-10. *Eur. J. Pharmacol.* **2013**, *699*, 112–117. [CrossRef]
11. Guimarães, A.G.; Scotti, L.; Scotti, M.T.; Mendonça, F.J.B.; Melo, N.S.R.; Alves, R.S.; De Lucca, W.; Bezerra, D.P.; Gelain, D.P.; Quintans, L.J. Evidence for the Involvement of Descending Pain-Inhibitory Mechanisms in the Attenuation of Cancer Pain by Carvacrol Aided through a Docking Study. *Life Sci.* **2014**, *116*, 8–15. [CrossRef] [PubMed]
12. Guimarães, A.G.; Oliveira, M.A.; Alves, R.D.S.; Menezes, P.D.P.; Serafini, M.R.; De Souza Araújo, A.A.; Bezerra, D.P.; Quintans, L.J. Encapsulation of Carvacrol, a Monoterpene Present in the Essential Oil of Oregano, with β-Cyclodextrin, Improves the Pharmacological Response on Cancer Pain Experimental Protocols. *Chem. Biol. Interact.* **2015**, *227*, 69–76. [CrossRef] [PubMed]
13. Arunasree, K.M. Anti-Proliferative Effects of Carvacrol on a Human Metastatic Breast Cancer Cell Line, MDA-MB 231. *Phytomedicine* **2010**, *17*, 581–588. [CrossRef] [PubMed]
14. Cavalcante Melo, F.H.; Rios, E.R.V.; Rocha, N.F.M.; Citõ, M.D.C.D.O.; Fernandes, M.L.; De Sousa, D.P.; De Vasconcelos, S.M.M.; De Sousa, F.C.F. Antinociceptive Activity of Carvacrol (5-Isopropyl-2-Methylphenol) in Mice. *J. Pharm. Pharmacol.* **2012**, *64*, 1722–1729. [CrossRef]
15. Guimarães, A.G.; Oliveira, G.F.; Melo, M.S.; Cavalcanti, S.C.H.; Antoniolli, A.R.; Bonjardim, L.R.; Silva, F.A.; Santos, J.P.A.; Rocha, R.F.; Moreira, J.C.F.; et al. Bioassay-Guided Evaluation of Antioxidant and Antinociceptive Activities of Carvacrol. *Basic Clin. Pharmacol. Toxicol.* **2010**, *107*, 949–957. [CrossRef]
16. Lee, B.; Yeom, M.; Shim, I.; Lee, H.; Hahm, D.H. Inhibitory Effect of Carvacrol on Lipopolysaccharide-Induced Memory Impairment in Rats. *Korean J. Physiol. Pharmacol.* **2019**, *24*, 27–37. [CrossRef]
17. Gunal, M.Y.; Heper, A.O.; Zaloglu, N. The Effects of Topical Carvacrol Application on Wound Healing Process in Male Rats. *Pharmacogn. J.* **2014**, *6*, 10–14. [CrossRef]
18. Alagawany, M. Biological Effects and Modes of Action of Carvacrol in Animal and Poultry Production and Health—A Review. *Adv. Anim. Vet. Sci.* **2015**, *3*, 73–84. [CrossRef]
19. Wolf, A.M.; Wolf, D.; Rumpold, H.; Enrich, B.; Tilg, H. Adiponectin Induces the Anti-Inflammatory Cytokines IL-10 and IL-1RA in Human Leukocytes. *Biochem. Biophys. Res. Commun.* **2004**, *323*, 630–635. [CrossRef]
20. Laothaweerungsawat, N.; Neimkhum, W.; Anuchapreeda, S.; Sirithunyalug, J.; Chaiyana, W. Transdermal Delivery Enhancement of Carvacrol from *Origanum Vulgare* L. Essential Oil by Microemulsion. *Int. J. Pharm.* **2020**, *579*, 119052. [CrossRef]
21. Mufamadi, M.S.; Pillay, V.; Choonara, Y.E.; Du Toit, L.C.; Modi, G.; Naidoo, D.; Ndesendo, V.M.K. A Review on Composite Liposomal Technologies for Specialized Drug Delivery. *J. Drug Deliv.* **2011**, *2011*, 939851. [CrossRef] [PubMed]
22. Hashemi Dehaghi, M.; Dadash Zadeh, S.; Keshvari, H.; Abasian, P. Preparation and Characterization of Dorzolamide HCl Loaded in Niosome in Order to Study the Amount of Its Realese. *Res. Pharm. Sci.* **2012**, *7*, 290.
23. Hashemi, M.; Omidi, M.; Mohammadi, J.; Shalbaf, M.; Shayeh, J.S.; Mohagheghi, M.A. Nanohybrid Platform of Functionalized Graphene Oxide for Chemo-Photothermal Therapy. *Basic Clin. Cancer Res.* **2018**, *10*, 1–8.
24. Mullaicharam, A.R.; Murthy, R.S.R. Lung Accumulation of Niosome-Entrapped Rifampicin Following Intravenous and Intratracheal Administration in the Rat. *J. Drug Deliv. Sci. Technol.* **2004**, *14*, 99–104. [CrossRef]
25. Akbarzadeh, I.; Tavakkoli Yaraki, M.; Bourbour, M.; Noorbazargan, H.; Lajevardi, A.; Sadat Shilsar, S.M.; Heidari, F.; Mousavian, S.M. Optimized Doxycycline-Loaded Niosomal Formulation for Treatment of Infection-Associated Prostate Cancer: An in-Vitro Investigation. *J. Drug Deliv. Sci. Technol.* **2020**, *57*, 101715. [CrossRef]
26. Dehaghi, M.H.; Haeri, A.; Keshvari, H.; Abbasian, Z.; Dadashzadeh, S. Dorzolamide Loaded Niosomal Vesicles: Comparison of Passive and Remote Loading Methods. *Iran. J. Pharm. Res. IJPR* **2017**, *16*, 413.

27. Budhiraja, A.; Dhingra, G. Development and Characterization of a Novel Antiacne Niosomal Gel of Rosmarinic Acid. *Drug Deliv.* **2014**, *22*, 723–730. [CrossRef]
28. Patel, J.; Ketkar, S.; Patil, S.; Fearnley, J.; Mahadik, K.R.; Paradkar, A.R. Potentiating Antimicrobial Efficacy of Propolis through Niosomal-Based System for Administration. *Integr. Med. Res.* **2015**, *4*, 94–101. [CrossRef]
29. Juneja, R.; Roy, I. Iron Oxide-Doped Niosomes as Drug Carriers for Magnetically Targeted Drug Delivery. *Int. J. Nanomed.* **2018**, *13*, 7–9. [CrossRef]
30. Ghafelehbashi, R.; Akbarzadeh, I.; Tavakkoli Yaraki, M.; Lajevardi, A.; Fatemizadeh, M.; Heidarpoor Saremi, L. Preparation, Physicochemical Properties, in Vitro Evaluation and Release Behavior of Cephalexin-Loaded Niosomes. *Int. J. Pharm.* **2019**, *569*, 118580. [CrossRef]
31. Khan, D.H.; Bashir, S.; Figueiredo, P.; Santos, H.A.; Khan, M.I.; Peltonen, L. Process Optimization of Ecological Probe Sonication Technique for Production of Rifampicin Loaded Niosomes. *J. Drug Deliv. Sci. Technol.* **2019**, *50*, 27–33. [CrossRef]
32. Wahyuni, S.T.; Rahmasari, D.; Nugroho, R.S.; Agusta, I.; Daminda, R.D.K.; Sundugesti, R.V.; Ermawati, D. Enhanced Antibacterial Activity of Piper Betle Extract Niosome Serum Gel and Its Irritation Effects. *KnE Med.* **2023**, *2023*, 178–188. [CrossRef]
33. Jufri, M.; Muthaharrah, M.; Humairah, E.; Purwaningsih, E.H. Stability of Anti-Acne Niosome Gels Containing Betel Leaf (*Piper Betle* L.) Essential Oil. *Int. J. Appl. Pharm.* **2017**, *9*, 130–134. [CrossRef] [PubMed]
34. García-Díaz, M.; Patiño, B.; Vázquez, C.; Gil-Serna, J. A Novel Niosome-Encapsulated Essential Oil Formulation to Prevent Aspergillus Flavus Growth and Aflatoxin Contamination of Maize Grains During Storage. *Toxins* **2019**, *11*, 646. [CrossRef] [PubMed]
35. Trinh, L.H.; Takzare, A.; Ghafoor, D.D.; Siddiqi, A.F.; Ravali, S.; Shalbaf, M.; Bakhtiar, M. Trachyspermum Copticum Essential Oil Incorporated Niosome for Cancer Treatment. *J. Drug Deliv. Sci. Technol.* **2019**, *52*, 818–824. [CrossRef]
36. Abdelhamed, F.M.; Abdeltawab, N.F.; ElRakaiby, M.T.; Shamma, R.N.; Moneib, N.A. Antibacterial and Anti-Inflammatory Activities of Thymus Vulgaris Essential Oil Nanoemulsion on Acne Vulgaris. *Microorganisms* **2022**, *10*, 1874. [CrossRef]
37. Shah, P.; Goodyear, B.; Haq, A.; Puri, V.; Michniak-Kohn, B. Evaluations of Quality by Design (QbD) Elements Impact for Developing Niosomes as a Promising Topical Drug Delivery Platform. *Pharmaceutics* **2020**, *12*, 246. [CrossRef]
38. Gurumukhi, V.C.; Bari, S.B. Fabrication of Efavirenz Loaded Nano-Formulation Using Quality by Design (QbD) Based Approach: Exploring Characterizations and in Vivo Safety. *J. Drug Deliv. Sci. Technol.* **2020**, *56*, 101545. [CrossRef]
39. Yaghoobian, M.; Haeri, A.; Bolourchian, N.; Shahhosseni, S.; Dadashzadeh, S. The Impact of Surfactant Composition and Surface Charge of Niosomes on the Oral Absorption of Repaglinide as a BCS II Model Drug. *Int. J. Nanomed.* **2020**, *15*, 8767–8781. [CrossRef]
40. Nadzir, M.M.; Fen, T.W.; Mohamed, A.R.; Hisham, S.F. Size and Stability of Curcumin Niosomes from Combinations of Tween 80 and Span 80. *Sains Malays.* **2020**, *46*, 2455–2460. [CrossRef]
41. Shekhawat, P.; Pokharkar, V. Risk Assessment and QbD Based Optimization of an Eprosartan Mesylate Nanosuspension: In-Vitro Characterization, PAMPA and in-Vivo Assessment. *Int. J. Pharm.* **2019**, *567*, 118415. [CrossRef] [PubMed]
42. Shehata, T.M.; Ibrahim, M.M.; Elsewedy, H.S. Curcumin Niosomes Prepared from Proniosomal Gels: In Vitro Skin Permeability, Kinetic and In Vivo Studies. *Polymers* **2021**, *13*, 791. [CrossRef]
43. Dabbagh Moghaddam, F.; Akbarzadeh, I.; Marzbankia, E.; Farid, M.; Khaledi, L.; Reihani, A.H.; Javidfar, M.; Mortazavi, P. Delivery of Melittin-Loaded Niosomes for Breast Cancer Treatment: An in Vitro and in Vivo Evaluation of Anti-Cancer Effect. *Cancer Nanotechnol.* **2021**, *12*, 1–35. [CrossRef]
44. Mohanty, D.; Rani, M.J.; Haque, M.A.; Bakshi, V.; Jahangir, M.A.; Imam, S.S.; Gilani, S.J. Preparation and Evaluation of Transdermal Naproxen Niosomes: Formulation Optimization to Preclinical Anti-Inflammatory Assessment on Murine Model. *J. Liposome Res.* **2019**, *30*, 377–387. [CrossRef] [PubMed]
45. Jadupati, M.; Kumar, N.A.; Amites, G. Transferosome: An Opportunistic Carrier for Transdermal. *Int. Res. J. Pharm.* **2012**, *3*, 35–38.
46. Hajimehdipoor, H.; Shekarchi, M.; Khanavi, M.; Adib, N.; Amri, M. A Validated High Performance Liquid Chromatography Method for the Analysis of Thymol and Carvacrol in *Thymus Vulgaris* L. Volatile Oil. *Pharmacogn. Mag.* **2010**, *6*, 154. [CrossRef]
47. Dey, S.; Pramanik, S.; Malgope, A. Formulation and Optimization of Sustained Release Stavudine Microspheres Using Response Surface Methodology. *ISRN Pharm.* **2011**, *2011*, 627623. [CrossRef] [PubMed]
48. Waghule, T.; Rapalli, V.K.; Singhvi, G.; Gorantla, S.; Khosa, A.; Dubey, S.K.; Saha, R.N. Design of Temozolomide-Loaded Proliposomes and Lipid Crystal Nanoparticles with Industrial Feasible Approaches: Comparative Assessment of Drug Loading, Entrapment Efficiency, and Stability at Plasma PH. *J. Liposome Res.* **2020**, *31*, 158–168. [CrossRef]
49. Das, P.; Das, M.K. Production and physicochemical characterization of nanocosmeceuticals. In *Nanocosmeceuticals*; Academic Press: Cambridge, MA, USA, 2022; pp. 95–138.
50. Mohammad, E. Anti-aging Effect of Free Curcumin and Niosome Entrapping Curcumin in H2O2-induced Aging in Human Fibroblast Cell Lines. *J. Adv. Phys.* **2019**, *16*, 237–246. [CrossRef]
51. Malatesta, M. Transmission Electron Microscopy for Nanomedicine: Novel Applications for Long-Established Techniques. *Eur. J. Histochem.* **2016**, *60*, 8–12. [CrossRef]
52. Kapoor, H.; Aqil, M.; Imam, S.S.; Sultana, Y.; Ali, A. Formulation of Amlodipine Nano Lipid Carrier: Formulation Design, Physicochemical and Transdermal Absorption Investigation. *J. Drug Deliv. Sci. Technol.* **2019**, *49*, 209–218. [CrossRef]

53. Balasubramaniam, A.; Kumar, V.A.; Pillai, K.S. Formulation and In Vivo Evaluation of Niosome-Encapsulated Daunorubicin Hydrochloride. *Drug Dev. Ind. Pharm.* **2002**, *28*, 1181–1193. [CrossRef]
54. Moolakkadath, T.; Aqil, M.; Ahad, A.; Imam, S.S.; Praveen, A.; Sultana, Y.; Mujeeb, M.; Iqbal, Z. Fisetin Loaded Binary Ethosomes for Management of Skin Cancer by Dermal Application on UV Exposed Mice. *Int. J. Pharm.* **2019**, *560*, 78–91. [CrossRef]
55. Jain, A.; Deveda, P.; Vyas, N.; Chauhan, J.; Khambete, H.; Jain, S. Development of Antifungal Emulsion Based Gel for Topical Fungal Infection. Available online: https://www.researchgate.net/publication/284789522_Development_of_antifungal_emulsion_based_gel_for_topical_fungal_infection (accessed on 6 April 2023).
56. Aodah, A.H.; Hashmi, S.; Akhtar, N.; Ullah, Z.; Zafar, A.; Zaki, R.M.; Khan, S.; Ansari, M.J.; Jawaid, T.; Alam, A.; et al. Formulation Development, Optimization by Box–Behnken Design, and In Vitro and Ex Vivo Characterization of Hexatriacontane-Loaded Transethosomal Gel for Antimicrobial Treatment for Skin Infections. *Gels* **2023**, *9*, 322. [CrossRef]
57. Jahan, S.; Aqil, M.; Ahad, A.; Imam, S.S.; Waheed, A.; Qadir, A.; Ali, A. Nanostructured Lipid Carrier for Transdermal Gliclazide Delivery: Development and Optimization by Box-Behnken Design. *Inorg. Nano-Met. Chem.* **2022**, *2022*, 1–14. [CrossRef]
58. Gupta, D.K.; Aqil, M.; Ahad, A.; Imam, S.S.; Waheed, A.; Qadir, A.; Iqubal, M.K.; Sultana, Y. Tailoring of Berberine Loaded Transniosomes for the Management of Skin Cancer in Mice. *J. Drug Deliv. Sci. Technol.* **2020**, *60*, 102051. [CrossRef]
59. Aryal, S.; Baniya, M.K.; Danekhu, K.; Kunwar, P.; Gurung, R.; Koirala, N. Total Phenolic Content, Flavonoid Content and Antioxidant Potential of Wild Vegetables from Western Nepal. *Plants* **2019**, *8*, 96. [CrossRef] [PubMed]
60. Loo, C.H.; Basri, M.; Ismail, R.; Lau, H.L.N.; Tejo, B.A.; Kanthimathi, M.S.; Hassan, H.A.; Choo, Y.M. Effect of Compositions in Nanostructured Lipid Carriers (NLC) on Skin Hydration and Occlusion. *Int. J. Nanomed.* **2013**, *8*, 13–22. [CrossRef]
61. Chaubey, P.; Patel, R.R.; Mishra, B. Development and Optimization of Curcumin-Loaded Mannosylated Chitosan Nanoparticles Using Response Surface Methodology in the Treatment of Visceral Leishmaniasis. *Expert Opin. Drug Deliv.* **2014**, *11*, 1163–1181. [CrossRef]
62. Jianxian, C.; Saleem, K.; Ijaz, M.; Ur-Rehman, M.; Murtaza, G.; Asim, M.H. Development and in vitro Evaluation of Gastro-protective Aceclofenac-loaded Self-emulsifying Drug Delivery System. *Int. J. Nanomed.* **2020**, *15*, 5217–5226. [CrossRef]
63. Sarheed, O.; Dibi, M.; Ramesh, K.V.R.N.S. Studies on the Effect of Oil and Surfactant on the Formation of Alginate-Based O/W Lidocaine Nanocarriers Using Nanoemulsion Template. *Pharmaceutics* **2020**, *12*, 1223. [CrossRef] [PubMed]
64. Iqubal, M.K.; Iqubal, A.; Imtiyaz, K.; Rizvi, M.M.A.; Gupta, M.M.; Ali, J.; Baboota, S. Combinatorial Lipid-Nanosystem for Dermal Delivery of 5-Fluorouracil and Resveratrol against Skin Cancer: Delineation of Improved Dermatokinetics and Epidermal Drug Deposition Enhancement Analysis. *Eur. J. Pharm. Biopharm.* **2021**, *163*, 223–239. [CrossRef] [PubMed]
65. Emami, J.; Yousefian, H.; Sadeghi, H. Targeted Nanostructured Lipid Carrier for Brain Delivery of Artemisinin: Design, Preparation, Characterization, Optimization and Cell Toxicity. *J. Pharm. Pharm. Sci.* **2018**, *21*, 225s–241s. [CrossRef]
66. Qamar, Z.; Ashhar, M.U.; Annu; Qizilibash, F.F.; Sahoo, P.K.; Ali, A.; Ali, J.; Baboota, S. Lipid Nanocarrier of Selegiline Augmented Anti-Parkinson's Effect via P-Gp Modulation Using Quercetin. *Int. J. Pharm.* **2021**, *609*, 121131. [CrossRef] [PubMed]
67. Pardakhty, A.; Varshosaz, J.; Rouholamini, A. In Vitro Study of Polyoxyethylene Alkyl Ether Niosomes for Delivery of Insulin. *Int. J. Pharm.* **2007**, *328*, 130–141. [CrossRef] [PubMed]
68. Farmoudeh, A.; Akbari, J.; Saeedi, M.; Ghasemi, M.; Asemi, N.; Nokhodchi, A. Methylene Blue-Loaded Niosome: Preparation, Physicochemical Characterization, and in Vivo Wound Healing Assessment. *Drug Deliv. Transl. Res.* **2020**, *10*, 1428–1441. [CrossRef]
69. Qadir, A.; Aqil, M.; Ali, A.; Warsi, M.H.; Mujeeb, M.; Ahmad, F.J.; Ahmad, S.; Beg, S. Nanostructured Lipidic Carriers for Dual Drug Delivery in the Management of Psoriasis: Systematic Optimization, Dermatokinetic and Preclinical Evaluation. *J. Drug Deliv. Sci. Technol.* **2020**, *57*, 101775. [CrossRef]
70. Pande, V.; Patel, S.; Patil, V.; Sonawane, R. Design Expert Assisted Formulation of Topical Bioadhesive Gel of Sertaconazole Nitrate. *Adv. Pharm. Bull.* **2014**, *4*, 121. [CrossRef]
71. Ahmad, N.; Afzali, R.; Neupane, Y.; Amin, S.; Kohli, K. Optimization of Gel Based System of Lercanidipine by Statistical Design for Transdermal Delivery; Histopathological Examination and Rheological Characterization. *J. Biopharm. Sci.* **2014**, *2*, 15–25.

gels MDPI

Article

Preparation and Characterization of Silymarin Gel: A Novel Topical Mucoadhesive Formulation for Potential Applicability in Oral Pathologies

Divyambika Catakapatri Venugopal [1,†], Reshma Devi Senthilnathan [2,†], Saba Maanvizhi [2], Yasasve Madhavan [1], Sathasivasubramanian Sankarapandian [1], Vijayalakshmi Ramshankar [3,*] and Mangathayaru Kalachaveedu [4,*]

1 Department of Oral Medicine and Radiology, Sri Ramachandra Institute of Higher Education and Research (DU), Porur, Chennai 600116, India
2 Department of Pharmacognosy, Sri Ramachandra Institute of Higher Education and Research (DU), Porur, Chennai 600116, India
3 Department of Preventive Oncology (Research), Cancer Institute (WIA), Adyar, Chennai 600020, India
4 Herb and Food Division, Isha Arogya, Coimbatore 641010, India
* Correspondence: r.vijayalakshmi@cancerinstitutewia.org (V.R.); mangathayaru.kalachaveedu@ishalife.com (M.K.)
† These authors contributed equally to this work.

Abstract: *Silybum marianum* has been used for centuries by herbalists and physicians to treat different forms of liver diseases. It contains flavonoid, which has antioxidant, anti-inflammatory, antifibrotic and anticancer properties. The objective of this research was to develop a silymarin-based mucoadhesive gel for prolonged release in oral mucosa and to evaluate the same by using *in vitro* drug release kinetic models and *ex vivo* methods for drug permeation using chicken buccal mucosa. The mucoadhesive gel was formulated in different trials by varying the concentration of silymarin and polymer. Out of 10 formulation trials, the F10 optimized trial was characterized for *in vitro* physicochemical parameters such as pH, homogeneity, viscosity, stability, drug content, *in vitro* drug release, *in vitro* antioxidant assay and *ex vivo* permeation study. Trial 10 was chosen as the best trial formulation among the other trials and was marked as an optimal trial. The physicochemical properties observed were pH to be 6.4 ± 0.01, the gel free of lumps, spreadability of 23.75 ± 0.03 and drug content of 32.77 ± 0.20 mg/g. It had no physiological changes such as color shift or fluid exudate segregation after 6 months of storage at room temperature. *In vitro* drug release established the presence of a non-fickian mechanism and demonstrated dose-dependent antioxidant activity. *Ex vivo* findings indicated 21.97 ± 0.18% release, proving that the gel can permeate through the oral mucosal membrane. Our future research will concentrate on expanding the therapeutic scope by developing the formulation trial F10 to a nanoformulation and conducting clinical trials for its potential use in various oral diseases.

Keywords: silymarin; mucoadhesive gel; topical formulation; drug release; bioavailability

Citation: Venugopal, D.C.; Senthilnathan, R.D.; Maanvizhi, S.; Madhavan, Y.; Sankarapandian, S.; Ramshankar, V.; Kalachaveedu, M. Preparation and Characterization of Silymarin Gel: A Novel Topical Mucoadhesive Formulation for Potential Applicability in Oral Pathologies. *Gels* **2023**, *9*, 139. https://doi.org/10.3390/gels9020139

Academic Editor: Shige Wang

Received: 31 December 2022
Revised: 27 January 2023
Accepted: 2 February 2023
Published: 7 February 2023

1. Introduction

Silymarin is a flavonolignan derived from milk thistle that contains silibinin, isosilibinin, silychristin, isosilychristin, and silydianin, as well as taxifolin. Amongst its various constituents, silybin with its two diastereoisomeric compounds, namely silybin A and silybin B, are present in a higher percentage (around 70%) and contribute to the biological effect exerted by silymarin [1]. Silymarin has been used as a therapeutic agent in an array of liver disorders such as chronic liver disorders, cirrhosis, hepatocellular carcinoma, alcohol abuse, non-alcoholic fatty liver disease, virus-related liver damage and end stages of different hepatopathies because of its inherent antiviral, antioxidant, anti-inflammatory, and antifibrotic properties [2]. Silymarin has also demonstrated chemopreventive action and antimetastatic activity in *in vitro* and *in vivo* conditions in various cancers, chiefly gastrointestinal cancers and also in other cancers such as colorectal and pancreatic cancers [3]. Silymarin-mediated apoptosis, through increased cleavage of caspase 8, has been demonstrated using *in vitro* studies using oral cancer cell lines [4]. Silymarin has been found to have a good safety profile without toxicity, even after prolonged administration [5]. Although there has been a long history of silymarin use in various systemic forms such as tablets, capsules, and suspension, currently randomized controlled trials (RCT) using a topical formulation of silymarin have been employed as a therapeutic option.

Gels offer certain advantages over other formulation types, including quicker drug release, smoother delivery, and better bio-compatibility and mucoadhesivity. Gels permit adherence to the oral mucosa at the site of the lesion and rapid elimination by regular catabolic routes [6]. A topical gel formulation for skin, prepared by Sampatrao and the team showed drug content, pH, and spreadability of the formulation to be 96.6%, 6.8 and 25.45 gm.cm/s, respectively, with a maximum drug release of 96.30% throughout 3 h. The gel demonstrated pseudoplastic flow properties without showing any acute skin irritancy [7]. A randomized, double-blinded, placebo-controlled clinical trial evaluated the preventive effect of a silymarin 1% gel in comparison with a placebo, on the occurrence of radiodermatitis in breast cancer patients. The results demonstrated a significant delay in radiodermatitis development and progression when treated with silymarin [8]. Unlike skin formulations, oral gels should possess good mucoadhesion, without getting easily washed off by saliva, high water content and low surface friction for better retention in the injured oral mucosa [9]. Although currently, research is ongoing on the utility of topical skin formulation of silymarin, there is no evidence of oral mucoadhesive gel formulation for therapeutic use in various oral diseases. The reason could be attributed to poor water solubility and low bioavailability [2]. However, with the introduction of complexing with phosphatidylcholine, which has better absorption, and new silibinin glyco-conjugates (gluco, manno, galacto, and lacto-conjugates), which both have a high solubility in the water, this may be overcome [10].

Silymarin as an oral topical formulation can aid in the treatment of many oral diseases, taking advantage of its antioxidant, anti-inflammatory, antifibrotic and anticancer activity. To the best of our knowledge, since there is no oral topical formulation available in the literature, the current study aims to design, formulate and evaluate the preclinical release studies of silymarin-based mucoadhesive oral topical gel.

2. Results

2.1. Characterization of Silymarin Gel Formulation

The following findings were obtained from the characterization studies, which gave an overview of the physical properties of the chosen silymarin gel formulation (F10) (Figure 1 and Table 1). The various gel formulations tested in various compositions have been listed in Table S1.

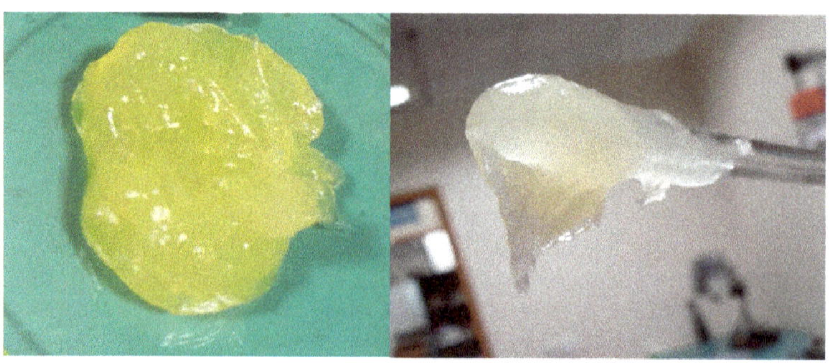

Figure 1. Silymarin mucoadhesive gel.

Table 1. Various parameters evaluated in the chosen silymarin gel formulation (F10).

S.No	Parameters	Observation
	Preformulation studies	
1	**Organoleptic characters**	Color: Yellow Odor: Characteristic Taste: Bitter
	Melting point	150 °C
	Linear regression analysis	It obeys Beers-Lamberts law
	FTIR spectroscopy	No interaction was observed
	Evaluation parameter for gel	
2	Organoleptic character	Color: Pale yellow color Odor: Mint odor Taste: Astringent taste
	pH	6.42
	Viscosity	3700 ± 0.98 to 7400 ± 0.32 cps
	Homogeneity	No visible particles are seen
	Spreadability	23.75
	Theoretical Drug content	32.77 ± 0.20 mg/g
	Stability	No change in physical properties was observed
	Drug release	Percentage cumulative drug release was found to be 2.6% and 3.25% for open-ended cylinder and diffusion cell, respectively, after 3 h
	Release kinetics	Zero order, first order, Higuchi kinetic
	In vitro antioxidant	Presence of antioxidant activity is confirmed
	Ex vivo diffusion study	In 3 h, it was found to be 3.03% and coefficient range was 0.9701

The scanning range ultraviolet spectrophotometric analysis was performed in distilled water, and 287 nm was used as the experimental value of maximum. The absorbance value of standard concentrations of 1–5 µg/mL was plotted, and linearity was observed for silymarin when analyzed at 287 nm with an R^2 = 0.9962.

2.2. FTIR Spectral Analysis

The FTIR spectra of the pure drug showed prominent peaks at 3566.7 cm^{-1} due to O-H stretch, 2925.48 cm^{-1} due to C-H stretch, 2.360.44 cm^{-1} due to absorption of carbon dioxide, 1716.34 cm^{-1} due to C=O stretch cyclic ketone, 1684.52 cm^{-1} due to C=C stretch aromatic, 1508.06 cm^{-1} due to C-H bending, 1281.47 cm^{-1} due to C-OH stretch, and 1166.72 cm^{-1} due to C-O stretch. From the spectra (Figure 2), it was observed that there was no significant change in the original peak of the drug and the polymer when compared with the spectra of the physical mixture of the formulated gel. This indicates that there was no interaction between drug, polymer and other excipients.

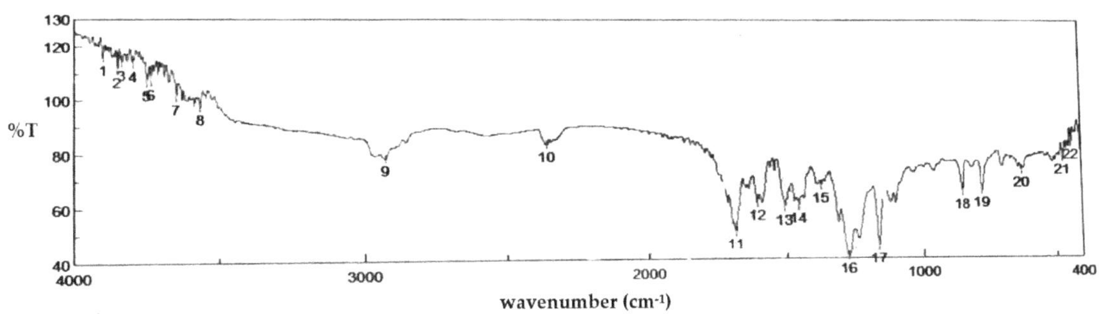

Figure 2. FTIR spectra of the formulation ingredients such as silymarin, carbopol and other excipients.

2.3. Evaluation Parameters

2.3.1. pH

The standard pH range of the oral mucosa is 6.0 to 7.0 and the gel pH values were found to be 6.4.

2.3.2. Homogeneity

There were no lumps or grittiness in any of the gel formulations that were made. (Table 2).

2.3.3. Spreadability

After 1 min, the spreadability of the gels ranged from 10.73 ± 0.01 mm to 23.75 ± 0.03 mm. The gel formulations were prepared with Carbopol 934, and hence demonstrated high spreadability and extrudability (Table 2).

2.3.4. Drug Content Uniformity

The percentage drug content of all formulations ranged from 0.2 ± 0.01% mg/g to 32.77 ± 0.20% mg/g (Table 2) The drug content was consistent across all gels.

2.3.5. Stability

Physiological changes such as change in color, or segregation of fluid exudates of the gels were not observed after 6 months of stability studies. The pH of all the gels was unaffected and ranged from 6.5 ± 0.03 to 7.0 ± 0.06. After 6 months, and the drug content ranged from 0.17 ± 0.19 to 28.00 ± 0.004% mg/g (Table 2). The results are expressed as mean ± SD, where n = 3 and SD stands for standard deviation.

2.4. In Vitro Drug Diffusion Studies

Cumulative percent drug release ranged from 21.97 ± 0.02% for the open-ended cylinder and cumulative percent drug release ranged from 65.71 ± 0.04% for the diffusion cell apparatus across all formulations (Figure 3). As a result, F10 was chosen as the optimal formulation.

2.5. Release Kinetics

Release kinetics of the optimized formulation F10 were investigated using zero-order, first-order, and Higuchi diffusion (Table 3 and Figure 4). Zero-order kinetics is defined as a constant amount of drug that is eliminated per unit of time, but the rate is independent of the concentration of the drug. First-order kinetics describes the constant proportion of the drug that is eliminated per unit of time. The rate of elimination is proportional to the amount of drug in the body. Higuchi's kinetics dominated drug release in the formulation and proves that the gel is a controlled drug delivery system. It also confirms a non-Fickian mechanism. Non-Fickian drug release refers to the drug being released from the gel through a diffusion mechanism as well as another method known as chain relaxation.

2.6. In Vitro Antioxidant Study

The optimized formulation F10 was subjected to anti-oxidant analysis. The antioxidant activity of samples were investigated at various concentrations and standards (ascorbic acid). By scavenging DPPH (free radical) and converting it to DPPH, the samples demonstrated strong antioxidant activity (Figure 5). Nitric oxide (NO) is a quite unstable species that further, under aerobic conditions, reacts with O_2 to obtain uniform nitrate and nitrite products using NO_2, N_2O_4 and N_3O_4 intermediates. In this study, the nitrite developed by the reaction mixture was limited.

Table 2. Physicochemical evaluation for various gel formulations.

Evaluation Parameters		F1	F2	F3	F4	F5	F6	F7	F8	F9	F10
pH		4.2 ± 0.02	4.9 ± 0.04	5.0 ± 0.01	5.5 ± 0.03	5.7 ± 0.05	5.8 ± 0.04	6.0 ± 0.01	6.2 ± 0.04	6.3 ± 0.03	6.4 ± 0.01
Homogeneity		good	Good	good	good	good	good	good	good	good	good
Spreadability		10.73 ± 0.01	15.98 ± 0.03	18.01 ± 0.02	18.95 ± 0.04	19.73 ± 0.09	19.89 ± 0.01	20.56 ± 0.03	21.64 ± 0.02	22.05 ± 0.04	23.75 ± 0.03
Viscosity (cpm)		3700 ± 0.98	3490 ± 0.07	5000 ± 0.56	5231 ± 0.02	5409 ± 0.34	6000 ± 0.51	6302 ± 0.33	6589 ± 0.54	7000 ± 0.94	7400 ± 0.32
Theoretical drug content (mg/g)		0.2 ± 0.01 (Silymarin:0.5 g)	0.5 ± 0.21 (Silymarin:0.5 g)	0.87 ± 0.78 (Silymarin:0.5 g)	0.95 ± 0.12 (Silymarin:0.5 g)	1.9 ± 0.10 (Silymarin:0.5 g)	10.72 ± 0.76 (Silymarin:1 g)	12.1 ± 0.80 (Silymarin:1 g)	17.20 ± 0.37 (Silymarin:1 g)	29.83 ± 0.25 (Silymarin:1 g)	32.77 ± 0.20 (Silymarin:1 g)
Stability (6 months at room temperature)	**pH**	6.5 ± 0.03	6.5 ± 0.12	6.5 ± 0.15	6.6 ± 0.04	6.8 ± 0.21	6.8 ± 0.40	6.8 ± 0.31	6.9 ± 0.01	7.0 ± 0.04	7.0 ± 0.06
	Drug content (mg/g)	0.17 ± 0.19	0.23 ± 0.01	0.30 ± 0.83	0.52 ± 0.06	0.55 ± 0.15	1.9 ± 0.34	5.34 ± 0.57	9.8 ± 0.93	15.34 ± 0.05	28.00 ± 0.004

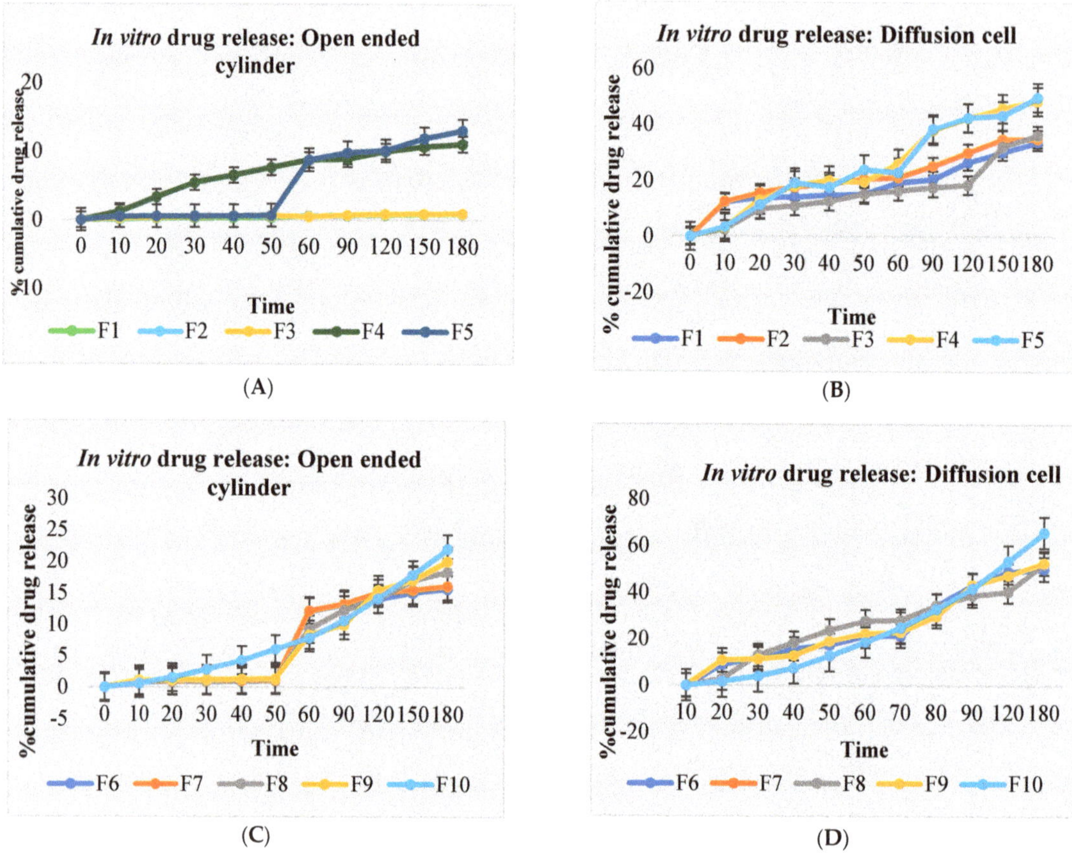

Figure 3. Percent cumulative drug release from gel formulations. (**A**) Open−ended cylinder (F1 to F5), (**B**) diffusion cell apparatus (F1 to F5), (**C**) Open−ended cylinder (F6 to F10), (**D**) diffusion cell apparatus (F6 to F10).

Table 3. Release kinetics for optimized formulation F10.

	Zero Order	First Order	Higuchi Diffusion	Korsmeyer Peppas
R^2	0.9475	0.9475	0.9869	0.9448

2.7. Ex Vivo Permeation Study

The gel can permeate easily across the mucosal membrane. The permeation correlation coefficient range was 0.9586 and the percentage was found to be 21.97% in 3 h. The result of the study is shown in (Table 4 and Figure 6).

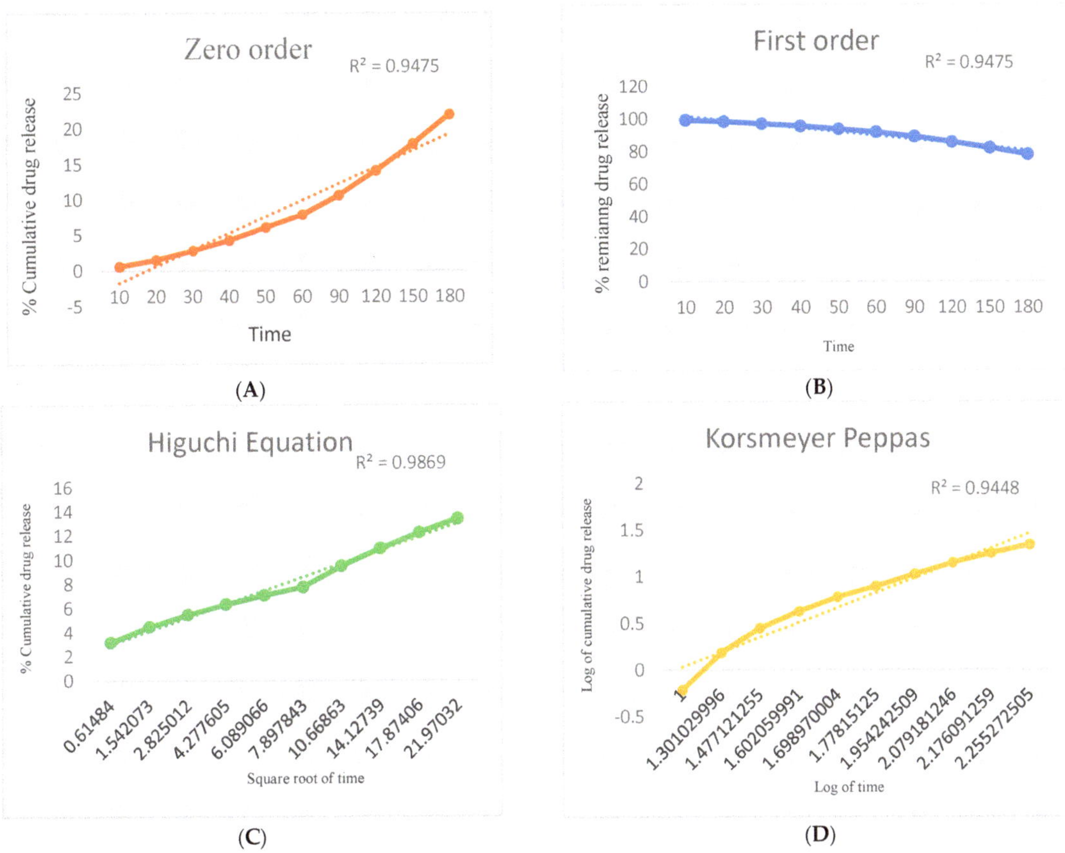

Figure 4. Higuchi's kinetics drug release in the formulation. (**A**) Zero order, (**B**) first order, (**C**) Higuchi equation, (**D**) Korsmeyer Peppas.

Table 4. Ex vivo diffusion study of silymarin.

S.NO	Time (mins)	% Cumulative Drug Release
1.	10	0.057 ± 0.0
2.	20	1.52 ± 0.03
3.	30	2.66 ± 0.05
4.	40	4.11 ± 0.09
5.	50	6.05 ± 0.02
6.	60	8.32 ± 0.06
7.	90	11.05 ± 0.03
8.	120	14.03 ± 0.09
9.	150	17.56 ± 0.04
10.	180	21.97 ±0.18

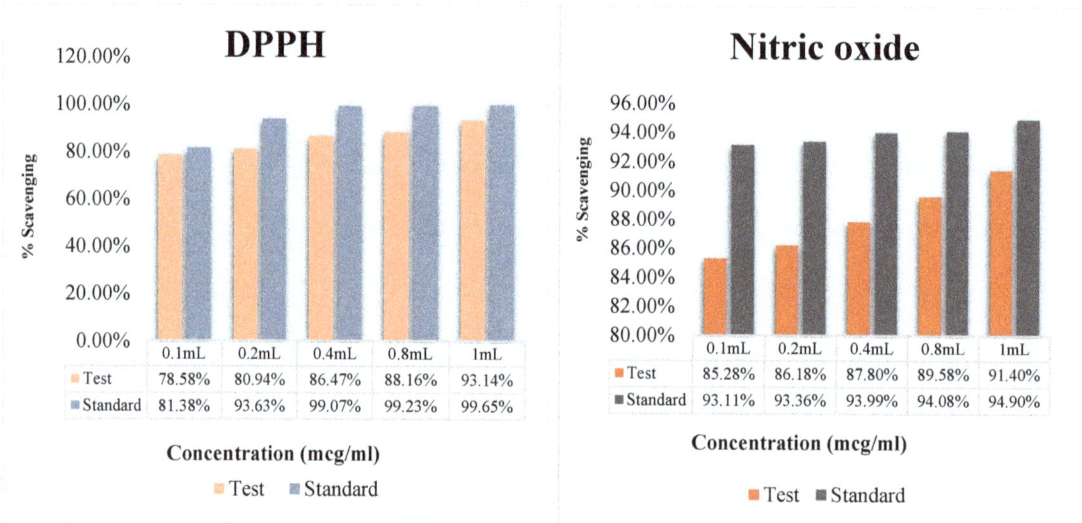

(A) (B)

Figure 5. In vitro antioxidant study for optimized gel F10. (**A**) DPPH activity, which shows that it is dose-dependent, (**B**) nitric oxide, the scavenging activity.

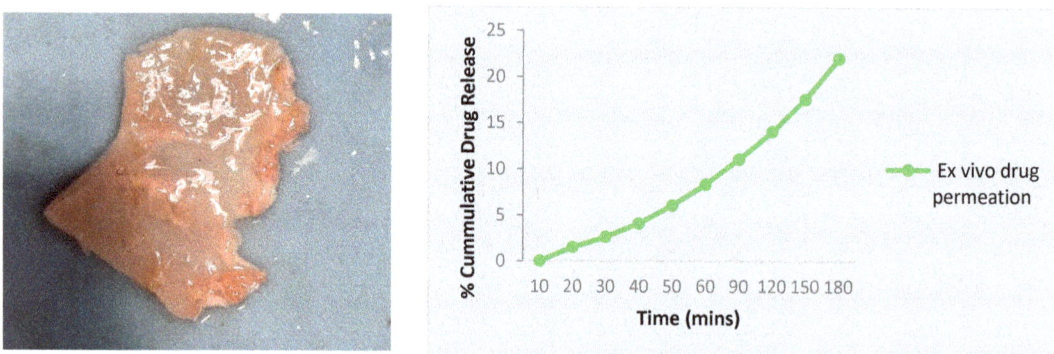

(A) (B)

Figure 6. Ex vivo permeation study for the optimized formulation F10. (**A**) Chicken buccal mucosa, (**B**) Cumulative drug release.

3. Discussion

Mucosal drug delivery systems have the superior advantage of intimate contact between the diseased mucosa and the drug, localizing the drug to the specific site and better patient compliance, compared to systemic drugs. Mucoadhesive forms also offer better plasma concentration and therapeutic efficacy [11]. The pre-formulation phase or study helps to lay down the foundation for transforming a new drug into a pharmaceutical formulation in such a way that it can be administered in the right manner and optimum dose. Moreover, pre-formulation studies provide better stability to the formulation by developing a proper design with adequate constituents, protecting the drug component from environmental conditions and evaluating the performance of the developed formulation to bring it to translational use in pharmacological practice [12]. Among various topical formulations such as ointments, creams, pastes, emulsions and gels, pastes and gels are the preferred topical formulations for the oral mucosa, because of their superior retentivity and mucoadhesive property. Gels, in particular, have higher water content and exhibit lower friction and hence are better retained in the injured sites of the oral cavity for a prolonged period, thereby improving the drug's therapeutic efficacy [9].

Silymarin is a commonly used therapeutic drug with proven anti-inflammatory, antioxidant, anticancer and anti-fibrotic activities. Hence, the silymarin-based topical oral mucoadhesive gel could be a promising medication for inflammatory diseases of oral mucosa such as oral ulcers and oral mucositis, fibrotic diseases such as oral submucous fibrosis and radiation-induced fibrosis. To the best of our knowledge, silymarin-based oral mucoadhesive gel is not available for therapeutic use. However, a study done using silymarin cream for skin application is available in the literature, where its application in radiodermatitis of breast cancer patients, as a double-blinded placebo-controlled clinical trial, showed a significant delay in radiodermatitis development and progression in the silymarin group, with absence of notable side effects on the skin [8]. A study conducted using topical skin formulations of silymarin in melasma showed good results with minimal side effects on prolonged application for three months [13]. Silymarin also exhibited a gastroprotective effect in a study done by Sharma et al., in 2018, where silymarin-loaded beads of chitosan-MMT exhibited good mucoadhesion and efficient release of the drug, and were found to be a promising drug for the treatment of gastric ulcers [14]. Silymarin is known to be a safe drug with very minimal side effects. No major side effects or life-threatening complications have been reported so far in the literature, especially on prolonged and high-dosage use [15,16]. In 2011, Becher-Schiebe et al., conducted an observational study on 101 women with breast cancer who had had breast-conserving surgery, where they used an alternate, open, nonrandomized schedule to assign patients to the silymarin and standard of care (SOC) groups. This trial exhibited silymarin-based cream to be a promising candidate for a safe prophylactic treatment option of radiodermatitis [17].The absence of significant side effects, even upon long-term use makes it an ideal candidate for therapeutic use in oral inflammatory and fibrotic conditions, which warrants a longer duration of treatment and follow-up.

According to the literature survey, there is no silymarin-based mucoadhesive gel for oral topical application. The major hurdle is solubility, and the same is predicted to be overcome with formulation trials using a combination of newer polymers such as carbopol. Carbopol constitutes acrylic acid-poly alkenyl ether/divinyl glycol cross-linked primary polymer particles of about 0.2 to 6.0 micron average diameter with high molecular weight. Demonstrated to create a tenacious bond with the mucus membrane resulting in strong bio-adhesion, several commercial oral and topical products available today and under investigation have been formulated with carbopol polymers. Along with excellent adhesion forces, they lower the concentration of active ingredients and provide patient compliance with increased bioavailability of certain drugs. In the present study, silymarin gel was prepared using carbopol, propylene glycol, methylparaben, propylparaben, triethanolamine and peppermint oil. The stability, spreadability and release studies, *in vitro* antioxidant activity and *ex vivo* diffusion study using a chick embryo Franz diffusion model showed good promising results. These results are in concurrence with the previous study of silymarin gel for skin preparation, using the same ingredients, which demonstrated to have good viscosity, spreadability and antifungal properties with pH within the range of skin pH. The skin formulation also showed stability for up to two months at 40 °C, and had no skin irritation in human volunteers [7]. Among the three gel formulations prepared using pomegranate flowers by Aslan and his colleagues, one was preferred as a superior formulation because of its proper appearance and uniformity, acceptable viscosity, mucoadhesiveness and stability at different temperatures [9]. The present study of silymarin gel formulation trials resulted in an optimal gel formulation comparable to the gel obtained from the above-mentioned study with acceptable physicochemical parameters as shown in the results. Post six months of storage, no physiological alterations or changes of pH were observed. Non-fickian drug release and 22% drug permeation established its oral mucosal membrane permeability. Some of the physical methods to improve drug permeation include nanoformulations, liposomes, ethosomes, microemulsion, hydrogels, etc, which can be taken up in future studies [18].

The pH adjustment in the F10 was optimized maximally to 6.4, with further adjustment bearing on its viscosity. The normal salivary pH is 6.2–7.6, varying to 6–7 in various oral pathologies. A study performed by Foglio-Bonda et al., in 2017, reported that the salivary pH of 88 patients with oral lesions was 6.7 and the mean pH of 80 patients without oral mucosal lesions was 6.95 [19]. Oral disease severity, due to smoking and poor oral hygiene habits, is also strongly correlated with low salivary pH values (6.25 in stage IV periodontal disease) [20]. Hence, the pH of the oral mucosa tends to vary with oral disease presentations and habits such as tobacco.

A study employing ultrasonication (for better drug permeation) for the preparation of a silymarin-loaded nanostructured lipid carrier (NLC) gel showed antiproliferative, antioxidant, anti-inflammatory and antitumor activity in ex vivo and in vivo studies of skin cancer models [21]. The establishment of clinical efficacy of the gel shall be the impetus for further expansion of its therapeutic index, possibly as nano gel. Plans of preparation of a nano-based silymarin gel could probably increase the release percentage and offer better drug availability for improved efficacy. This formulation has set the pace for the development of the needed therapeutic option for oral diseases such as oral ulcers (aphthous ulcers and traumatic ulcers), oral mucositis and potentially malignant disorders such as oral submucous fibrosis, in the form of optimally bioavailable silymarin, thus potentializing its antifibrotic, anti-oxidant, anti-inflammatory and cytoprotective properties. Hence, multi-centric clinical trials on the topical application of silymarin mucoadhesive gel for different oral diseases can pave the way for identifying its therapeutic efficacy in different oral diseases.

4. Conclusions

Silymarin mucoadhesive gel was prepared using a carbopol base and was found to show optimum results with pre-formulation studies. The prepared silymarin oral mucoadhesive gels, formulation F10, had improved mucoadhesive properties, drug delivery and bioavailability to the oral mucosa, according to the physicochemical evaluation results. Our *ex vivo* analysis confirms formulation permeation and investigated the potential for oral mucosal delivery. The next step in the research process is to develop a nano gel using poly(lactic-co-glycolic acid) nanoparticles and conduct clinical trials in different oral pathologies.

5. Materials and Methods

The Serum Institute of India Pvt. Ltd. Pune, India supplied silymarin. Carbopol 934, methyl paraben and propyl paraben were purchased from Loba Chemie Pvt. Ltd. Mumbai, India. Propylene glycol and triethanolamine were obtained from SRL Chemicals Pvt. Ltd. Chennai, India.

5.1. Preparation of Silymarin-Based Mucoadhesive Gel

Silymarin, which is the bioflavanoid present in *Silybum marianum*, has poor water solubility. Because of its low bioavailability, mucoadhesive gels with optimized polymer combinations were planned as an improved silymarin formulation for potential use in oral mucosa as a topical preparation.

The gel was prepared by dispersing carbopol into distilled water to which propylene glycol had previously been added. Methylparaben and propylparaben were added to this precisely weighed quantity. Then, silymarin, which was previously solubilized in propylene glycol, was slowly added with gentle stirring in the carbopol gel base. Triethanolamine was added to neutralize the mixture and peppermint oil was added as a flavoring agent.

5.2. Pre-Formulation Studies

The color, odor, taste, and melting point of silymarin were determined. According to the certificate of analysis on the label, the purity was found to be 70% total silymarin. The standard stock solution was prepared in distilled water, and 100 mg of silymarin was accurately weighed and transferred to a 100 mL standard volumetric flask, which was then filled with distilled water to a volume of 100 mL. A sufficient aliquot of the standard stock solution (10 mL) was transferred to a 100 mL standard volumetric flask; the volume was then made up to 100 mL with distilled water. 1 mL, 2 mL, 4 mL and 5 mL was pipetted out, and the final volume was made up to 100 mL with distilled water to achieve concentrations of 1–5 μg/mL solutions. The absorbance was measured against distilled water at 287 nm (λ max silymarin) and the standard curve of concentration versus absorbance was plotted. The Fourier-transform infrared spectroscopy (FT-IR) spectra matching system was used to identify any potential chemical reactions between the drugs and polymers. It was scanned between 4000 and 600 cm^{-1}.

5.3. Evaluation Parameters

5.3.1. Viscosity

The measurement of the viscosity of the formulated gel was determined by a Brookfield Viscometer.

5.3.2. pH

The pH of the gel formulation was determined by using a digital pH meter [22].

5.3.3. Homogeneity

Visual inspection was used to verify the homogeneity of the formed gel formulation. They were examined for the existence of aggregates and the appearance of aggregates [23].

5.3.4. Spreadability

The therapeutic effectiveness of the formulations was also determined by its spreadability. It was expressed in terms of the time it takes two slides to slip off the gel that is put between the slides, under the direction of a certain load. It was calculated using the following formula: S = M*L/t, where M denotes the weight attached to the upper slide, L denotes the length of the glass slide, and T denotes the time required to separate the slides [24,25].

5.3.5. Drug Content

A specified amount of formulated gel was dissolved in 100 mL of pH 6.8 phosphate buffer. This solution was filtered and spectrophotometrically estimated at 287.0 nm with phosphate buffer (pH 6.8) as a blank.

5.3.6. Stability Study

As per the I.C.H. guidelines, the shelf-life of the formulation was determined and any incompatibility within the formulation was identified as changes in appearance and drug content of the stored mucoadhesive gel, which was observed after 1, 2, 3, 4, 5 and 6 months [26].

5.3.7. Drug Release Study by Open-Ended Cylinder

A glass cylinder measuring 10 cm in height, 3.7 cm in diameter, and 3.1 cm in diameter with both ends open were used. A glycerol-soaked cellophane membrane was attached to one end of the cylinder. It kept 5 g of the gel under investigation. At the receptor compartment, a beaker containing 100 mL of 6.8 pH buffer solution was used. The sample was submerged to a depth sufficient to keep it below the medium's surface in the receptor compartment. While at 37 °C, the medium in the compartment was agitated with a magnetic stirrer. Every 10 min, 5 mL of the sample was removed and measured at 287 nm for 3 h. Every time, the volume extracted was replaced by an equivalent amount of medium [27]. The investigations were carried out in triplicate, and the standard deviation (SD) was determined.

5.3.8. Drug Release Study by Diffusion Cell Apparatus

Phosphate buffer was used as the dissolution medium (pH 6.8). The membrane (DURAPORE HVLP 45 μm) was positioned over the receptor compartment. The receptor compartment was packed with 12 mL of dissolution medium and held at 37 °C. The dissolution medium was stirred with a magnetic stirrer at 50 rpm. At various intervals, 2 mL aliquots were removed and replaced with an equal volume of receptor medium [28]. The Electrolab diffusion cell apparatus was used for this study. The trials were carried out in triplicate, and the standard deviation (SD) was estimated.

5.3.9. Release Kinetics

Dissolution and release of drugs are important phenomena for solid dosage forms such as tablets, capsules, and semisolid dosage forms such as creams, ointments, and implants, which deliver the drugs over the intended period ranging from hours to weeks and years. It is also applicable to the design and optimization of all kinds of modified release dosage forms such as sustained, delayed, controlled release dosage forms and novel drug delivery systems. For the evaluation of *in vitro* release kinetics to investigate the mechanism of release, the data were analyzed with the following mathematical models [29].

To investigate the mechanism of release, the data were analyzed with the following mathematical models: zero-order kinetic (Equation (1)), first-order kinetic (Equation (2)) and Higuchi kinetic (Equation (3)) Korsmeyer-Peppas model (Equation (4)), and Hixson-Crowell cube root law (Equation (5)) [30,31].

$$Q_t = K_0 t \tag{1}$$

$$\ln Q_t = \ln Q_0 - K_1 t \tag{2}$$

$$Q_t = K_h t^{1/2} \tag{3}$$

$$Mt/M\alpha = K_p t^n \tag{4}$$

$$Q_0^{\ 1/3} - Q_t^{\ 1/3} = K_{HC} t \tag{5}$$

5.4. *In Vitro Antioxidant Studies for Prepared Gel*

5.4.1. 2,2-Diphenylpicrylhydrazyl (DPPH) Free Radical Scavenging Test

Various concentrations of standard ascorbic acid and samples, namely 100, 200, 400, 800 and 1000 μg/mL, were prepared in distilled water. An equal volume of different concentrations of standards and DPPH were mixed separately in a clean and labelled tube and the tubes were incubated at room temperature in the dark for 30 min. The absorbance was measured at 517 nm using a UV-VIS spectrophotometer [32]. The percentage inhibition was calculated using the formula:

$$\text{Percentage inhibition} = (\text{Abs control} - \text{Abs sample})/\text{Abs control} \times 100$$

5.4.2. Nitric Oxide Scavenging Assay

A 3 mL reaction mixture containing sodium nitroprusside (10 mM in phosphate-buffered saline) and various concentrations of standard ascorbic acid and samples, namely 100, 200, 400, 800 and 1000 μg/mL, was prepared in distilled water and kept in incubated at 37 °C for 4 h. To the incubation solution, 0.5 mL of Griess reagent was added and the absorbance was read at 546 nm [33]. The percentage inhibition was calculated using the formula:

$$\text{Percentage inhibition} = (\text{Abs control} - \text{Abs sample})/\text{Abs control} \times 100$$

5.5. Ex Vivo Permeation Study

An earlier study performed by El-Samaligy et al., in 2006, proposed that silymarin liposomal buccal delivery showed steady-state permeation through a chicken buccal mucosa for 6 hrs, and was higher than free drug powder. Hence, the drug permeation study was carried out through chicken buccal mucosa using a Franz diffusion cell. The chicken buccal mucosa was surgically removed from the underlying muscular layer by cutting the loose connective fibers, then cleaned from surface fats and washed with isotonic phosphate buffer at pH 6.8. The buccal mucosa, with the mucosal surface facing up, was placed between the donor and receptor compartments. The receptor compartment was filled with phosphate buffer (pH 6.8) and the diffusion cells were placed on hot plate-multiple magnetic stirrers at 37 °C and 150 rpm for 3 h. Samples were withdrawn at different intervals, filtered and replaced by fresh PB, and analyzed spectrophotometrically at wavelength 287 nm. The investigations were carried out in triplicate, and the standard deviation (SD) was determined [18,34].

Supplementary Materials: The following supporting information can be downloaded at: https://www.mdpi.com/article/10.3390/gels9020139/s1, Table S1: Composition of formulated silymarin mucoadhesive gel.

Author Contributions: Conceptualization, D.C.V., V.R. and M.K.; data curation, R.D.S., Y.M. and S.S.; formal analysis, R.D.S., S.M., S.S. and M.K.; investigation, D.C.V., R.D.S., Y.M. and M.K.; methodology, D.C.V. and R.D.S.; project administration, V.R. and M.K.; resources, S.M.; software, S.M.; supervision, V.R. and M.K.; validation, V.R.; writing—original draft, D.C.V., R.D.S. and Y.M.; writing—review and editing, V.R. and M.K. All authors have read and agreed to the published version of the manuscript.

Funding: This research received no external funding.

Data Availability Statement: The data associated with the manuscript are available from the first and corresponding authors.

Acknowledgments: The authors would like to thank Microlabs Pvt. Ltd., Bangalore and Serum Institute of India Pvt Ltd., Pune for providing silymarin.

Conflicts of Interest: The authors declare no conflict of interest.

References

1. Loguercio, C.; Festi, D. Silybin and the Liver: From Basic Research to Clinical Practice. *World J Gastroenterol.* **2011**, *17*, 2288–2301. [CrossRef] [PubMed]
2. Federico, A.; Dallio, M.; Loguercio, C. Silymarin/Silybin and Chronic Liver Disease: A Marriage of Many Years. *Molecules* **2017**, *22*, 191. [CrossRef] [PubMed]
3. Fallah, M.; Davoodvandi, A.; Nikmanzar, S.; Aghili, S.; Mirazimi, S.M.A.; Aschner, M.; Rashidian, A.; Hamblin, M.R.; Chamanara, M.; Naghsh, N.; et al. Silymarin (Milk Thistle Extract) as a Therapeutic Agent in Gastrointestinal Cancer. *Biomed. Pharmacother.* **2021**, *142*, 112024. [CrossRef] [PubMed]
4. Won, D.-H.; Kim, L.-H.; Jang, B.; Yang, I.-H.; Kwon, H.-J.; Jin, B.; Oh, S.H.; Kang, J.-H.; Hong, S.-D.; Shin, J.-A.; et al. In Vitro and in Vivo Anti-Cancer Activity of Silymarin on Oral Cancer. *Tumour Biol.* **2018**, *40*, 1010428318776170. [CrossRef] [PubMed]
5. Koltai, T.; Fliegel, L. Role of Silymarin in Cancer Treatment: Facts, Hypotheses, and Questions. *J. Evid. Based Integr. Med.* **2022**, *27*, 2515690X211068826. [CrossRef] [PubMed]
6. Fini, A.; Bergamante, V.; Ceschel, G.C. Mucoadhesive Gels Designed for the Controlled Release of Chlorhexidine in the Oral Cavity. *Pharmaceutics* **2011**, *3*, 665–679. [CrossRef] [PubMed]
7. Sampatrao, K.A.; Bhalchandra, K.J.; Ravindra, J.; Manohar, P. Formulation Development and Evaluation of Silymarin Gel. *Res. J. Pharm. Technol.* **2011**, *4*, 1633–1636.
8. Karbasforooshan, H.; Hosseini, S.; Elyasi, S.; Fani Pakdel, A.; Karimi, G. Topical Silymarin Administration for Prevention of Acute Radiodermatitis in Breast Cancer Patients: A Randomized, Double-Blind, Placebo-Controlled Clinical Trial. *Phytother. Res.* **2019**, *33*, 379–386. [CrossRef]
9. Aslani, A.; Zolfaghari, B.; Davoodvandi, F. Design, Formulation and Evaluation of an Oral Gel from Punica Granatum Flower Extract for the Treatment of Recurrent Aphthous Stomatitis. *Adv. Pharm. Bull.* **2016**, *6*, 391–398. [CrossRef]
10. Zarrelli, A.; Romanucci, V.; Tuccillo, C.; Federico, A.; Loguercio, C.; Gravante, R.; Di Fabio, G. New Silibinin Glyco-Conjugates: Synthesis and Evaluation of Antioxidant Properties. *Bioorg. Med. Chem. Lett.* **2014**, *24*, 5147–5149. [CrossRef]

11. Boddupalli, B.M.; Mohammed, Z.N.K.; Nath, R.A.; Banji, D. Mucoadhesive Drug Delivery System: An Overview. *J. Adv. Pharm. Technol. Res.* **2010**, *1*, 381–387. [CrossRef] [PubMed]
12. Patel, P. *Preformulation Studies: An Integral Part of Formulation Design*; IntechOpen: London, UK, 2019; ISBN 978-1-78985-839-6.
13. Nofal, A.; Ibrahim, A.-S.M.; Nofal, E.; Gamal, N.; Osman, S. Topical Silymarin versus Hydroquinone in the Treatment of Melasma: A Comparative Study. *J. Cosmet. Derm.* **2019**, *18*, 263–270. [CrossRef] [PubMed]
14. Sharma, A.; Puri, V.; Kakkar, V.; Singh, I. Formulation and Evaluation of Silymarin-Loaded Chitosan-Montmorilloite Microbeads for the Potential Treatment of Gastric Ulcers. *J. Funct. Biomater.* **2018**, *9*, 52. [CrossRef] [PubMed]
15. Dunnick, J.K.; Singh, B.; Nyska, A.; Peckham, J.; Kissling, G.E.; Sanders, J.M. Investigating the Potential for Toxicity from Long-Term Use of the Herbal Products, Goldenseal and Milk Thistle. *Toxicol. Pathol.* **2011**, *39*, 398–409. [CrossRef] [PubMed]
16. Saller, R.; Brignoli, R.; Melzer, J.; Meier, R. An Updated Systematic Review with Meta-Analysis for the Clinical Evidence of Silymarin. *Komplementmed* **2008**, *15*, 9–20. [CrossRef]
17. Becker-Schiebe, M.; Mengs, U.; Schaefer, M.; Bulitta, M.; Hoffmann, W. Topical Use of a Silymarin-Based Preparation to Prevent Radiodermatitis. *Strahlenther. Onkol.* **2011**, *187*, 485. [CrossRef]
18. Gao, Y.; Du, L.; Li, Q.; Li, Q.; Zhu, L.; Yang, M.; Wang, X.; Zhao, B.; Ma, S. How Physical Techniques Improve the Transdermal Permeation of Therapeutics: A Review. *Medicine* **2022**, *101*, e29314. [CrossRef]
19. Foglio-Bonda, P.L.; Brilli, K.; Pattarino, F.; Foglio-Bonda, A. Salivary Flow Rate and PH in Patients with Oral Pathologies. *Eur. Rev. Med. Pharm. Sci.* **2017**, *21*, 369–374.
20. Lăzureanu, P.C.; Popescu, F.; Tudor, A.; Stef, L.; Negru, A.G.; Mihăilă, R. Saliva PH and Flow Rate in Patients with Periodontal Disease and Associated Cardiovascular Disease. *Med. Sci. Monit.* **2021**, *27*, e931362-1–e931362-13. [CrossRef]
21. Iqbal, B.; Ali, J.; Ganguli, M.; Mishra, S.; Baboota, S. Silymarin-Loaded Nanostructured Lipid Carrier Gel for the Treatment of Skin Cancer. *Nanomedicine* **2019**, *14*, 1077–1093. [CrossRef]
22. Shukla, R.; Tiwari, G.; Tiwari, R.; Rai, A.K. Formulation and Evaluation of the Topical Ethosomal Gel of Melatonin to Prevent UV Radiation. *J. Cosmet. Dermatol.* **2020**, *19*, 2093–2104. [CrossRef] [PubMed]
23. Shaikh, S.; Shete, A.; Doijad, R. Formulation and Evaluation Pharmaceutical Aqueous Gel of Powdered Guava Leaves for Mouth Ulcer Treatment. *Pharmatutor* **2018**, *6*, 32. [CrossRef]
24. Jain, N.; Verma, A.; Jain, N. Formulation and Investigation of Pilocarpine Hydrochloride Niosomal Gels for the Treatment of Glaucoma: Intraocular Pressure Measurement in White Albino Rabbits. *Drug Deliv.* **2020**, *27*, 888–899. [CrossRef] [PubMed]
25. Shanmugam, D.K.; Madhavan, Y.; Manimaran, A.; Kaliaraj, G.S.; Mohanraj, K.G.; Kandhasamy, N.; Amirtharaj Mosas, K.K. Efficacy of Graphene-Based Nanocomposite Gels as a Promising Wound Healing Biomaterial. *Gels* **2023**, *9*, 22. [CrossRef]
26. Uhljar, L.É.; Kan, S.Y.; Radacsi, N.; Koutsos, V.; Szabó-Révész, P.; Ambrus, R. In Vitro Drug Release, Permeability, and Structural Test of Ciprofloxacin-Loaded Nanofibers. *Pharmaceutics* **2021**, *13*, 556. [CrossRef]
27. Amatya, S.; Park, E.J.; Park, J.; Kim, J.; Seol, E.; Lee, H.; Choi, H.; Shin, Y.; Na, D.H. Drug Release Testing Methods of Polymeric Particulate Drug Formulations. *J. Pharm. Investig.* **2013**, *43*, 259–266. [CrossRef]
28. Rahman, Z.; Zidan, A.; Khan, S.; Reddy, I.; Khan, M. Cholorpheniramine Tannate Complexes: Physicochemical, Chemometric, and Taste Masking Evaluation. *Int. J. Pharm.* **2012**, *436*, 582–592. [CrossRef]
29. Lee, P.I. Kinetics of Drug Release from Hydrogel Matrices. *J. Control. Release* **1985**, *2*, 277–288. [CrossRef]
30. Paarakh, M.P.; Jose, P.A.N.I.; Setty, C.M.; Peter, G. V Release Kinetics–Concepts and Applications. *Int. J. Pharm. Res. Technol.* **2019**, *8*, 12–20. [CrossRef]
31. Dhana lekshmi, U.M.; Poovi, G.; Kishore, N.; Reddy, P.N. In Vitro Characterization and in Vivo Toxicity Study of Repaglinide Loaded Poly (Methyl Methacrylate) Nanoparticles. *Int. J. Pharm.* **2010**, *396*, 194–203. [CrossRef]
32. Baliyan, S.; Mukherjee, R.; Priyadarshini, A.; Vibhuti, A.; Gupta, A.; Pandey, R.P.; Chang, C.-M. Determination of Antioxidants by DPPH Radical Scavenging Activity and Quantitative Phytochemical Analysis of Ficus Religiosa. *Molecules* **2022**, *27*, 1326. [CrossRef] [PubMed]
33. Habu, J.B.; Ibeh, B.O. In Vitro Antioxidant Capacity and Free Radical Scavenging Evaluation of Active Metabolite Constituents of Newbouldia Laevis Ethanolic Leaf Extract. *Biol. Res.* **2015**, *48*, 16. [CrossRef] [PubMed]
34. El-Nahas, A.E.; Allam, A.N.; El-Kamel, A.H. Mucoadhesive Buccal Tablets Containing Silymarin Eudragit-Loaded Nanoparticles: Formulation, Characterisation and Ex Vivo Permeation. *J. Microencapsul.* **2017**, *34*, 463–474. [CrossRef] [PubMed]

gels

Communication

Delivery of *B. subtilis* into Animal Intestine Using Chitosan-Derived Bioresorbable Gel Carrier: Preliminary Results

Alexander Samokhin [1,*], Anastasia Korel [1], Elena Blinova [1], Alexander Pestov [2], Galina Kalmykova [1], Nadezhda Akulova [1], Viktoria Betz [1], Vadim Tkachenko [3] and Ekaterina Litvinova [1,4]

1 Novosibirsk State Technical University, 630073 Novosibirsk, Russia
2 Institute of Organic Synthesis n.a. I. Ya. Postovsky UB RAS, 620137 Ekaterinburg, Russia
3 Institute of Nuclear Physics SB RAS, 630090 Novosibirsk, Russia
4 Scientific Research Institute of Neurosciences and Medicine SB RAS, 630117 Novosibirsk, Russia
* Correspondence: motorist@inbox.ru

Abstract: The oral delivery of bacteria in the human intestine is of great interest because of its potential to correct the gut microbiota and treat inflammatory bowel diseases. The aim of this study was to evaluate sodium N-(2-sulfonatoethyl)chitosan gel cross-linked with glutaraldehyde as a delivery carrier for probiotic bacteria to the gut using in vitro and in vivo experiments. The bacterial test strain was *B. subtilis 20*. The cytotoxicity of the gel was evaluated via cell culture using flow cytometry and light microscopy. The gel as a delivery system was assessed by the dye release in medium with different pH levels in vitro, and by bacterial titer monitoring in mouse feces using the microbiology method in vivo. Results of an in vitro experiment showed that tested gel has no cytotoxicity. The use of gel as a carrier for bacterial delivery into the intestine was more effective than oral gavage of bacterial suspension. Therefore, gel delivery of bacteria decreased the titer level by up to two times. However, a gavage of bacterial suspension decreased the titer level by over 200 times. Tested gel has the potential to be a carrier for the safe delivery of bacteria to the intestine through the stomach, reducing the rate of the elimination of probiotic bacteria from the intestine.

Keywords: bacteria; bacillus subtilis; intestine; carrier; sodium N-(2-sulfonatoethyl)chitosan; gel; release; probiotics; cytotoxicity; animal

Citation: Samokhin, A.; Korel, A.; Blinova, E.; Pestov, A.; Kalmykova, G.; Akulova, N.; Betz, V.; Tkachenko, V.; Litvinova, E. Delivery of *B. subtilis* into Animal Intestine Using Chitosan-Derived Bioresorbable Gel Carrier: Preliminary Results. *Gels* **2023**, *9*, 120. https://doi.org/10.3390/gels9020120

Academic Editor: Shige Wang

Received: 15 December 2022
Revised: 26 January 2023
Accepted: 26 January 2023
Published: 1 February 2023

1. Introduction

Numerous articles about microbiota–host interactions published in the past two decades have confirmed that the gut microbiota has a significant impact on host homeostasis. There is interest in ways to correct the imbalance in the microbiota to treat intestinal inflammation [1].

The pathogenesis of inflammatory bowel diseases (IBD) related to microbiota occurs in both animals and humans. The microbiota plays important roles in the pathophysiology of IBD by impacting the immune system, the host metabolism and gastrointestinal function [1–3].

The treatment of IBD involves restoring the microbiota by first decreasing the amount of existing pathobionts and then delivering probiotics, therapeutic agents, or microbiota of from a healthy donor. There are two ways to deliver bacteria in IBD patients: transoral delivery and fecal transplantation. Both ways have their drawbacks. For transoral delivery, the beneficial bacteria need to be protected from damage by the acidic environment of the stomach [4,5].

The *Bacillus* species are probiotic bacteria that support a healthy gut microbiome and promote digestion and immunity. They include *B. subtilis*, *B. coagulans*, *B. licheniformis* and *B. clausii* [6–8]. However, the lowest favorable pH for the *B. subtilis* strains is 5.0; furthermore, this bacterium prefers neutral pH environments [9–11]. An optimal pH is crucial in order for the bacteria to survive during the stomach passage, which has an aggressively acidic pH, until it reaches the treatment site.

Different delivery systems for bacterial safety are used to avoid their destruction in acidic pHs, including biodegradable gel carriers based on alginate, chitosan or gelatin [12–15]. At the same time, the carrier should provide not only protection for the bacteria from the gastric acid, but also should be able to effectively release it when it reaches the alkaline pH of the intestine.

Despite there being a wide selection of delivery systems for all proven probiotics (*Lactobacillus, Bifidobacterium, Bacillus, Streptococcus* and others), most studies are still at the in vitro stage. Carriers require bacteria delivery capabilities to be tested in vivo [15].

It is known that sulfated polysaccharides are ubiquitous in living systems. These polysaccharides are involved in such biological functions as organism development, cell proliferation and differentiation, tissue homeostasis and host defense. Engineered sulfated polysaccharides are structural derivatives not found in nature. They are generated through the chemical and enzymatic modification of natural polysaccharides, as well as chemically synthesized oligo- and polysaccharides. Engineered sulfated polysaccharides exhibit novel and augmented biological properties compared with their unmodified counterparts, mainly through facilitating interactions with other macromolecules [16]. However, data on the usage of carrier gels made from such modified sulfated chitosan in living systems to deliver bacteria into the gut are scarce.

In this paper, we present a short communication of our pilot study, which aimed to identify the properties of a bioresorbable sodium N-(2-sulfonatoethyl)chitosan (SEC) gel crosslinked with glutaraldehyde as a potential carrier for the oral delivery of probiotics through the gastrointestinal tract in an animal model using *B. subtilis 20* as a test strain.

2. Results

2.1. Release of Dye from the Gel in Acidic and Alkaline Media

In an acidic medium (pH = 2.2), the dye was not released from the SEC gel. The level of dye was constantly in a range of 0.098–0.106 optical density units (ODU). However, the dye was successfully released from the SEC gel in the alkaline medium (pH = 7.4). There was a more effective release in the first 15 min, which amounted to 1.008 ODU, followed by a slow release of dye until reaching the lower limit of detection: 0.595 ODU at 30 min, 0.321 ODU at 45 min, 0.181 ODU at 60 min and 0.103 at 105 min (Figure 1).

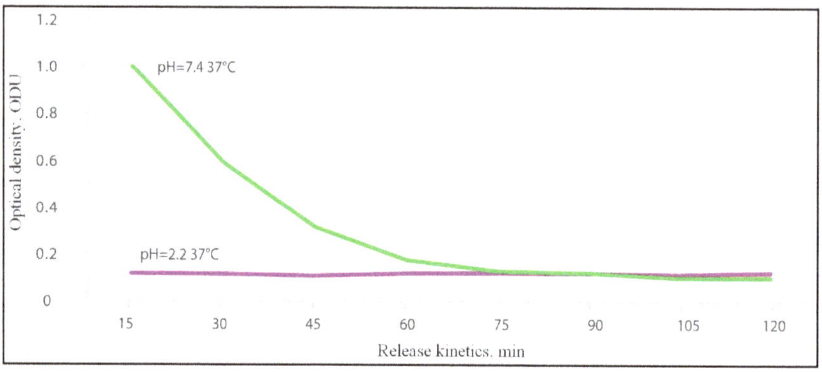

Figure 1. Kinetics of dye release from the sodium N-(2-sulfonatoethyl)chitosan gel in two different pH mediums at 37 °C.

2.2. Cell Viability/Cytotoxicity of Fibroblast Culture with Gel

Microscopy of the cell culture showed no signs of cytotoxicity after incubation with the gel sample: fibroblasts had a typical morphology and proliferated and formed a monolayer for three days. There were no rounded cells that are specific for dead cells (Figure 2A).

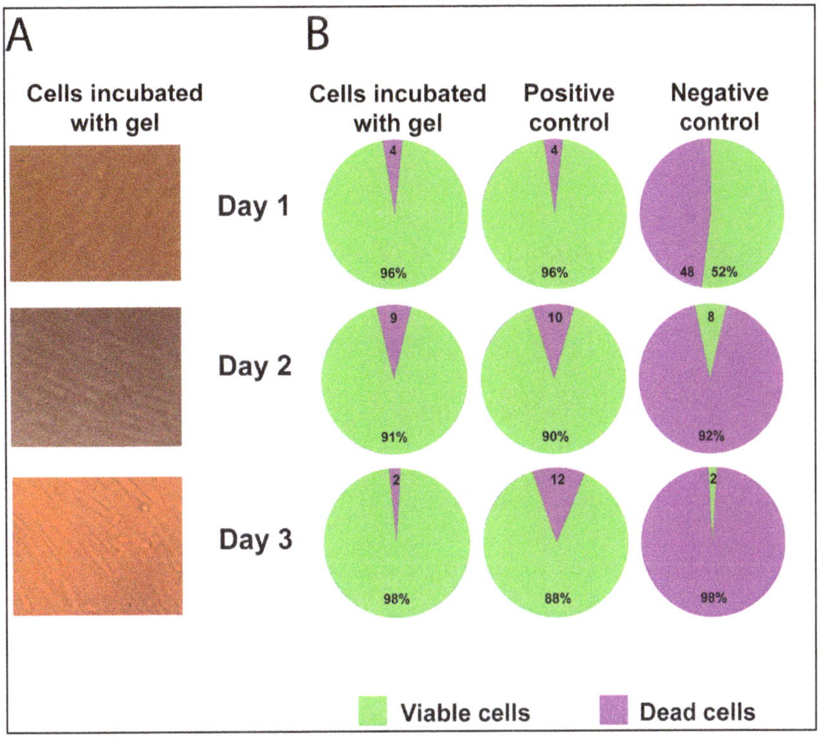

Figure 2. Cell viability/cytotoxicity of fibroblasts, observed during three days of culture with the gel. (**A**). Cell culture microscopy images. (**B**). Live/dead cells ratio (flow cytometry with Calcein-AM and 7-AAD dyes).

The percent of viable cells on Day 1 did not differ from the positive control—it was 96% in both groups (Fisher exact test = 1.0). The percent of viable cells in the negative control (with DMSO) was two times lower than in the positive control (48% vs. 96%) (Chi-square = 57.1; $p < 0.00001$).

On Day 2, 91% of the cells incubated with gel were alive, and 90% were still living in the positive control, which was almost equal to the percentage on Day 1 (Chi-square = 2.1 and Chi-square = 2.8, $p > 0.05$, respectively). However, in the negative control, the number of live cells on Day 2 was reduced drastically to 8% (Chi-square = 39.7; $p < 0.0001$).

On Day 3, there was no significant decrease in the number of live cells compared to Day 2 in the positive control—88% (Chi-square = 0.2; $p > 0.05$). In the negative control, the number of live cells decreased by four times to 2% (Chi-square = 3.8; $p = 0.05$). In the cells incubate with gel, after three days, the number of live cells was increased to 98% in comparison with Day 2's values (Chi-square = 4.7; $p = 0.03$).

2.3. Test Strain of *B. subtillus 20* Transit in the Gastrointestinal Tract of the Mice

Microbiological analysis of the feces of mice fed a standard diet showed the absence of the test strain *B. subtillus 20* on Day 0. After five days of administering the gel loaded with test strain *B. subtillus 20* during feeding, the number of bacteria increased, and the number was only two times less than in the loaded gel (baseline titer). After the gel additive was withdrawn on Day 5, the number of bacteria decreased by 10 and 40 times on Days 6 and 7, respectively (Table 1).

Table 1. Titer of test strain *B. subtilis 20* in mice feces after gel carrier's passage through the gastrointestinal tract.

Days of the Experiment	Number of Mice	Titer Value, CFU/g
Bacteria-loaded gel carrier (baseline titer)	1	8.75×10^8
	2	8.75×10^8
Day 0 (before test strain feeding)	1	0.0
	2	0.0
Day 5 after feeding every day with gel carrier additive (end of test strain feeding)	1	4.65×10^8
	2	3.75×10^8
Day 6 (feeding with standard ration only)	1	0.43×10^8
	2	0.89×10^8
Day 7 (feeding with standard ration only)	1	0.10×10^8
	2	0.09×10^8

At the same time, the titer of the test strain *B. subtilis 20* in the feces of mice fed a diet with an additive suspension of *B. subtilis 20* (concentration of 10^9 spores/g of diet) decreased by more than 200 times during feeding.

3. Discussion

The use of synthetic sulfated derivatives of chitosan as carriers has been discussed in various articles over the last five years. Basically, these materials are described as sorbents for divalent metal ions with controlled sorption characteristics [17–21]. N-Sulfonatoethylation decreases the sorption ability of chitosan but expands the sorption pH and shifts the optimal pH value toward higher acidities. Such materials exhibit higher sorption rates than standard synthetic ion exchangers do. The degree of sulfonatoethylation of the material changes its sorption ability [22,23].

Artificial gastric juice in environmental models of the stomach can show the impact of gastric juice on probiotic bacteria. The viability of probiotic bacteria is reduced in such stomach models, e.g., by 8.5 logs for *L. casei* and by 11 logs for *B. bifidum* after 2 h [24]. The proteolysis of bacteria by pepsin in the stomach also has an effect on the bacteria's viability [25].

In this regard, the use of protective shells, as well as transport carriers for the delivery of bacteria into the intestine, seems to be quite justified.

Studies have shown that *B. subtilis* is able to self-coats with biofilm, which could prolong its retention time in the intestine to compared with uncoated *B. subtilis* [26–28].

Furthermore, a lipid-membrane-encapsulated oral delivery system for probiotics was constructed by Cao Z. et al. as an IBD treatment [29]. This system has demonstrated that the delivered bacteria could reside in the intestine for 4 h after oral administration. Five days of treatment with lipid-membrane-coated bacteria significantly alleviated the damage to the colonic mucosa and epithelium and the infiltration of immune cells in a murine colitis model [29].

Our results showed that oral delivery using an SEC gel carrier prolonged the residence of test strain bacteria in the animal intestine. The delivery of *B. subtilis 20* using a gel carrier decreased the amount of the test strain in the gut by only two times, compared with a gavage of bacterial suspension that decreased the amount of *B. subtilis* by 200 times. Similar results were established by Khosravi Zanjani MA et al. [30] and Krasaekoopt et al. [31];

they showed that encapsulation in chitosan coatings considerably increased the survival rates of the probiotics.

Despite the *B. subtilis* used in our pilot experiment being stable in acidic environments, it has been shown that the safety of the bacteria of the test strain was increased 100 times as a result of being carried by the SEC gel. A certain contribution to the result could be explained by more effective bacteria adhesion to the intestinal mucosal surface due to prolonged intestinal transit in the gel carrier and a reduced rate of the elimination of probiotic bacteria from the intestine. This gel may be applied in the delivery of other probiotic bacteria or phages that have worse abilities than *B. subtilis* to withstand aggressive external influences, such as hydrochloric acid and gastrointestinal enzymes in the gastrointestinal tract.

The possibility of using gel for the delivery of bacteria in the gastrointestinal tract is also confirmed by our results of in vitro cytotoxicity. In vitro, the gel showed no toxic effect on the cell culture, and when it was used to feed mice we did not observe any adverse reactions in the animals, be they behavioral or allergic.

Our results in vitro of the dye release rate from the gel in an alkaline medium showed that gel demonstrates the first order kinetics, with an initial rapid dose of dye, followed by a slow release and gel degradation in an environment similar to the intestine. The absence of toxic effects on experimental animals after consuming the gel confirmed the safety of its use, since gel degradation and utilization do not lead to the manifestation of intoxication in mice.

In the pilot study, we did not perform a histological assessment of the mice's tissue after eating the gel, but there others have studied the biocompatibility of sulfated chitosans in vivo. A reparative effect was shown for O,N-(2-sulfoethyl)chitosan on the blood vessel walls of rabbits [32].

Chitosan gel may, therefore, have a therapeutic effect when used as a carrier for bacteria to repair the inflamed intestine in IBD. The results obtained by Ding, K. et al. show that 6-O-sulfated chitosan can act as a supplement aiding cell differentiation [33].

Thus, our results confirmed the possibility of using sodium N-(2-sulfonatoethyl)chitosan gel for the delivery of bacteria into the gastrointestinal tract through the acidic environment in the stomach and for reducing the rate of the elimination of probiotic bacteria from the intestine. The gel was also shown to be non-toxic, and could be used in animals.

4. Materials and Methods

4.1. *Bioresorbable Gel Carrier*

Samples of sodium N-(2-sulfonatoethyl)chitosan gel crosslinked with glutaraldehyde were synthesized using a method described previously [34]. The composition and structure of the gel were confirmed by elemental analysis using "CHN PE 2400" (Perkin Elmer, Waltham, MA, USA), and FTIR-spectroscopy using a "Spectrum Two" spectrometer (Perkin Elmer, Waltham, MA, USA) (see Figure 3).

All gel samples were sterilized before the experiment, using electron beam sterilization with an absorbed dose 25 kGy [35].

4.2. *Gel Loading Procedure*

Dry sterile gel samples (0.5 g each) under sterile conditions were loaded with a sterile solution of bacteria in a laminar flow hood. Samples were placed in Petri dishes and filled with a PBS solution containing *B. subtilis 20* bacteria in the titer indicated in Table 1. Next, the dishes were stored at +4 °C for 24 h, and then swollen samples were cut into pieces 0.5 × 0.5 cm in size the next day. Each cut sample was transferred into a separate well of a 24-well plate.

Then, all bacteria-loaded gel samples were encapsulated in 1% agar (to prevent gel desiccation in air and as a taste modifier for animal feeding) at +37 °C into the same plate wells. After solidification, the resulting gel–agar complexes were removed from the wells (Figure 4), placed in portions (one portion per one feeding) and stored at +4 °C before use.

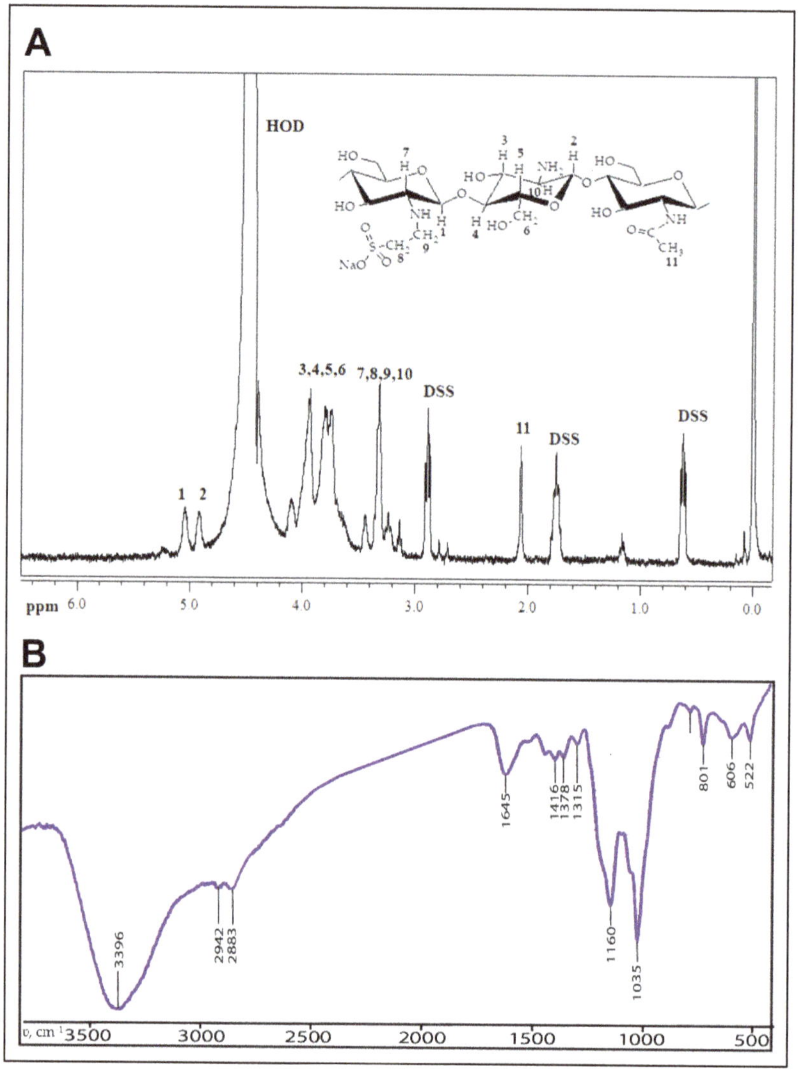

Figure 3. (**A**). The structural units and the 400 MHz ^{1}H NMR spectrum of sodium N-(2-sulfonatoethyl)chitosan with degrees of substitution 1.0 (D_2O/DCl) (DSS–2,2-dimethyl-2-silapentane-5-sulfonate sodium salt was used as an internal ^{1}H NMR standard; ppm—parts per million); (**B**). FT-IR spectrum of crosslinked N-(2-sulfoethyl)chitosan with glutaraldehyde.

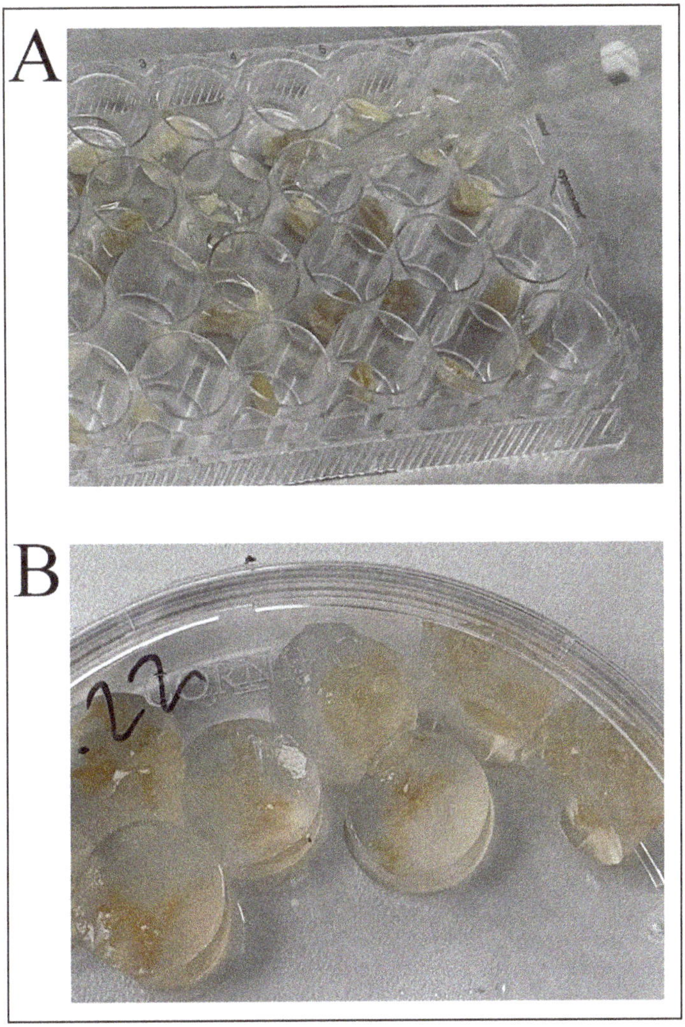

Figure 4. Overview of bacteria-loaded gel samples encapsulated in agar. (**A**). Preparation of gel–agar complexes; (**B**). Final solidified gel–agar complexes.

4.3. Animals

Four 14–16 week-old adult CD1 outbred female mice (with confirmed SPF status according to [36]) were housed in individually ventilated cages in a 12/12 h light/dark photoperiod under standard conditions. Food and water were provided ad libitum. Animals were divided into two groups (study and control) in equal proportions, based on feed additives; the study group (n = 2) was fed with standard ration with the test strain-loaded gel carrier samples added once per day for five days, and the control group (n = 2) was fed with standard ration with the test strain bacteria solution added using a gavage.

All the procedures were conducted according to the 3R principle [37] (including small sample size for the pilot experiment), standards of Good Laboratory Practice, institutional Ethics Committee guidelines and the European Convention for the Protection of Vertebrate Animals.

4.4. Bacterial Strain and Growth Conditions

The kanamycin-resistant strain *Bacillus subtilis 20* was used to feed the animals. Strain selection for the experiment was substantiated by its ability to grow on selective kanamycin-added medium that allows for precisely calculating CFU for that strain. Bacteria were cultured on plates of Dextrose Casein–peptone agar (Merck, Darmstadt, Germany) at +37 °C for 72–96 h until complete spore formation. Spores of the bacteria were resuspended in sterile phosphate-buffered saline (PBS) solution.

4.5. Method of Characterization of Bacillus subtilis 20 Colony Forming Units

The aliquots of the gel (1 g) or mouse feces (100 mg) were placed into 9.0 or 0.9 mL sterile PBS, respectively, and homogenized using a sterile glass rod. A decimal dilution series was prepared in PBS, and 100 μL of the appropriate dilutions were plated on Dextrose Casein–peptone agar containing 100 mg/mL kanamycin. The plates were incubated at +37 °C for 48 h.

4.6. Gel Release Model in Different pH Mediums

A pH value equal to 2.2 is equivalent to the stomach pH, and for the intestine pH environment the pH value is equal to 7.4 [38]. The corresponding pH buffers were prepared based on the protocol in [39]. Experimental setup was mounted from polypropylene containers (5 mL), made from 50 mL tubes, and gel placed into the buffer with regard to pH. To maintain the flow around the gel and the required temperature (37 °C), an ImmunoChem-2200 plate shaker-incubator (HTI, North Attleboro, MA, USA) was used.

To detect the release of dye from the gels, spectrophotometry was used. The dye's (acridoneacetic acid) (Polisan, St. Petersburg, Russia) kinetic release from the gel carrier was visualized based on changes in the optical density of the solution at 355 nm [40].

After loading the gels with dye, the gel samples were washed once in a 1× phosphate-buffered saline (PBS) and placed in the containers (5 mL) with an appropriate pH buffer.

Gel samples prepared for dye release were placed in an appropriate (alkaline or acidic) pH buffer (n = 6 for each medium) and stirred for 2 h at 37 °C.

Every 15 min, 200 μL aliquots of buffer from all containers were sampled, and gels were placed in a new portion of buffer. The optical density of each aliquot was analyzed in black 96-well plates with a translucent bottom on a TriStar LB 941 spectrophotometer (Berthold Technologies, Bad Wildbad, Germany).

4.7. Cytotoxicity Evaluation

The effect of the gel on the viability of human fibroblasts was evaluated after 24, 48 and 72 h of incubation with cells (confluent monolayer) in 12-well plates (TPP, Switzerland). Cells were seeded in plates (at least 6×10^4 cells per cm^2) using the medium, which consisted of DMEM (Gibco, Carlsbad, CA, USA) with 10% fetal serum (StemCell Technologies, Vancouver, Canada), 2 mM L-glutamine (StemCell Technologies, Canada) and combined antibiotics of 100 IU/mL of penicillin and 100 μg/mL of streptomycin (Biolot, St. Petersburg, Russia). The plates were incubated in a CO_2 incubator at +37 °C in air with 5% CO_2. After reaching the confluent monolayer, sterile gel samples (approximately 3 mm^3 in size) were placed on top of the cell monolayer (study group, n = 4). Cultivation of samples with cells was carried out in three replicates.

Fibroblasts cultured without gel samples and any other medium components were used as a positive control (n = 7). Fibroblasts cultured in addition to 10% DMSO (Biolot, St. Petersburg, Russia) were used as negative controls (n = 4). At the end of the incubation period, the samples were removed and the cells were washed with Dulbecco's phosphate–saline solution (DPBS) (Gibco, Carlsbad, CA, USA). After that, the cells were detached with a 1:1 trypsin: Versen solution (Biolot, St. Petersburg, Russia) and centrifuged for 10 min at 1100 rpm. The pellet was resuspended in the complete medium. Next, the cells were centrifuged for 5 min at 1100 rpm, then washed with 1 mL of PBS. After that, the cell pellet was added to 500 μL of DPBS (Gibco, Carlsbad, CA, USA) containing 100 nM of fluorescent dye Calcein AM Viability Dye (eBioscience Invitrogen, San Diego, CA, USA) for visualization of living cells, followed by incubation for 15 min in the dark at room temperature. Then, 5 μL of fluorescent dye 7-AAD Viability Staining Solution (eBioscience Invitrogen, San Diego, CA, USA) was added to the solution to visualize dead cells and stained for 5 min in the dark at room temperature. For detection and quantitative analysis of living and dead cells, samples stained with the above dyes were analyzed on a FACS Canto II flow cytometer (Becton Dickinson, Franklin Lakes, NJ, USA) using FACS Diva 6.0 software (Becton Dickinson, Franklin Lakes, NJ, USA).

The cell culture was also examined using inverted light microscopy of culture plates to visualize cell growth conditions.

4.8. Statistics

Statistical analysis was performed using the IBM SPSS Statistics software (version 25.0) and R software (version 4.1.2). For frequency variables, proportions and confidence intervals were calculated using the method of Wilson E.B. Comparisons were performed by the Pearson Chi-square method or Fisher's exact method, if applicable (with Bonferroni correction for multiple comparisons). Significance levels are indicated either as absolute values or (in the case of exponential values) as $p < 0.0001$.

Author Contributions: Conceptualization, E.L., A.S. and A.K.; methodology, E.L., A.K., G.K., E.B. and A.P.; software, A.S.; validation, A.S., G.K. and E.L.; formal analysis, G.K., N.A., A.S. and E.B.; investigation, E.L., A.K., G.K., E.B., A.S., N.A. and V.B.; resources, E.L., A.K., V.T., A.P. and N.A.; data curation, A.S., G.K. and E.L.; writing—original draft preparation, A.S.; writing—review and editing, A.S, E.L.; visualization, A.S. and A.K.; supervision, E.L.; project administration, E.L.; funding acquisition, E.L. All authors have read and agreed to the published version of the manuscript.

Funding: This research was supported by the "Priority-2030" program in Novosibirsk State Technical University. The work with mice (acquisition and maintenance) was supported by budgetary funding for basic scientific research (grant number 122042700001-9) to E.L.

Institutional Review Board Statement: The animal study protocol was approved by the Ethics Committee of the Scientific Research Institute of Neurosciences and Medicine (protocol #3 dated 03.11.2021) for studies involving animals.

Informed Consent Statement: Not applicable.

Data Availability Statement: Not applicable.

Conflicts of Interest: The authors declare no conflict of interest.

References

1. Dixit, K.; Chaudhari, D.; Dhotre, D.; Shouche, Y.; Saroj, S. Restoration of Dysbiotic Human Gut Microbiome for Homeostasis. *Life Sci.* **2021**, *278*, 119622. [CrossRef] [PubMed]
2. Richard, M.L.; Sokol, H. The Gut Mycobiota: Insights into Analysis, Environmental Interactions and Role in Gastrointestinal Diseases. *Nat. Rev. Gastroenterol. Hepatol.* **2019**, *16*, 331–345. [CrossRef]
3. Guan, Q. A Comprehensive Review and Update on the Pathogenesis of Inflammatory Bowel Disease. *J. Immunol. Res.* **2019**, *2019*, 7247238. [CrossRef]
4. Ding, W.K.; Shah, N.P. Effect of Various Encapsulating Materials on the Stability of Probiotic Bacteria. *J. Food Sci.* **2009**, *74*, M100–M107. [CrossRef] [PubMed]
5. Weingarden, A.R.; Vaughn, B.P. Intestinal Microbiota, Fecal Microbiota Transplantation, and Inflammatory Bowel Disease. *Gut Microbes* **2017**, *8*, 238–252. [CrossRef]
6. Maher, M. Tolerance and Effect of a Probiotic Supplement Delivered in Capsule Form. *FNS* **2019**, *10*, 626–634. [CrossRef]
7. Cuentas, A.M.; Deaton, J.; Khan, S.; Davidson, J.; Ardita, C. The Effect of Bacillus Subtilis DE111 on the Daily Bowel Movement Profile for People with Occasional Gastrointestinal Irregularity. *J. Prob. Health* **2017**, *5*, 4. [CrossRef]
8. Paytuví-Gallart, A.; Sanseverino, W.; Winger, A.M. Daily Intake of Probiotic Strain Bacillus Subtilis DE111 Supports a Healthy Microbiome in Children Attending Day-Care. *Benef. Microbes* **2020**, *11*, 611–620. [CrossRef]
9. Tohidifar, P.; Plutz, M.J.; Ordal, G.W.; Rao, C.V. The Mechanism of Bidirectional PH Taxis in Bacillus Subtilis. *J. Bacteriol.* **2020**, *202*, e00491-19. [CrossRef]
10. Gauvry, E.; Mathot, A.-G.; Couvert, O.; Leguérinel, I.; Coroller, L. Effects of Temperature, PH and Water Activity on the Growth and the Sporulation Abilities of Bacillus Subtilis BSB1. *Int. J. Food Microbiol.* **2021**, *337*, 108915. [CrossRef]

11. Jiménez-Delgadillo, R.; Valdés-Rodríguez, S.E.; Olalde-Portugal, V.; Abraham-Juárez, R.; García-Hernández, J.L. Efecto Del PH y Temperatura Sobre El Crecimiento y Actividad Antagónica de Bacillus Subtilis Sobre Rhizoctonia Solani. *RMF* **2018**, *36*, 256–275. [CrossRef]
12. Mortazavian, A.; Razavi, S.; Ehsani, M. Principles and Methods of Microencapsulation of Probiotic Microorganisms. *Iran J. Microbiol.* **2007**, *5*, 1–18. Available online: http://www.ijbiotech.com/article_7032.html. (accessed on 10 December 2022).
13. Yao, M.; Wu, J.; Li, B. Microencapsulation of Lactobacillus Salivarious Li01 for Enhanced Storage Viability and Targeted Delivery to Gut Microbiota. *Food Hydrocoll.* **2017**, *72*, 228–236. [CrossRef]
14. Silva, K.; Cezarino, E.; Michelon, M. Symbiotic Microencapsulation to Enhance Lactobacillus Acidophilus Survival. *LWT-Food Sci. Technol.* **2018**, *89*, 503–509. [CrossRef]
15. Liu, J.; Li, X.; Zhang, X.; Wang, R.; Han, S.; Qin, P.; Xing, X.-H.; Zhang, C. Gut Lumen-Targeted Oral Delivery System for Bioactive Agents to Regulate Gut Microbiome. *J. Future Foods* **2022**, *2*, 307–325. [CrossRef]
16. Arlov, Ø.; Rütsche, D.; Asadi Korayem, M.; Öztürk, E.; Zenobi-Wong, M. Engineered Sulfated Polysaccharides for Biomedical Applications. *Adv. Funct. Mater.* **2021**, *31*, 2010732. [CrossRef]
17. Petrova, Y.S.; Bukharova, A.V.; Neudachina, L.K.; Adamova, L.V.; Koryakova, O.V.; Pestov, A.V. Chemical Properties of N-2-Sulfoethylchitosan with a Medium Degree of Substitution. *Polym. Sci. Ser. B* **2014**, *56*, 487–493. [CrossRef]
18. Kapitanova, E.I.; Ibragimova, A.A.; Petrova, Y.S.; Pestov, A.V.; Neudachina, L.K. Influence of the Degree of Chitosan Sulfoethylation on the Sorption of Palladium(II) Chloride Complexes from Multicomponent Solutions. *Russ. J. Appl. Chem.* **2018**, *91*, 297–303. [CrossRef]
19. Petrova, Y.S.; Neudachina, L.K.; Oseeva, M.Y.; Pestov, A.V. Effect of Complex-Former Ion Concentration on the Selectivity of Metal Ion Sorption on Cross-Linked N-2-Sulfoethylchitosan. *Russ. J. Inorg. Chem.* **2018**, *63*, 400–405. [CrossRef]
20. Petrova, Y.S.; Pestov, A.V.; Alifkhanova, L.M.K.; Neudachina, L.K. Effect of the Degree of Cross-Linking of N-2-Sulfoethylchitosan on the Sorption Selectivity of Copper(II) and Silver(I). *Russ. J. Appl. Chem.* **2015**, *88*, 1434–1439. [CrossRef]
21. Petrova, Y.S.; Kapitanova, E.I.; Neudachina, L.K.; Pestov, A.V. Sorption Isotherms of Metal Ions onto an N-(2-Sulfoethyl)Chitosan-Based Material from Single- and Multi-Component Solutions. *Sep. Sci. Technol.* **2017**, *52*, 2385–2394. [CrossRef]
22. Evdokimova, O.V.; Pestov, A.V.; Pechishcheva, N.V.; Shunyaev, K.Y. Preparation of Sorbents Based on Chitosan and Its Sulfoethyl Derivative for Sorption of Perrhenate Ions. *Russ. J. Appl. Chem.* **2016**, *89*, 1377–1382. [CrossRef]
23. Petrova, Y.S.; Pestov, A.V.; Usoltseva, M.K.; Kapitanova, E.I.; Neudachina, L.K. Methods for Correction of Selectivity of *N* -(2-Sulfoethyl)Chitosan-Based Materials towards Platinum(IV) and Palladium(II) Ions. *Sep. Sci. Technol.* **2019**, *54*, 42–50. [CrossRef]
24. Khosravi Zanjani, M.A.; Ghiassi Tarzi, B.; Sharifan, A.; Mohammadi, N. Microencapsulation of Probiotics by Calcium Alginate-Gelatinized Starch with Chitosan Coating and Evaluation of Survival in Simulated Human Gastro-Intestinal Condition. *Iran J. Pharm. Res.* **2014**, *13*, 843–852. [PubMed]
25. Zhu, H.; Hart, C.A.; Sales, D.; Roberts, N.B. Bacterial Killing in Gastric Juice–Effect of PH and Pepsin on Escherichia Coli and Helicobacter Pylori. *J. Med. Microbiol.* **2006**, *55*, 1265–1270. [CrossRef]
26. Higgins, D.; Dworkin, J. Recent Progress in *Bacillus Subtilis* Sporulation. *FEMS Microbiol. Rev.* **2012**, *36*, 131–148. [CrossRef]
27. Wang, X.; Cao, Z.; Zhang, M.; Meng, L.; Ming, Z.; Liu, J. Bioinspired Oral Delivery of Gut Microbiota by Self-Coating with Biofilms. *Sci. Adv.* **2020**, *6*, eabb1952. [CrossRef] [PubMed]
28. Tallawi, M.; Opitz, M.; Lieleg, O. Modulation of the Mechanical Properties of Bacterial Biofilms in Response to Environmental Challenges. *Biomater. Sci.* **2017**, *5*, 887–900. [CrossRef] [PubMed]
29. Cao, Z.; Wang, X.; Pang, Y.; Cheng, S.; Liu, J. Biointerfacial Self-Assembly Generates Lipid Membrane Coated Bacteria for Enhanced Oral Delivery and Treatment. *Nat. Commun.* **2019**, *10*, 5783. [CrossRef] [PubMed]
30. Khosravi Zanjani, M.A.; Ehsani, M.R.; Ghiassi Tarzi, B.; Sharifan, A. Promoting Probiotics Survival by Microencapsualtion with Hylon Starch and Genipin Cross-Linked Coatings in Simulated Gastro-Intestinal Condition and Heat Treatment. *Iran J. Pharm. Res.* **2018**, *17*, 753–766. [PubMed]
31. Krasaekoopt, W.; Bhandari, B.; Deeth, H. The Influence of Coating Materials on Some Properties of Alginate Beads and Survivability of Microencapsulated Probiotic Bacteria. *Int. Dairy J.* **2004**, *14*, 737–743. [CrossRef]
32. Bolshakov, I.N.; Gornostaev, L.M.; Fominykh, O.I.; Svetlakov, A.V. Synthesis, Chemical and Biomedical Aspects of the Use of Sulfated Chitosan. *Polymers* **2022**, *14*, 3431. [CrossRef]
33. Ding, K.; Wang, Y.; Wang, H.; Yuan, L.; Tan, M.; Shi, X.; Lyu, Z.; Liu, Y.; Chen, H. 6- *O* -Sulfated Chitosan Promoting the Neural Differentiation of Mouse Embryonic Stem Cells. *ACS Appl. Mater. Interfaces* **2014**, *6*, 20043–20050. [CrossRef] [PubMed]
34. Pestov, A.V.; Petrova, Y.S.; Bukharova, A.V.; Neudachina, L.K.; Koryakova, O.V.; Matochkina, E.G.; Kodess, M.I.; Yatluk, Y.G. Synthesis in a Gel and Sorption Properties of N-2-Sulfoethyl Chitosan. *Russ. J. Appl. Chem.* **2013**, *86*, 269–272. [CrossRef]
35. Samokhin, A.G.; Tkachenko, V.O.; Kuznetsov, V.A.; Zemlyakova, E.O.; Nesterov, D.V.; Larionov, P.M.; Pestov, A.V. Selection of an Optimal Method for Sterilization of the Medical Grade Biodegradable Polymers. In Proceedings of the "2nd International Conference Modern Synthetic Methodologies for Creating Drugs and Functional Materials (MOSM2018)" (15–17 November 2018), published in AIP Conference Proceedings, Kuantan, Malaysia, 11 January 2019; p. 030017. [CrossRef]
36. FELASA Working Group on Revision of Guidelines for Health Monitoring of Rodents and Rabbits; Mähler, M.; Berard, M.; Feinstein, R.; Gallagher, A.; Illgen-Wilcke, B.; Pritchett-Corning, K.; Raspa, M. FELASA Recommendations for the Health Monitoring of Mouse, Rat, Hamster, Guinea Pig and Rabbit Colonies in Breeding and Experimental Units. *Lab. Anim.* **2014**, *48*, 178–192. [CrossRef]

37. Hubrecht, R.C.; Carter, E. The 3Rs and Humane Experimental Technique: Implementing Change. *Animals* **2019**, *9*, 754. [CrossRef] [PubMed]
38. Pokrovskii, V.M.; Korot'ko, G.F. *Human Physiology: Text-book*, 3rd ed.; Revised and Complete; JSC Publishing House "Medicina": Moscow, Russia, 2011; p. 656, ISBN 5-225-04729-7.
39. McIlvaine, T.C. A Buffer Solution For Colorimetric Comparison. *J. Biol. Chem.* **1921**, *49*, 183–186. [CrossRef]
40. Starostenko, A.A.; Zaikin, P.A.; Troitskij, A.V.; Bystrova, T.N. Method of Producing Oxidised Dextran, Suitable For Its Visualization In Blood Serum. Abstract of Invention RU2690380. Bulletin No. 16. 3 June 2019. Available online: https://yandex.ru/patents/doc/RU2690380C1_20190603 (accessed on 10 December 2022).

gels

MDPI

Review

State of the Art of Hydrogel Wound Dressings Developed by Ionizing Radiation

Maria Demeter, Anca Scărișoreanu * and Ion Călina *

National Institute for Lasers Plasma and Radiation Physics, 409 Atomiștilor, 077125 Măgurele, Romania
* Correspondence: anca.scarisoreanu@inflpr.ro (A.S.); calina.cosmin@inflpr.ro (I.C.)

Abstract: The development of an ideal hydrogel wound dressing with excellent characteristics is currently a significant demand in wound therapy. The ideal hydrogel wound dressing must provide a moist environment between the wound and the dressing, promote wound healing, absorb excess exudate and toxins, be completely sterile, and not adhere to the wound. The evolution and current status of research on hydrogel wound dressings obtained exclusively through production by ionizing radiation are discussed in this paper review, along with the preparation methods, properties, standard characterization techniques, and their applications in wound dressing. First, we described the methods for synthesizing hydrogel wound dressings with ionizing radiation. Then, standard methods of characterization of hydrogel wound dressings such as gel fraction, swelling degree, sol–gel analysis, rheological properties, morphology, moisture retention capability, and water vapor transmission rate have been investigated. In the end, specific attention was paid to the drug release, antibacterial performance, and cytotoxicity of hydrogels. Moreover, the application of hydrogel in regenerative medicine as wound healing dressing was covered.

Keywords: hydrogel; ionizing radiation; wound dressings; crosslinking; healing

check for updates

Citation: Demeter, M.; Scărișoreanu, A.; Călina, I. State of the Art of Hydrogel Wound Dressings Developed by Ionizing Radiation. *Gels* **2023**, *9*, 55. https://doi.org/10.3390/gels9010055

Academic Editor: Shige Wang

Received: 19 December 2022
Revised: 5 January 2023
Accepted: 9 January 2023
Published: 10 January 2023

1. Introduction

The largest organ of the human body is the skin, which represents the main defence barrier against various pathogens commonly encountered in the environment. Wounds on the skin break these barriers and form entry gates for external pathogens [1]. To overcome the exposure of the wounds to the external environment, it is recommended to immediately cover them with suitable dressings depending on the severity of the wound. Dry dressings generally used in medical practice provide only physical protection, they do not help the healing of the wound.

In recent decades, hydrogels based on natural/synthetic polymers have become a modern dressing alternative for wound care, due to the excellent properties of polymers that absorb a large volume of water through the hydrophilic functional groups. Hydrogels derived from polymers can be obtained through several methods that ensure the polymer's crosslinking. Physical crosslinking can occur through ionic interactions [2], hydrogen bonds [3], crystallization [4], hydrophobic interaction [5], or protein interaction [6]. Chemical crosslinking methods include free radical polymerization [7] and enzymatic reaction [8]. Since physical methods do not ensure homogeneous crosslinking in the hydrogel structure, and in the case of chemical crosslinking, the inclusion of crosslinking agents leads to a decrease in the biocompatibility of the hydrogel, the best crosslinking method that removes the disadvantages of the methods above mentioned is radiation crosslinking.

The ionizing radiation technology applied to hydrogel synthesis was first reported in 1958 when γ-radiation was used for PVA crosslinking [9]. In this period, the first studies on the crosslinking of poly(vinyl alcohol) (PVA) and poly(N-vinyl pyrrolidone) (PVP) by irradiating aqueous solutions with γ-radiation were also achieved [10,11]. Charlesby and Alexander were the first researchers who reported the crosslinking of PVP with ionizing radiation. Over the years, various other water-soluble polymers have been treated or crosslinked with ionizing radiation, and this treatment is often also a step in the preparation of new polymeric biomaterials. Thus, the radiation crosslinking technique has become one of the standard methods for obtaining hydrogels [12].

In the following decades, the study of hydrogels remained focused on networks with a relatively simple structure obtained by chemical crosslinking of synthetic polymers, with applications in medicine (ophthalmology and controlled drug release). Hydrogels with a simple network structure offered the possibility of characterizing some of their physicochemical properties, such as crosslinking density and the elucidation of diffusion processes.

The synthesis of hydrogels for biomedical applications by irradiation with ionizing radiation has seen constant development since the late 1960s, as mentioned in the papers and patents published by Sakurada and Ikada [13], Kaetsu [14,15], and Hoffman [16].

A substantial contribution in this field was made by Professor Janusz M. Rosiak and his collaborators from the Technical University of Lodz, Poland, the main inventor of the technology for obtaining hydrogel dressings by irradiation. In the 1980s, Rosiak et al. obtained PAAM hydrogels by crosslinking with γ-radiation. The obtained hydrogel was intended to be used as a dressing for skin drug release [17].

Later, in 1989, Janusz M. Rosiak and his collaborators patented the first commercial hydrogel for biomedical applications based on PVP known under the trade name (AquaGel®), currently marketed in Poland and some Central European countries. This hydrogel has bacterial barrier properties eliminating wound superinfection, inhibiting the loss of body fluids, allowing oxygen access to the wound, and, in general, accelerating the healing process. The hydrogels developed by Rosiak consist of water, poly(N-vinyl-2-pyrrolidone) (PVP), poly(ethylene glycol) (PEG), and agar. In the end-use form, they are supplied as transparent sheets with a thickness of a few millimeters, containing over 90% of water [18].

Another aspect related to the production of hydrogels using irradiation as the main production technology is that recently, worldwide, natural and synthetic polymers that are soluble in water, biodegradable, and biocompatible are used more and more, to the detriment of synthetic ones derived from petroleum. The second aspect is related to the promotion of non-polluting technologies and processing with ionizing radiation, especially because irradiation with electron accelerators is a non-polluting technology. Due to the diverse and numerous applications, the number of scientific studies on the preparation of hydrogels by irradiation with ionizing radiation has increased rapidly, especially after the year 2000.

Since 2003, at the Institute of Energy and Nuclear Research (IPEN) in Brazil, there has been a fully automatic production line for the production of dressings in the form of hydrogels commercially named Biogel®. Important contributions were made by Benamer, Darwis et al., who reported obtaining pure and stable PVP hydrogels using γ-radiation, at concentrations of 7–10% PVP and doses of 2–3 kGy, in saturated solutions in the environment of inert Ar or N_2 [19,20]. Can et al. studied the effect of γ-irradiation on aqueous solutions of PVP in the presence of persulfate anions [21]. Ozurek, Dafader et al. studied hydrogels based on PVP/tartaric acid and agar [22,23]. Dergunov [24,25], Sohail [26], and Abd Alla [27] studied hydrogels sensitive to pH variations obtained from PVP/chitosan or PVP/AAc. Gottlieb [28] studied hydrogels obtained from PVP and PVME poly(vinyl methyl ether) sensitive to temperature variations used to obtain controlled drug release systems. Awadallah [29] studied hydrogels obtained by γ-irradiation from PVA/PVP/PAA, used in meniscus replacement treatment.

The most important PVP-based hydrogels obtained using γ-radiation crosslinking for biomedical applications (dressings, controlled drug release systems) were made by Razzak et al. [30]. By adding PVA, they obtained a hydrogel with high resistance to *E. coli*. Himly et al., by adding PEG, obtained elastic, transparent, flexible hydrogels, impermeable to bacteria, which absorb a large amount of water and are easy to remove [31]. Lugao et al., by irradiation of aqueous solutions of PVP (6%) and PEO (3%), obtained hydrogels with increased elasticity [32].

Pekel et al. synthesized biodegradable hydroxypropylmethylcellulose (HPMC) in a paste-like condition using e-beam irradiation and demonstrated that the HPMC with a higher degree of substitution gave a higher gel content at each concentration and irradiation dose above 20 kGy [33]. Rashid et al. evaluated the effect of γ-irradiation on chitosan (CS) samples. They showed that the degree of deacetylation, swelling degree, and antimicrobial activity of CS improved after irradiation [34].

Wang et al., by adding CMC, obtained hydrogels with better mechanical strength than that of pure CMC hydrogel, and with better flexibility and equilibrium swelling than that of pure PVP hydrogel. These hydrogels show a similar moisture retention capacity to that of a commercial hydrogel-based dressing [35]. Park and Nho, by adding Aloe Vera to the hydrogel obtained from PVP and PVA, obtained dressings with superior qualities [36]. Nho and Park, by adding chitosan to the hydrogel obtained from PVP and PVA, manufactured dressings with superior qualities for rapid healing and stopping bleeding [37]. Singh et al., by adding alginate and Ag nanoparticles, obtained a hydrogel with a strong antimicrobial effect [38]. Song et al. fabricated a GEL/alginate/CMC hydrogel by radiation-induced crosslinking with good biocompatibility and accelerated wound healing [39]. El-Rehim et al., by adding PAA, obtained a hydrogel used as a polymer matrix for the release of drugs in the gastric environment where the solubility of PVP is extremely low [40]. Kadlubowski et al., by adding PAA, obtained a hydrogel very sensitive to pH variations [41].

Until now, most of the scientific works related to hydrogels manufactured to be used as wound dressings present mostly studies, which are based either on the development of new innovative products, with multiple functionalities, or are focused on the characterization of newly developed materials, using methods of very complex analysis. Other studies are also focused on the detailed description of the materials used for the manufacture of hydrogels and their properties.

However, we have identified that in the specialized literature, there is a need for a new perspective on the properties a hydrogel must fulfill to be used successfully in wound dressing applications. The current review aims to include the physicochemical and biological properties required for hydrogel dressings for wound healing.

In general, a hydrogel designed for wound dressings must demonstrate a series of physicochemical, biological, and mechanical properties: be insoluble (gel fraction properties), have the power to absorb liquids (be absorbent), be physically and chemically stable in different pH environments, be non-toxic, be biocompatible, biodegradable, be elastic from a mechanical point of view, allow the transmission of gases, keep its moisture, not adhere to the skin, improve the rapid healing of wounds, and be cheap to manufacture [42].

Until 2022, about 157 publications on hydrogel wound dressing obtained by ionizing radiation were published, with a total of 6144 citations in the last 20 years (Figure 1). The most relevant papers were published in polymer science topics as well as in nuclear science technology.

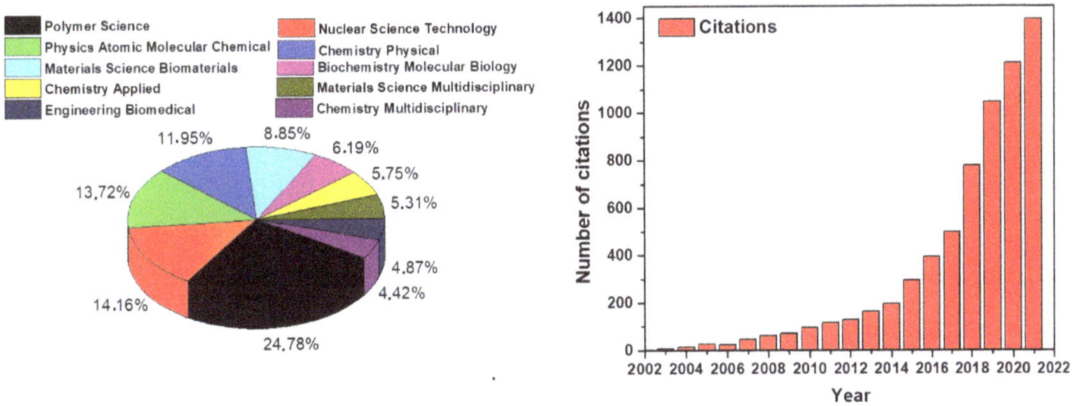

Figure 1. Papers published on hydrogel wound dressings prepared by ionizing radiation in various disciplines and the number of citations per year.

2. Wound Dressing Hydrogel Crosslinking Using Irradiation Technology

High-energy irradiation of aqueous solutions of polymers leads to the formation of radicals on the polymer chain resulting in the scission of the C–H bonds. In addition, the hydroxyl radicals resulting from the radiolysis of water molecules are attached to the polymer chains, forming macroradicals. The crosslinked structure of hydrogels is given by the recombination of macroradicals [43]. Using irradiation technology to crosslink natural/synthetic polymers, hydrogels based on poly(acrylic acid), poly(ethylene glycol), poly(vinyl alcohol), or poly(methyl vinyl ether), were obtained [44–46].

This crosslinking method has several advantages: it is carried out in aqueous solutions, at physiological pH and room temperature, does not require other special conditions, and the irradiated polymers change their physical and chemical properties at the same time during irradiation [47]. Moreover, no chemical initiators are required for hydrogel crosslinking [48]; this is beneficial for biomedical applications because the resulting medical devices are also sterilized during irradiation.

Internationally, we have identified several companies that produce dressings in the form of hydrogels obtained by ionizing radiation. Some of these are produced by e-beam crosslinking and marketed by the KikGel company (Ujazd, Poland), under various trade names: BurnTec®, HydroAid®, and Neoheal®, and their main use is the treatment of burns and various procedures in plastic surgery (http://kikgel.com.pl, accessed on 12 December 2022). Nexgel (Langhorne, PA, USA) produces hydrogel based on PEO, PEG, or PVP by crosslinking with e-beam for wound healing care, cosmetics, and other applications (https://nexgel.com, accessed on 12 December 2022). Alliqua Inc. (Langhorne, PA, USA) produces SilverSeal® and Hydress® products, which are hydrogels for wound dressings made of polymers crosslinked by e-beam and are used for superficial wounds, abrasions, burns, and dermal/leg ulcers (https://www.prnewswire.com/news-releases/alliqua-signs-distribution-agreement-for-silverseal-and-hydress-wound-dressings-228603251.html, accessed on 12 December 2022).

Besides the radiation simultaneously crosslinking and sterilizing in one step, easy control of hydrogel physical properties can be obtained by controlling the absorbed dose and polymer composition. The biomaterials, methods of preparation, and irradiation dose used to develop hydrogel wound dressings are summarized in Table 1.

Table 1. Classification of preparation methods and irradiation dose.

Hydrogel Composition	Radiation Crosslinking Types	Irradiation Dose	References
Agarose	e-beam	5–30 kGy	[49]
BC/AA	e-beam	35–50 kGy	[50]
Collagen	e-beam	5–100 kGy	[51]
CS/PVP/PEG/AA	e-beam	15–25 kGy	[52]
PEO/PVA	e-beam	20–100 kGy	[53]
PVA/AA	e-beam	10–100 kGy	[54]
PVA/PVP/Carr/silk	e-beam	10–100 kGy	[55]
PVP/PEG/Agar	e-beam	20–50 kGy	[56–58]
PVP/PEG/HEMA	e-beam	20 kGy	[59]
PVP/PEG/Starch	e-beam	25 kGy	[60]
AA/PLST/MMT/CS	γ-irradiation	15 kGy	[61]
$AgNO_3$/GEL/CM-CS	γ-irradiation	30 kGy	[62]
AgNP/GEL/PVA	γ-irradiation	30–50 kGy	[63]
AgNPs/PHEMA	γ-irradiation	12 kGy	[64]
AgNPs/PVP	γ-irradiation	25–45 kGy	[65]
CHS/SA	γ-irradiation	25 kGy	[66]
CIP/PLGA/PVA	γ-irradiation	25 kGy	[67]
CM-CS/GEL	γ-irradiation	30 kGy	[68]
CS/GEL/PVA	γ-irradiation	40 kGy	[69]
CS/PVA/TA	γ-irradiation	30 kGy	[70]
DMAEM/PEO/ZnS	γ-irradiation	10–30 kGy	[71]
ELT/PVP	γ-irradiation	25 kGy	[72]
GA/TG//PVA/PVP	γ-irradiation	27.3 kGy	[73]
GEL/γ-PGA	γ-irradiation	3–100 kGy	[74]
GEL/PVP	γ-irradiation	5–30 kGy	[75]
HACC/PVA/PEO	γ-irradiation	20–60 kGy	[76]
Lyophilized collagen	γ-irradiation	10–100 kGy	[77]
MOG/Carbopol	γ-irradiation	13.68–41.04 kGy	[78]
$PAA/ZnCl_2$	γ-irradiation	25–75 kGy	[79]
PEGDA/CS	γ-irradiation	5–25 kGy	[80]
PHEMA/ITA	γ-irradiation	25 kGy	[81]
PVA/Agar/Carr	γ-irradiation	25–30 kGy	[82]
PVA/AgNPs; PVP/AgNPs	γ-irradiation	25 kGy	[83]
PVA/Alginate	γ-irradiation	25–50 kGy	[84]
PVA/CA/GEL/AgNPs	γ-irradiation	15–25 kGy	[85]
PVA/CM-CS/honey	γ-irradiation	25–40 kGy	[86]
PVA/CS	γ-irradiation	20–50 kGy	[87,88]
PVA/CS/Gly	γ-irradiation	40 kGy	[89]
PVA/Gly	γ-irradiation	25 kGy	[90]
PVP/ITA/CS/ZnO	γ-irradiation	30 kGy	[91]
PVA/PVP	γ-irradiation	5–60 kGy	[92]
PVA/PVP/AgNPs	γ-irradiation	15–40 kGy	[93]
PVA/PVP/aloe vera	γ-irradiation	25–50 kGy	[94]
PVA/PVP/clay	γ-irradiation	10–40 kGy	[95]
PVA/PVP/CPC	γ-irradiation	50 kGy	[96]
PVA/PVP/CS	γ-irradiation	15–70 kGy	[97]
PVA/PVP/Gly	γ-irradiation	25 kGy	[98]
PVA/PVP/Gly/CS	γ-irradiation	25–60 kGy	[99]
PVA/ws-CS	γ-irradiation	30 kGy	[100]
PVA/ws-CS/Gly	γ-irradiation	30–70 kGy	[101]
PVP/Alginate/AgNPs	γ-irradiation	25 kGy	[38]
PVP/Carr/AgNPs	γ-irradiation	25 kGy	[102]
PVP/Carr/PEG	γ-irradiation	25 kGy	[103]
PVP/CMC	γ-irradiation	5–60 kGy	[35]
PVP/k-Carr	γ-irradiation	15–35 kGy	[104]
PVP/PEG	γ-irradiation	25–30 kGy	[105]
PVP/PEG/Agar	γ-irradiation	25 kGy	[106]
PVP/PEG/Alkhydin	γ-irradiation	25–75 kGy	[107]
SF/CS/PVA/AgNPs	γ-irradiation	20–60 kGy	[108]
SSD/SA/PG	γ-irradiation	2.5–120 kGy	[109]
Sterculia gum/PVA/PVP	γ-irradiation	8.42–50.54 kGy	[110]
TG/PVA/PVP	γ-irradiation	9–45.4 kGy	[111]
TG/SA/PAAM	γ-irradiation	34.7 kGy	[112]
TG/SA/PVA	γ-irradiation	9.1–36.3 kGy	[113]
XG/PVA/ZnO	γ-irradiation	10–30 kGy	[114]

3. Characterization of Hydrogel Wound Dressings Prepared by Ionizing Radiation

3.1. Gel Fraction and Swelling Degree

A mandatory step in the evaluation of hydrogel dressing produced by radiation crosslinking is the quantification of the hydrogel formation, in other words, the formation of insoluble material [115]. This step can be easily done using a classical gravimetrical method and can be performed as well using a Soxhlet extraction procedure [116].

The crosslinked hydrogel content (gel fraction), GF, is expressed as the mass of the insoluble gel sample in the dry state (m_d) divided by the initial mass (m_i) of the sample in the dry state, after immersion in water for a standardized period: 16 h [117], or 48 h [118].

One of the characteristic features of hydrogel is its swelling response when subjected to various environmental variations such as water, pH, and ionic strength. Hydrogels' good absorption properties are the most important characteristics, especially when intended to be used for biomedical applications. The swelling of a hydrogel is measured in terms of the swelling ratio of swollen gel to dry gel. One of the most crucial physical aspects of hydrogels is their soluble nature. This can be determined by calculating the insoluble component of a dried sample following immersion in water for 48 h at room temperature.

The body's blood contains about 90% water, and the excellent swelling properties of the hydrogels make the gel absorb a large amount of water in the blood favoring the bleeding to stop. Table 2 includes the values for gel fraction, swelling degree in water, and PBS of hydrogel wound dressing irradiated between 25–30 kGy.

Table 2. Gel fraction, the swelling degree in water, and PBS of hydrogel wound dressings irradiated between 25–30 kGy.

Hydrogel Composition	Gel Fraction (%)	Swelling Degree (%) in Water	Swelling Degree (%) in PBS	References
Agarose	-	163	-	[49]
BC/AA	-	5704–9903	~3000	[50]
CIP/PLGA/PVA	89–91	-	~250	[67]
Collagen	60	-	260	[51]
Collagen/PVP	84	9000	1450	[119]
GE/PVA/AgNPs	64–78	-	119–231	[63]
GE/IPN/PVP	82–83.4	485–507	-	[75]
GEL/γ-PGA	52.4	~3500	367	[74]
HACC/PVA/PEO	68	3800	-	[76]
PAA/$ZnCl_2$	78–84	-	5000–6000	[79]
PEO-PVA	62	1400–2000	-	[53]
PEGDA/CS	95–97		500–600	[80]
PVA/Alginate	80–98	-	700–1000	[84]
PVA/AgNPs	86	240	-	[83]
PVA/CM-CS/honey	55–57	1254–2551	-	[86]
PVA/CS	74–88	50–130		[87]
PVA/Gly	88–93.5	250–350	-	[90]
PVA/PVP	78	42–168	-	[92]
PVA/PVP/AgNPs	86	1400	-	[93]
PVA/PVP/aloe vera	18–35	2600–4600	-	[94]
PVA/PVP/Carr/silk	60–79	-	-	[55]
PVA/PVP/clay	70	210	-	[95]
PVA/PVP/CS	55–90	450–550	-	[97]
PVA/PVP/Gly	36–63	1190–1776	-	[98]
PVA/PVP/Gly/CS	30–52	4500–5400	-	[99]
PVA/ws-CS	61–80	-	-	[100]
PVP/Alginate/AgNPs	77–82	2144–2697	-	[38]
PVP/Agar/CS/LA	80.2–82.8	-	-	[120]
PVP/AgNPs	70–80	2100–3000	-	[65]
PVP/AgNPs	82	110	-	[83]
PVP/Carr/PEG	-	679	-	[103]
PVP/CMC	78–81	81–429	-	[35]
PVP/CS/PEG/AA	87	-	1250	[52]
PVP/ITA/CS/ZnO	54–66	1490–2607	-	[91]
PVP/PEG/Agar	71	1250		[59]
PVP/PEG/Alkhydin	33–71	400–1200	-	[106]
PVP/PEG/HEMA	89	355–385	-	[107]
SSD/SA/PG	-	150	164.5	[56]
XG/PVA/ZnO	83–89	1280–1950	550–892	[114]

3.2. Sol–Gel Analysis

For the hydrogels obtained by irradiation technology, sol–gel analysis is a useful tool to determine the insoluble gel and the radiochemical yield of the crosslinking. The radiochemical yield, (G) represents the number of units (molecules, atoms, ions) transformed by ionizing radiation following the absorption of a quantity of radiant energy of 100 eV and is expressed in $mol \cdot J^{-1}$ [121]. The sol–gel analysis method is applied in the case of polymer mixtures above the gelation dose and can estimate also the scission yield. The radiochemical yield of scission under standard irradiation conditions was found to be close to zero for most non-ionic polymers such as PVP, PVA, or PAAM [122]. The main parameters specific to irradiation technology are determined with the Charlesby–Rosiak equation, which is the basis of the numerical analyses corresponding to the gel/absorbed dose curves which are integrated into a calculation software called GelSol95 (made at Lodz University of Technology, Division of Applied Radiation Chemistry, Poland) [123]. The sol–gel parameters determined below using GelSol95 are the degradation/crosslinking ratio (p_0/q_0), the gelation dose (Dg), and the virtual dose (Dv). When the (p_0/q_0) is less than 2, for polymer mixtures that form insoluble fractions after irradiation, it can be said that crosslinking prevails more than degradation [124].

To determine the changes that take place in polymer solutions after irradiation with ionizing radiation, first are determined the parameters: radiochemical yield of the crosslinking G(X) and the radiochemical yield of the scission G(S), which give information about the number of new bonds formed per unit of absorbed energy, and the number of broken chains, respectively. The two parameters change simultaneously, and the final result of the irradiation is given by the value of their yields [125]. Considering that the unit of absorbed dose is Gy, this corresponds to 1 Joule per 1 kg of irradiated material. For the crosslinking process, the most important is the radiochemical yields of the radiolysis products of water (HO·, H·, $e^-_{aq.}$, H_2O_2, H_2, and H^+), which are very well established (2.8, 0.6, 2.7, 0.7, 0.5, and 2.7×10^{-7} mol J^{-1}, respectively). The radicals HO· and H·, formed in the process of radiolysis of water, are the main species responsible for the formation of macroradicals. The HO· radical extracts a hydrogen atom, and forms a water molecule, and the hydrogen radical extracts a hydrogen atom from the polymer molecule, resulting in an H_2 molecule, which accumulates in the form of gas bubbles during the gelation process. When a polymer–water system is exposed to radiation, a series of macroradicals are formed, which participate in recombination reactions [126].

To determine the crosslinking parameters specific to collagen and PVP mixtures, Demeter et al. used sol–gel analysis. They observed that in some mixtures the gelation started before the chosen irradiation dose, depending on the amount of collagen in the mixture [127]. Depending on the composition of the six analyzed hydrogels, the ratio was between 0.09–0.79, which indicates that the crosslinking process prevailed and the degradation is negligible. The same thing is observed in the case of the values obtained for G(X) and G(S), i.e., G(X) > G(S), which means that the polymers were predominantly crosslinked [119]. In another study, the sol–gel analysis method was used for PAAM hydrogels in which poly (L-ascorbic acid) was incorporated. It was observed that the values of the degradation/crosslinking ratios ranged between 0.03 and 0.30 and predominantly indicated the crosslinking of polymer chains compared to degradation. They also have higher values when the content of poly (L-ascorbic acid) increases in the polymer mixture [128].

In the case of quaternary hydrogels from CS, PVP, PEG, and PAA obtained by e-beam irradiation, the ratio (p_0/q_0) had a value of 0.28. Additionally, the values for G(X) and G(S) when the hydrogel was irradiated with 25 kGy were equal to 0.23 μmol/J and 0.13 μmol/J, respectively, which suggests that the crosslinking yield was higher than the degradation yield. For the other hydrogel (made of CS, PVP, and PEO), the (p_0/q_0) ratio was 0.51, which means that the hydrogel showed polymer chain scissions in a moderate proportion and the values for G(X) and G(S) after irradiation were equal to 0.22 μmol/J and 0.25 μmol/J respectively, which means that the two yields are almost equal $G(X) \approx G(S)$ [52]. G(S) was determined for CS in solid state and aqueous solution after being irradiated with γ-irradiation and subjected to sonochemical degradation treatment [129]. The values obtained were equal to 0.6 μmol/J for CS in solid state and 0.19 μmol/J for CS in aqueous solution, values close to those reported in other studies. The crosslinking yield has lower values for polymers in the solid state than in the aqueous solution, due to reactions of transient species resulting from water irradiation [130]. The gelatin (GEL) hydrogels were crosslinked at 30 kGy and showed that the radiochemical yield ratios and Dg increased with the initial concentration of GEL [75]. The sol–gel analysis is a useful and necessary method for determining the characteristics of polymeric hydrogels to understand their behavior when used as biomedical devices.

3.3. Rheology and Mechanical Properties

The elasticity or stiffness properties of a hydrogel can be highlighted by a series of mechanical parameters [131]. The most used techniques for the quantitative description of the flow resistance (viscosity) and the deformation resistance of polymer solutions are rheology/viscometry. Rheology is a technique that provides valuable information both on the molecular structure (determining the size of the macromolecular chains and the crosslinking density), on the behavior of the hydrogel, as well as on the self-assembly mechanisms.

For polymer hydrogels, rheology is probably the most important physicochemical characterization technique [132]. When a hydrogel sample is inserted between two horizontal disks (or cylinders), and after applying an oscillating stress (or force) of a certain value to one of the disks, the motion induced in the other disk is decomposed into two components (in-phase and out-of-phase). The storage modulus (G′, elastic component) and loss modulus (G″, viscous component) can be measured as a function of applied stress or oscillation frequency. In general, to identify the formation of a hydrogel by rheology two experiments are commonly used: the linear response of the modules (G′ and G″) to a fixed stress of small amplitude applied by the frequency variation, and the second one refers to the nonlinear response of the modules at a fixed frequency by changing the shear force. For a typical hydrogel, G′ should be invariant with frequency up to yield stress corresponding to the transition from the gel state to the ground state and should exceed G″ by at least an order of magnitude. If the G′ decreases rapidly by applying a force, this indicates the breaking of the macromolecular network. The behavior and magnitude of the G′ and G″, as well as the flow stresses depending on the applied force can be mathematically modeled and lead to conclusions regarding the gel structure [133]. The values of the G′ and G″ moduli can be related to the crosslinking density (Ve) by using the rubber elasticity theory. In the reference [134], the equation for crosslinking density determination for hydrogels produced by radiation synthesis is described very well. Concerning the rheological demands for ideal wound dressing, the hydrogel must have adequate stretch and compression properties, an elastic modulus comparable to human skin, and good adhesion, but not excessively adhesive, to support deformation [135]. Similarly, in further paragraphs, we emphasize the rheological and mechanical properties of hydrogel wound dressings produced at the legally accepted sterilization dose.

Dynamic mechanical analysis performed on gels fabricated of PHEMA and P(HEMA/Itaconic Acid) showed improvement in the mechanical strength of the polymeric network if the itaconic acid content is not higher than 3.5 mol%. The moduli values were determined in the relaxation state, meaning after γ-irradiation crosslinking was between 4.73–18.4 kPa at 25 kGy [81]. PVP-k-carrageenan hydrogels wound dressings with the composition based on 10% *w*/*w* PVP, 2% *w*/*w* KC, 0.3% *w*/*w* KCl, and 1.5% *w*/*w* PEG showed a fracture strength of 1750 kPa and a toughness 83.6 mJ for the 25 kGy sample by comparison with irradiated ones. As the authors described, these results confirm the influence of radiation action on the crosslinking density of hydrogels. The hydrogels in question, besides their mechanical properties, exhibit fluid absorbency, elasticity, and transparency, and are easy to remove [104]. PVA-agar-carrageenan presented a drastic increase in elongation and tensile strength by adding 1% agar. These dressings were synthesized using γ-irradiation in a single irradiation stage, so the gel formation and sterilization have been performed at

25–30 kGy [82]. Lugão et al. have investigated the rheological properties of PVP hydrogels and the influence of some additives such as PEG, PEO, and glycerol on the gel properties using γ-irradiation synthesis. The hydrogel formulation containing PVP 6% and PEO 3% showed an elastic modulus of 33,000 Pa and the formation of a transparent fluid gel with high elasticity [32]. The PVA/cellulose acetate/gelatin hydrogels composites crosslinked by γ-irradiation at 25 kGy and loaded with 20 mM $AgNO_3$ showed an improved tensile strength of 34,000 Pa [85]. The nanosilver/GEL/CM-chitosan hydrogels possessed interconnected porous structures and had a compressive modulus of 54,000 Pa at 1 mM $AgNO_3$, which slightly reduced with the increase of nanoparticles. The highest compressive strength was obtained at 10 mM $AgNO_3$. The hydrogels were synthesized using the radiation-induced reduction of silver ions and crosslinking in the same fabrication step [62]. The CS, GEL, PVA, and polyacrylamide sterile hydrogel wound dressings produced by γ-irradiation at 45 kGy showed a tensile strength of 6.68 MPa and 2.45 MPa. The results indicated that hydrogel films can be stretched, folded, and are resilient; moreover, these dressings can provide mechanical protection to the wound and can mimic artificial skin function [136].

PVA/CM-CS/honey hydrogels were prepared by irradiation at 25–40 kGy and by freeze-thawing; this approach allowed for obtaining dressings with higher mechanical strength [86]. Hydrogel wound dressings prepared from PVP, PEG, and agar showed constant values of stress at break within 25–35 kGy of about ~4000 Pa and much higher at 50 kGy due to increased crosslinking density. The higher radiation dose decreased the strain at break [106]. E-beam synthesis of PVP/PEG and PVP/PEG/starch wound dressings have been performed also at the legal sterilization dose of 25 kGy. The PVP content increased the tensile strength above 25 MPa, while the PEG content increased the elongation at break. The irradiation dose has a defining role in the mechanical properties, mainly by increasing them [60]. Hydrogels based on polyglutamic acid and GEL, showed at 25 kGy a strong tensile strength of 1 MPa and an elongation at break of 377%. Higher radiation doses of 40 kGy produced an excess of crosslinking and led to a fragile and brittle hydrogel [74]. Quaternary ammonium CS/PVA/PEO hydrogels prepared using γ-irradiation reached a tensile strength of ~34 MPa at 40 kGy. This study proves that a higher irradiation dose induced radiation degradation effects causing a decrease in tensile strength, while a higher content of quaternary ammonium chitosan decreased the elongation [76]. The tensile strength of 2.2 MPa was obtained for hydrogel wound dressings of CS/GEL/PVA crosslinked by γ-irradiation at 40 kGy. Their mechanical properties were mainly influenced by the polymer's weight ratios [69]. PVP/PEG/agar hydrogel wound dressings produced at pilot scale using γ-irradiation crosslinking and sterilization demonstrated the same quality from batch to batch. Following the compression tests, a slow decrease of the compression rate of 20% until 7 h and about 70% at 24 h was observed [105]. The PVA/PVP/CS hydrogels crosslinked at 25 kGy presented a tensile strength between 0.38 to 0.72 MPa and an elongation at the break between 217% to 277%. The same hydrogels were treated by the freeze-thaw procedure, which suggested that a higher dose of 25 kGy affects the mechanical properties, decreasing the strength of hydrogel at 0.37 MPa and elongation at 178% [97]. The hydrogels obtained from PVA and PVP at 20 kGy presented a very low tensile strength of 147 Pa and an elongation of 175%. Advanced decreasing of mechanical properties was found after irradiation above 20 kGy [92]. PVA/ws-CS hydrogels made using γ-irradiation (30 kGy) exhibit almost no temperature dependence due to the stable chemical crosslinking structure. By adding 1% chitosan, the elastic modulus (G') was found to be ~5000 Pa [100]. Storage (G') and loss (G'') moduli of e-beam crosslinked agarose hydrogels corresponded to 12.8 kPa. After e-beam irradiation with 30 kGy, the viscoelastic properties underwent a reduction of more than 20% compared to the initial moduli [49]. Thermally resistant hydrogels were made from PVA by e-beam irradiation and PVA acetalization followed by additional grafting of AA on the final dressing. The hydrogel gained the maximum tensile strength of 20 MPa at a very high radiation dose of 100 kGy. After steam autoclave sterilization, the tensile strength decreased to 10 MPa and the elongation at break increased to 300%. The hydrogel adhered to the burning surface and healed the burns faster compared with a typical dressing [54]. The hydrogel wound dressings formed by radiation-induced crosslinking using moringa oleifera gum/carbomer were developed to be used as a drug carrier for levofloxacin release. The optimal γ-irradiation dose was set at 13.68 kGy. The principal mechanical properties were tensile strength = 2.68 MPa, elongation at break = 64.49%, burst strength = 3.66 N, stress–relaxation = 47.10, percentage resilience = 28.25, and folding endurance = 234. The hydrogel dressings were hemocompatible, antioxidant, and bioadhesive [78]. CS/PVA hydrogels crosslinked using γ-irradiation showed a tensile strength of ~0.4–0.5 MPa at 20–50 kGy. In these conditions, the elongation at break showed a systematic decrease while increasing the PVA content, as well as with increasing the irradiation dose [87]. Antibacterial AgNP/GEL/PVA hydrogels were produced using γ-irradiation at 30 kGy using a weight ratio of gelatin: PVA of 60:40. The higher AgNP content revealed a negative impact on the mechanical properties of these hydrogels. At 0.25% AgNP content, the hydrogels presented a tensile strength of 0.0256 MPa and an elongation percentage of ~180% [63]. PVA/alginate hydrogels were produced using γ-irradiation. The tensile strength of the PVA/alginate hydrogels increased as the concentration of alginate was lower and as the radiation dose increased. The hydrogel swollen samples showed a tensile strength of 14 MPa and an elongation of over 70% [84]. Hydrogels that exhibit antibiotic activity containing PVA and cetylpyridinium chloride were developed using γ-radiation synthesis. Their mechanical properties were assessed using dynamic viscoelastic measurements. The results indicated a G' value of 10,500 Pa, this value being affected by increasing the antibiotic active ingredient. The authors observed that hydrogels prepared only from PVA had a G' of 7000 Pa, and were very soft and strong adhesives [96]. Hydrogels based on PEO and PVA containing high PVA content (20% and 30%) were produced by e-beam synthesis at 60 kGy and showed high strength of 0.2 to 0.35 MPa and elasticity of 500%. The tensile strength and elongation decreased with increasing doses because of the increase in crosslinking [53]. Collagen-PVP superabsorbent hydrogels were obtained using e-beam irradiation. The hydrogels prepared without crosslinking agents showed very low elastic moduli, below 66.5 Pa; in comparison, the hydrogels prepared with 0.5% crosslinking agent and irradiated showed maximum elastic modulus of 14,000 Pa. Higher irradiation dose above 25 kGy, decreased the G' value considerably [119]. Gelatin-PVP hydrogels functionalized with gallic acid were developed to obtain hydrogels with antioxidant properties. The microindentation assay demonstrates the formation of hydrogels with an elastic modulus between ~15,800 to ~44,800 Pa. A dose of 15 kGy was selected to obtain mechanically stable materials having adequate stability at manipulation [75]. PVP/kappa-carrageenan/PEG hydrogel dressing produced at 25 kGy using γ-irradiation showed a higher tensile strength of 0.05 MPa than commercial hydrogels based on PVP and PVA (0.04 MPa). The elongation at a break of 138% was almost similar to that of PVP commercial hydrogels (142%) [103]. PVP/PEG hydrogel membranes reinforced with MMA and PP fibers were prepared using e-beam irradiation at 20 kGy. The tensile strength and elongation depended on the grafting degree and varied between 0.02 and 0.16 MPa and ~50–70%, respectively [56]. Hydrogel dressings based on PVA, NVLC, and HEA presented at 25 kGy a tensile strength of 547 Pa, elongation at break of 115%, and elastic modulus of 3.6 kPa. Elastic modulus and tensile strength increased as a consequence of sterilization dose, while elongation decreased [137]. A sterile hydrogel film made from 3.75% tragacanth gum, 3.125% sodium alginate, and 5% PVA was obtained at 18.2 kGy using γ-irradiation. The films showed a mechanical strength of 9.56 N, which classified them as adequate for wound dressing applications [113]. The same authors reported the fabrication of hydrogels containing 2.5% tragacanth gum, 4% SA, and 1.12×10^{-2} mol/L Aam at a total dose of 34.7 kGy. These hydrogels were characterized by a tensile strenght of 2.45 MPa and elongation of 40.15% [112]. The e-beam crosslinking of the collagen gels leads to a strong increase in G' modulus up to 69.3 kPa at 100 kGy. At 25 kGy, the collagen hydrogel had a G' of 6174 Pa. Higher irradiation doses induce the formation of a compact network that restricts the movement of collagen chains and prevents water from being absorbed [51]. Antibacterial composite hydrogels prepared from CS/PVA/TA at

30 kGy using γ-irradiation have a tensile strength of 200 kPa and 424% elongation at break. The mechanical properties varied as a function of the CS molecular weight [70]. Bacterial cellulose (BC)/AA hydrogel wound dressing was synthesized using e-beam irradiation at 35 and 50 kGy. Hydrogels fabricated with 40% AA and 60% BC showed a tensile strength of ~ 1.5 MPa and elongation of about 300% [50]. The e-beam synthesis of PVP/PEG/agar wound dressing at 20 and 25 kGy produced a decrease in tensile strength and elongation with increasing irradiation dose. The tensile strenght and elongation varied from 17 to 27 kPa and 120–160%, respectively. These parameters were strongly influenced also by the %PVP and %PEG, as well as by the PVP molecular weight [58]. The PVA/GEL bilayer hydrogel achieving both wound dressing and antiadhesion capabilities was developed using γ-irradiation. At the dose of 30 kGy, the PVA side showed a modulus of elasticity of about ~12 kPa, and the other side with GEL had ~14 kPa [138].

3.4. Network Parameters

To tailor the molecular weight between two successive crosslinking points (M_C), crosslinking density (Ve), and mesh size (ξ) of an e-beam/gamma crosslinked hydrogel to be used as a wound dressing, the most important role is played by the action of radiation. The Ve is a parameter that describes the number of crosslinking points formed between different polymeric chains (inter-molecular crosslinking) or inside of the same polymeric chain (intra-molecular crosslinking) upon irradiation. The M_C is a measure of the degree of hydrogel crosslinking, and ξ is defined as the linear distance between consecutive crosslinking points and the space available between the macromolecular chains [139]. In general, the network parameters depend on many factors, such as composition, nature, and temperature of swelling medium and irradiation dose [140].

Mohamad et al. investigated the biocompatibility of crosslinked hydrogels for wound dressings at different irradiation doses and AA concentrations. They obtained that the Ve significantly increased with the increasing amount of AA in the hydrogel (from 0.41 mol/cm^3 to 3.06 mol/cm^3), but the Ve was also influenced by the irradiation dose [50].

Demeter et al. obtained a collagen hydrogel by e-beam crosslinking and demonstrated that Mc depended on the absorbed dose. When the absorbed dose is more than 25 kGy, the radiation effect is almost directly proportional to the G' modulus and the crosslinking density [51]. In another study, collagen/PVP hydrogels were prepared with the addition of a crosslinking agent. With decreasing crosslinking agent concentration, collagen/PVP systems can be crosslinked via e-beam irradiation with a moderate radiation dose, close to the legally accepted sterilization dose (25 kGy) to obtain the best swelling and optimum rheological properties [119]. Calina et al. demonstrated that the network parameters (M_C, Ve, and ξ) indicated the formation of a crosslinked structure with a nanostructured mesh size with a typical dimension of 11–67 nm [52].

The sterile and pure TG/PVA/SA wound dressing hydrogels which were prepared through the radiation method showed a value for M_C of 77,7939 g/mol, Ve of 2.0×10^{-6} mol/cm^3, and ξ of 38.83 nm [113]. The same authors prepared sterile TG/PVA/PVP hydrogel dressings for simultaneous care of wound infection and wound pain. The TG/PVA/PVP hydrogel dressings showed an Mc value of 43,927 g/mol, Ve of 0.35×10^4 mol/cm^3, and ξ of 28.05 nm in PBS at 37 °C [111].

The radiation-induced synthesis of novel PHEMA/ITA copolymer hydrogels with different ITA content and the investigation of their potential use in biomedical applications were reported by Tomić et al. They showed that an increase in ITA content gives rise to the enhancement of hydrophilicity. Due to that, polymer–fluid interactions in hydrogels are strengthened, Mc increases, and Ve decreases [81].

3.5. Morphology

Scanning electron microscopy (SEM) is used to provide information on sample surface topography and composition. This technique is used to capture the characteristics of the network structure in hydrogels. An important characteristic of hydrogel is the investigation of the structural morphology [141].

The porosity of the dressing matrix is an essential requirement for wound dressing. The porous hydrogel wound dressings are permeable to water vapor and wound exudates. The slow delivery of the drug and the absorption of fluid in the wound depend on the architecture of the hydrogel. The porous structure of polymer films helps absorb wound fluids, keeps the wound dry, and helps supply oxygen to the injury. Cryo-SEM explores the hydrated microstructures of biological specimens. [142].

The TG/PVA/SA hydrogel wound dressing produced by radiation synthesis showed the presence of a porous network, which facilitates the application of the hydrogels in drug delivery and increases the swelling properties of the hydrogels. The hydrogel film had interconnected pores, thus demonstrating crosslinking during synthesis [113]. Cryo-SEM images of the TG/PVA/PVP hydrogel wound dressings depicted the porous surface morphology and crosslinked network structure, which is important for wound fluid absorption and slow drug delivery [111]. In another work, the GA/TG/PVA/PVP hydrogel was incorporated with gentamicin for wound dressing applications. The hydrogel films showed an interconnected, heterogeneous, and highly porous structure, which can lead to a high degree of swelling, permeability, and may support cellular growth [73]. Four years later, they developed a hydrogel dressing to enhance wound healing. The cryo-SEM images of MOG/Carbopol films demonstrated a porous crosslinked network structure. This crosslinked network structure promotes wound healing by enhancing fibroblast adhesion, proliferation, and tissue regeneration [78]. Recently, the cryo-SEM images of TG/SA/PAAM hydrogel in the swollen state showed a highly porous structure in the hydrogel. Furthermore, the porous structure in the hydrogel dressing controls drug loading and drug release. Drug release is dependent on porosity, so by adjusting the pore size, drug release from hydrogels can be controlled. They then showed that the lower pore size led to the slow release of the drug up to a limited period of 9 h as compared to hydrogels with larger pore sizes [112]. The morphology of wound dressing hydrogels was examined containing different concentrations of AA and radiation doses. For this study, the hydrogels exhibited a highly porous mesh network and the pore size of the hydrogels decreased from 61.88 μm (35 kGy dose) to 5.4 μm (50 kGy dose), indicating that the pore size was influenced by AA content as well as e-beam irradiation dose. A smaller pore size is due to the increase in the crosslinking density of the hydrogels at higher AA content and e-beam irradiation dose [50]. Guo et al. showed that the microstructures of PVA hydrogel had a large pore size with uniformity.

After radiation treatment and TA immersion, the pore size was reduced and a dense network was formed, and the lamellar structure disappeared, due to strong hydrogen bonding interactions between TA and CS/PVA leading to the decrease of the pore size [70]. De Silva et al. developed a wound dressing hydrogel using PVP/Carr/PEG by γ-irradiation, and from the SEM analysis, it was observed that the top surfaces of the unirradiated and irradiated hydrogel showed a increased difference. The surface of the unirradiated hydrogel showed cracks in a few places, while the irradiated hydrogel surface did not show cracks [103]. The structure of the collagen hydrogel varied with the irradiation dose, from a more compact structure to a more macroporous structure. It was observed that the crosslinked collagen hydrogel at the dose of 25 kGy has a macroporous state with large pores (50 μm) [51]. The hydrogels have porous network structures and the approximate pore diameter was between 1 and 5 μm [63]. The hydrogels of PVA developed by irradiation followed by freeze-thawing showed a morphology with smaller pores, while those developed by freeze-thawing and freeze-thawing followed by irradiation have a morphology with larger pores. Smaller pores obtained in hydrogels prepared by irradiation and irradiation followed by freeze-thawing allow slower evaporation of water [100]. The HACC/PVA/PEO hydrogel with significant antibacterial activity irradiated at 30 kGy had many micropores, and the pores were uniformly distributed on the surface of the hydrogel. Moreover, it was shown that the pore size of the hydrogels decreased with increasing radiation dose, and the density of the pores on the surface of the hydrogels increased with the increase of the radiation dose [76]. Then in another work, they prepared CS/GEL/PVA hydrogels by the γ-irradiation method for use in wound dressing applications and showed that the microstructure of the hydrogels exhibited many micropores, and the pores were distributed uniformly on the surface of the materials [69]. The PVP/PVA/clay nanocomposite produced by γ-irradiation as a wound dressing showed that the clay particles were finely dispersed through the interconnecting pores of the hydrogel. Moreover, an increase in the clay content leads to a decrease in the pore size [95]. It was shown that the pore size of the PVP/PEG/Agar polymeric hydrogel mainly depends on the radiation dose and the copolymer composition [105]. The XG/PVA/ZnO dressing hydrogels exhibited interconnected porous structures that varied extensively depending on ZnO content. An increased amount of ZnO favors a denser network with a smaller pore size [114]. Zhou et al. found the porous crosslinked network structures with smooth surface morphology in all $AgNO_3$/GEL/CM-CS hydrogels. The pore diameter distributions were between 200–600 mm, which were expected to be insufficient for the permeation of nutrients and oxygen from the cells [62]. The PVA/PVP hydrogel has a relatively wider pore structure but a distinguished aggregation of the silver ions is observed as somewhat of a dark particle in the case of PVA/PVP hydrogel-loaded $AgNO_3$ [93]. From the microstructure of the pure and PVA/CA/GEL/AgNP composite gels, many micropores were observed on the surface of the materials. The pure and AgNP hydrogels possessed similar micropore structures and showed a tendency to agglomerate with an increase of AgNPs in the gel matrix [85]. The surface of the morphology of the DMAEM/PEO/ZnS nanocomposite at pH 7 (pore size about 5–15 μm) had a relatively uniform pore structure compared to that at pH 4. Moreover, the SEM results revealed that DMAEM/PEO/ZnS nanocomposite exhibited uniform structure and antimicrobial properties [71]. The SEM analysis revealed a PVP/CS/ITA copolymer with a rough surface, full of polymer aggregates with some cavities and hills due to the inhomogeneous effect of radiation during crosslinking processes. After loading the ZnO nanoparticles into the PVP/CS/ITA hydrogel, they obtained a hydrogel with a very rough surface [91]. The PVP/CMC hydrogel produced by irradiation at 30 kGy showed a porous structure and the pore size was 25 μm [35]. Varshney demonstrated that the presence of polysaccharides in aqueous solutions forms a gel network structure that affects the properties of the gel formed upon irradiation. The gels have a highly porous structure and pores of 1–3 μm size can be seen within the porous structure. The well-defined pore structure could be due to the presence of gel-forming polysaccharides and the evaporation of water from them, which form the basic structure before irradiation [82]. The smart PHEMA/ITA copolymer, with ITA content up to 5 mol% was studied upon γ-irradiation. For the PHEMA/ITA gel sample that was not lyophilized, the surface was compact, smooth, and dense, and for the lyophilized PHEMA/ITA gel sample, a heterogeneous distribution of pores in the sample structure was observed. Thus, the authors obtained intelligent hydrogels for the active treatment of skin and wounds [81].

In another article, the authors observed that all the PHEMA/ITA/AgNPs hydrogels with pore size less than 30 nm have relatively good dispersibility. They also proved that AgNPs formed on the surfaces of the hydrogels and were uniformly distributed [64]. The PVP/AgNP hydrogels irradiated at 25 kGy confirmed by SEM analysis that they were well distributed and have a spherical shape in the range of 4–10 nm [65]. Alcântara et al. obtained an antibacterial hydrogel dressing with AgNPs crosslinked by radiation and SEM analysis revealed a uniform distribution of AgNPs on the surface of the hydrogel dressing [83]. The ELT/PVP hydrogel exhibited a porous structure, showing interconnected pores (16 ± 8.42 μm) [72].

3.6. Moisture Retention Capability and Water Vapor Transmission Rate

Moisture retention capacity (MRC (%)) is an important factor for wound dressing hydrogels because a moist environment helps to heal the wound faster [143].

The MRC (%) of PVP/PEG/Agar wound dressing hydrogels after 5 h was 60% at room temperature and 40% at 37 °C [105]. The XG/PVA/ZnO nanocomposite retained approximately 50–65% of its water content after exposure to air for 6 h [114]. Experimental investigation showed that hydrogels from CS/PVP/PEG retain after 24 h between 80.52 and 82% humidity [52]. The MRC (%) in 4 h was 69% for the PVP/CMC hydrogels, having a similar moisture retention capacity, regardless of their composition and crosslinking density [35].

An ideal wound dressing must control the water loss from a wound at an optimal rate. The hydrogel prepared as a wound dressing must decelerate the loss of body liquid from the wound and at the same time, maintain a suitable humidity in the wound area. The water vapor permeability of a wound dressing hydrogel should prevent excessive dehydration or edema. An extremely high water vapor transmission rate (WVTR) may lead to wound dehydration, whereas a low WVTR may cause the accumulation of wound exudates. Hence, a dressing with a suitable WVTR is required to provide a moist environment for establishing the best milieu for natural healing [144]. The WVTR values of the hydrogel wound dressings obtained by irradiation are presented in Table 3.

Table 3. WVTR of hydrogel wound dressings.

Hydrogel	WVTR (g/m^2 Day)	References
BC/AA	2105–2666	[50]
CS/GEL/PVA	472.32 ± 133.22	[136]
CS/GEL/PVA/PAAM	541.81 ± 54.05	
CS/PVP/PEG	272.67	[52]
GA/TG	188–287	[73]
GEL/PVA/$AgNO_3$	4200–4600	[63]
MOG/Carbopol	2461.14 ± 39.16	[78]
PAAM/LAA	106–185	[128]
PVA/CS	40–73.35	[87]
PVA/PVP	89–173.97	[92]
PVP/AgNPs	783.52–940.16	[65]
PVP/Alginate/AgNPs	278.44	[38]
PVP/Carr/$AgNO_3$	481–536	[102]
TG/PVA/PVP	582.23 ± 86.55	[111]
TG/SA/PVA	197.39 ± 25.34	[113]
XG/PVA/ZnO	164.89–184.57	[114]

The water vapor transmission rate (WVTR) for normal skin is 204 ± 12 g/m^2 day, which for injured skin can range from 279 ± 26 g/m^2 day for a first-degree burn to 5138 ± 202 g/m^2 day for a granulating wound [145]. Commercial wound dressings demonstrate WVTR values between 76 and 9360 g/m^2 day [146]. When selecting dressings for wound management, they have to decide which type of dressing can maintain moisture for each type of wound. Therefore, WVTR characterization is important for wound dressing testing.

3.7. Drug Release

Wound dressings loaded with drugs can be used as a potential wound dressing with improved healing effects in wound care. The release of the drug-loaded in hydrogels occurs after water penetrates the polymeric networks, and this is followed by diffusion along the aqueous medium. The drug release is related to the swelling characteristics of hydrogels. The release profile of moxifloxacin hydrochloride from the drug-loaded TG/SA/PVA hydrogels showed that the amount of drug released in simulated wound fluid was higher than the release in PBS, pH 2.2 buffer, and distilled water [113]. The release rate of gentamicin from the TG hydrogel was found to be quite fast in both water and simulated wound fluid at 37 °C. The antibiotic release was observed more in water compared to simulated wound fluid [73]. The release profile of the antibiotic drug levofloxacin from the MOG/Carbopol polymer films in pH 2.2 buffer, PBS, and simulated wound fluid was studied. In pH 2.2 the drug release was more pronounced as compared to the other swelling media, thus levofloxacin showed pH-dependent solubility. Moreover, the solubility of the drug was the controlling factor for the release of the drug from the polymer matrix [78]. The drug-loaded TG/SA/PAAM hydrogels in different media indicated that the drug release was higher in PBS and simulated wound fluid as compared to pH 2.2 buffers. They obtained significant release values for lidocaine (34.86%) and amikacin (64.75%) [112].

For the AA/PLST/MMT/CS hydrogels with 10 and 20% of CS, the release of Sulfanilamide reaches equilibrium after 200 min [61]. The hydrogels containing ciprofloxacin hydrochloride (CIP)-loaded PLGA nanoparticles as antibacterial materials using γ-radiation were prepared. After four days, the amounts of CIP released from the PLGA nanoparticles were between 43.38 and 94.63 $\pm$ 2.91% [67]. PVA/PVP/CS hydrogels containing antibiotic ciprofloxacin lactate showed that the total amounts of the drug released were 85 and 65% for initial drug contents of 2.0 and 1.0 mg/mL, respectively [97]. The ibuprofen loaded in CS/PVP/PEG hydrogels was investigated at pH = 7.4 and pH = 9.4 (an infected wound). It was observed that about 30–45% of ibuprofen was released in the first 5 h, equivalent to around 30 mg. In the weak alkaline media, the ibuprofen release was lower, reaching 25–35% in the first 5 h [52]. The drug release from DMAEM/PEO/ZnS nanocomposite hydrogel loaded with gentamicin (G), colistin (C), and neomycin (N) at pH 4 and pH 7 was investigated. The neomycin and colistin drugs were more highly released at pH 4 than at pH 7, only the gentamicin release increased from 40% at pH 4 to 96% at pH 7 after 330 min [71]. The amoxicillin drug release on PVP/ITA/CS/ZnO nanocomposite hydrogels was achieved at pH 2.1 and 7.4. The release profile of amoxicillin at pH 2.1 was prolonged and the total amount released after 240 min was 33.5%. At pH 7.4, nearly 67% of the loaded drug was released in the first 60 min [91]. Ciprofloxacin was used as a model antibiotic to study the release properties of the PEDGA/CS hydrogels obtained by γ-irradiation. At 150 min, the cumulative release of PEDGA/ciprofloxacin and PEDGA/CS/ciprofloxacin hydrogels at an irradiation dose of 25 kGy were 87.96 $\pm$ 0.83% and 97.34 $\pm$ 0.41%, respectively [80]. In vitro cumulative release of sericin from the PVA hydrogels in PBS solution at 37 °C demonstrated that the sericin (50–70%) was released from the hydrogels within 6 h [147]. The TG/PVA/PVP hydrogels were prepared at an irradiation dose of 27.3 kGy. The lidocaine and gentamicin were released from the hydrogel in 360 min and showed different release profiles due to solubility and swelling in different simulated environments. Moreover, hydrogels loaded with these drugs can be useful as dressings in wound care [111]. The PVP/PEO/Agar γ-irradiated hydrogels loaded with neomycin drug showed a neomycin release of about 40% in 48 h [148].

3.8. *In Vitro Biological Properties*

3.8.1. Antimicrobial Studies

Antibacterial hydrogel wound dressings have been developed with the consideration of the fact that moist, warm, and nutritious environments in wound beds provide ideal conditions for microbial growth.

The hydrogels based on CS/PVA/TA presented significant antibacterial properties against *E. coli* and *S. aureus* with an inhibition rate between 95 and 97% [70]. The antimicrobial activity of SSD-alginate hydrogel and SSD-alginate-PG hydrogel produced using γ-irradiation at 2.5 kGy was determined against most encountered bacterial strains in burn wound infection such as *E. coli*, *K. pneumoniae*, *P. aeruginosa*, *A. baumannii*, *E. aerogenes*, and *S. aureus*. The irradiation treatment improved the performance of the formulated hydrogels and the highest inhibiting bacterial growth was observed for SSD-hydrogel loaded with prodigiosin known as an active ingredient with important antifungal, antibacterial, antimalarial, and antineoplastic properties [109]. CPC/PVA hydrogels demonstrated antibacterial activity on the *E. coli* strain, but this property is due to the action of cetylpyridinium chloride [96]. Regarding the antimicrobial effects of gelatin/PVA hydrogel in which various concentrations of AgNP were added, it was demonstrated that a proper concentration of 0.25% AgNP is efficient against *E. coli* and *S. aureus*, as well as MRSA. These values are considered a balance between biological and mechanical properties [63]. The CIP/PLGA-loaded PVA hydrogels denote important antibacterial activity in the *E. coli* and *S. aureus* strains [67]. The hydrogels fabricated from PVA and ws-chitosan showed efficiency only on *E. coli* due to their capability to bind Gram-negative bacteria to the chitosan amino groups [100]. PVP/carrageenan hydrogels incorporated with nanosilver prepared at 25 kGy by in situ reduction of Ag^+ and hydrogel crosslinking using γ-irradiation showed antimicrobial activity against the common contaminants isolated from burn patients, *P. aeruginosa*, *S. aureus*, *E. coli*, and *C. albicans*. The most effective hydrogels against bacterial strains were those loaded with 100 ppm nanosilver after 3–6 h [102]. PVA/PVP/glycerin/antibacterial hydrogels produced by irradiation at 25 kGy and freeze-thawing were verified for their efficiency against the three bacterial strains *E. coli*, *S. aureus*, and *S. aeruginosa*. In this case, the antibacterial effect generally decreased after irradiation. The hydrogels containing chloramine-T showed an accentuated decrease of antibacterial activity due to its weak resistance at irradiation. Upon irradiation can be generated radiolysis product, which reduces its bactericidal effect. It must be specified that the antibacterial properties of hydrogels containing sulfadiazine sodium salt were not affected by the γ-irradiation [98]. Quaternary ammonium chitosan hydrogel blended with PVA and PEO performed great antibacterial activity against *S. aureus* and *E. coli* [76]. The quaternary ammonium chitosan exhibits important antibacterial activity due to its polycationic structure which is attracted to the surface of bacteria and leads to the alteration of their cell membrane, releasing a significant amount of protein material or cytoplasmatic constituents out from the bacterial cell, finally producing the death of the bacteria [149]. Xanthan-based wound dressing hydrogels containing ZnO nanoparticles crosslinked at 30 kGy were tested to demonstrate the antibacterial activity against *S. aureus*, *E. coli*, and *C. albicans*. The results indicated a broad antimicrobial spectrum, both antifungal and antibacterial, and were correlated with the increase of ZnO nanoparticles included within xanthan dressings [114]. The nanosilver/gelatin/CM-chitosan hydrogels clearly showed antibacterial ability against *E. coli*. The zone of inhibition was wider at 10 mM of nanosilver [62]. The radiation synthesized of 2-(dimethylamino)ethyl methacrylate/PEO/ZnS nanocomposite hydrogel at 20 kGy loaded with antibiotics (colistin, gentamicin, and neomycin) was tested against both bacterial strains (*E. coli*, *P. aeruginosa*, *S. aureus*, and *B. subtilis*) and unicellular fungi (*A. niger*, *A. terreus*, *A. flavus*, and *A. fumigatus*). The presence of ZnS nanoparticles indicates the increase of antimicrobial and antifungal activity on all strains tested, especially those hydrogels loaded with neomycin [71]. PVP/chitosan/itaconic acid/ZnO nanocomposite hydrogel was efficient against *S. aureus* and *E. coli*. The incorporation of amoxicillin into these hydrogels increased significantly the antimicrobial activity [91]. Strong antimicrobial effect and complete inhibition of *P. aeruginosa*, *S. aureus*, and *C. albicans* were observed in the case of PVP/alginate hydrogel dressings containing 70 ppm nanosilver crosslinked by γ-irradiation at 25 kGy [38]. Smart P(HEMA/IA/silver NP) hydrogels were produced in the same step with the reduction of Ag+ by the γ-irradiation method at 12 kGy. The Ag/P(HEMA/IA) hydrogel demonstrated high antibacterial effects on *E. coli*, *S. aureus*, and *C. albicans* with effectiveness even at small silver concentrations [64]. AA/$ZnCl_2$ crosslinked hydrogels at 25 kGy indicated that they have good antimicrobial activity against *S. aureus* and *E. coli* strains. The antibacterial effects of AA/$ZnCl_2$ hydrogels increased as a function of the $ZnCl_2$ concentration [79]. Antimicrobial hydrogel wound dressing was developed, where they included CS dissolved in lactic acid to the original composition of PVP/Agar hydrogel dressing before irradiation. The Gram-negative bacteria *E. coli* growth was not inhibited and the test involving *S. aureus* showed that Gram-positive bacteria growth was hindered in the presence of hydrogel dressings containing CS as compared with the original hydrogel dressing [120]. The synthesized PVP/AgNPs hydrogel showed antimicrobial activity against *P. aeruginosa* and *S. aureus* and the PVA/AgNPs hydrogel indicated that it only had antibacterial activity against *P. aeruginosa* [83]. In another work, the pure PVP hydrogels had no inhibition zone, while the PVP/AgNPs hydrogels (5 mM) presented inhibition zones against all tested bacteria: *B. cereus*, *S. aureus*, *S. epidermidis*, *S. pyogenes*, *A. iwoffii*, *E. coli*, and *P. aeruginosa*. The PVP/AgNPs hydrogels with different concentrations of AgNPs irradiated at 25 kGy were investigated for bactericidal efficacy against *S. aureus*, it was proven that the rate of bacterial reduction increased with the increase of AgNPs content up to 99%, over 6 h (5 mM AgNPs) and 12 h (1 mM AgNPs) [65]. The Psyllium/Carbopol hydrogel dressings with frankincense essential oil developed using γ-irradiation showed efficient antimicrobial activities against *S. aureus*, *E. coli*, and *C. albicans* [150].

The antibacterial activity data for different hydrogel wound dressings are presented in Table 4. The in vitro antimicrobial tests of hydrogels were explored using microorganisms on Gram-positive bacteria (*S. aureus*, *C. albicans*) and Gram-negative bacteria (*P. aeruginosa*, *E. coli*).

Table 4. Antibacterial activity data of the hydrogel wound dressings.

Hydrogel	Antibacterial Agent	Inhibition Zone Diameter (mm)				References
		S. aureus	*C. albicans*	*P. aeruginosa*	*E. coli*	
CS/PEG/ZnO	-	-	-	-	20	[151]
CS/PEG/ZnO	GM	-	-	-	30	
DMAEM/PEO/ZnS	CL	15.2	8.5	13.2	12.5	[71]
DMAEM/PEO/ZnS	GM	15.1	8.4	13.1	12.4	
DMAEM/PEO/ZnS	NEO	15.3	8.5	13.2	12.5	
DMAEM/PEO/ZnS	AM	29.5	11.3	21	15.3	
GEL/CM-CS	AgNO3	-	-	-	13.5–19.6	[62]
GEL/PVA	-	15	-	-	15	[63]
GEL/PVA	AgNPs	15–26	-	-	15–27	
PVA/CA/GEL	AgNPs	14–23.2	14.2–17.3	11–23.7	9–24.9	[85]
PVA/CS	TA	10.5 ± 0.3	-	-	10.4 ± 1.1	[70]
PVA/PVP/Gly	AgNO3	19	-	21	17	[98]
PVA/PVP/Gly	SSD	20	-	26	21	
PVA/PVP/Gly	SD-Na	37	-	43	41	
PVA/PVP/Gly	Chloramine T	11	-	11	11	
PVP/Carr	AgNPs	0.8–4.87	-	1.9–9.56	1.2–8.31	[102]
PVP/ITA/CS/ZnO	-	0.4	-	-	0.8	[91]
PVP//ITA/CS/ZnO	AM	2.8	-	-	2.8	
PVP/PEG/Agar	NEO	22 ± 1	-	-	-	[148]
SF/CS/PVA	AgNPs	1.46–1.67	-	1.44–1.59	-	[108]
Alginate	SSD	16 ± 0.3	-	16 ± 0.6	16	[109]
Alginate/PG	SSD	15	16 ± 0.3	16 ± 0.3	16	
XG/PVA	ZnO	25–40	-	-	15–25	[114]

3.8.2. Cytotoxicity Studies

It is important to underline that from our study it emerged that the most used polymer for the development of hydrogels by radiation crosslinking for dressing applications is poly(vinyl alcohol). Cytotoxicity studies showed γ-crosslinked PVA hydrogel exhibited partial toxic effects on the L929 fibroblasts by comparison with PVA hydrogels crosslinked with glutaric aldehyde. This behavior has been associated with the partial crosslinking of the PVA polymer and some degradation products resulting after the irradiation process, or it is due to the high radiation dose used for the crosslinking reaction [152].

Considering that wound dressing hydrogels generally have as their main application the healing of skin wounds, this review takes into account the works that refer to the testing of these materials from the point of view of their cytotoxicity on skin-specific cells. The in vitro biological properties of BC/AA hydrogels were assessed by cytotoxicity test on human fibroblast cells. After a 24 h incubation, the hydrogels formulation was found nontoxic and showed cell viability above 88% [50]. The extract of PVA/ws-chitosan/glycerol hydrogels was nontoxic towards L929 mouse fibroblasts; moreover, they showed an accelerated wound healing process [89]. The cytotoxicity test using the NCTC Clone 929 cell line on the PVP hydrogels reinforced with PP showed cell viability above 75%. The percent of cell viability depended on the extract concentration [56]. PVA/alginate hydrogels, after incubation for 48 h, showed a considerable reduction in toxicity on L929 fibroblasts. All hydrogels tested showed a cell viability greater than 85% [84]. The cytotoxicity of AgNP/Gelatin/PVA hydrogels was investigated on normal human dermal fibroblasts showing 80% cell viability. The cell viability was little influenced by the increase in AgNP concentration and the incubation period [63]. The toxicity of CIP/PLGA-loaded PVA hydrogel prepared by γ-irradiation at 25 kGy was tested on human dermal fibroblasts. The studies indicated the non-toxicity of 244 mg CIP/PLGA-loaded PVA hydrogels [67]. In vitro cytotoxicity studies of gelatin/poly (γ-glutamic acid) hydrogels performed on the L929 fibroblasts revealed no cytotoxicity [74]. PVA/ws-chitosan/glycerol hydrogels produced by combining γ-irradiation at 40 kGy with a freeze-thawing cycle demonstrated non-toxicity towards L929 mouse fibroblasts. Moreover, these hydrogels showed cell viability higher than 100% after 24 h of incubation [101]. The ELT/PVA hydrogel demonstrated high cell viability (99.4%) on calcein and ethidium homodimer fibroblasts at lower concentrations (0.10% *w*/*v* ELT), whereas higher concentrations of ELT (0.15–0.25% *w*/*v*) revealed low values of viability [72]. The CM-CS/PEDGA hydrogels did not show significant cytotoxicity on the mouse fibroblast cell line L929, and even an increase in the number of cells was observed at lower extract concentrations (84.3%) [153]. PVA/sericin hydrogels were evaluated to determine the cytotoxicity of an indirect method on L929 mouse fibroblast cells. The cells cultured with the extracts of PVA/sericin hydrogels containing a higher sericin content (25% *w*/*v*) had proliferated more than cells cultured with the extracts of hydrogels containing a lower sericin content [147]. The PVP/AgNPs presented a low toxicity to human fibroblast cell line HaCaT of approx. 10%. For the PVA/AgNPs, cellular death was observed in up to just over 20% of the tested fibroblasts, so the hydrogel could be considered cytotoxic. Therefore, the PVA and PVP hydrogels with AgNPs were shown to be biocompatible, and they were considered nontoxic [83]. The effectiveness of the PVP/AgNPs hydrogels as a wound dressing was investigated for their cytotoxicity using mouse fibroblast cells (L929) and these hydrogels presented cell viability greater than 80% [65]. Cell viabilities for PVA/GEL bilayer hydrogel using 3T3-Swiss albino mouse embryonic fibroblasts were 98% (GEL) and 88% (PVA), respectively, thus showing good biocompatibility [138].

4. General Remarks and Future Perspectives

The scientific literature presents hydrogel wound dressings as promising candidates for improving traditional methods in various biomedical applications. Different advanced concepts have been developed in the last twenty years to prepare unique and biocompatible hydrogel wound dressings mainly using ionizing radiation.

This review paper includes methods of characterizing polymer hydrogels crosslinked with ionizing radiation. Furthermore, the paper includes general results published by different research groups that have studied the behavior of hydrogels used as wound dressings.

Each characterization method is essential and helps to obtain a hydrogel with properties that mimic human skin and its healing/regeneration capacity as faithfully as possible. In addition to crosslinking, the irradiation technology ensures dressing sterilization, a fundamental requirement for medical devices that come into contact with skin wounds. Additionally, this review paper includes studies on the antimicrobial and cytotoxicity properties of various hydrogels. Cytotoxicity studies represent an efficient method of testing cell proliferation and are discussed by several authors.

This review paper demonstrates that the hydrogels developed by radiation technologies are excellent dressing materials for wound healing. The crosslinked hydrogels in this way can absorb wound fluids, have good antimicrobial properties, improved functional adhesion, are cost-effective, and are easy to remove.

Hence, the polymers, antibacterial agents, drugs, and radiation synthesis used significantly improved the physicochemical properties of the prepared hydrogel dressings. Since e-beam and γ-irradiation are common sterilization methods, the properties of the irradiated hydrogels at doses of 25–30 kGy were investigated.

In the specialized scientific literature, a small number of clinical studies used hydrogel wound dressings. Therefore, it is necessary to carry out new such studies considering the useful properties of hydrogel dressings obtained by irradiation. The future development of hydrogel wound dressings will lead to much more precise and intelligent properties, biocompatibility, and well-regulated antibacterial activity.

Author Contributions: Conceptualization, M.D., I.C. and A.S.; data curation, M.D., I.C. and A.S.; writing—original draft preparation, M.D., I.C. and A.S.; writing—review and editing, M.D., I.C. and A.S. All authors have read and agreed to the published version of the manuscript.

Funding: This work was supported by a grant from the Ministry of Research, Innovation and Digitization, CNCS—UEFISCDI, project number PN-III-P1-1.1-PD-2021-0552, within PNCDI III, as well as from the Romanian National Core Program LAPLAS VII–contract no. 30N/2023.

Institutional Review Board Statement: Not applicable.

Informed Consent Statement: Not applicable.

Data Availability Statement: No new data were created or analyzed in this study. Data sharing is not applicable to this article.

Conflicts of Interest: The authors declare no conflict of interest.

Abbreviations

AA	Acrylic acid
AM	Amoxicillin
AgNPs	Ag nanoparticles
BC	Bacterial cellulose
CA	Cellulose acetate
CIP	Ciprofloxacin hydrochloride
CHS	Chondroitin sulfate
CMC	Carboxymethyl cellulose
CM-CS	Carboxymethyl-chitosan
CPC	Cetylpyridinium chloride
CS	Chitosan
CL	Colistin
DMAEM	2-(Dimethyl amino)ethyl methacrylate
Dv	virtual dose
Dg	gelation dose
E-beam	electron beam
ELT	Elastin
G	Radiochemical yield
GA	Gum acacia
GEL	Gelatin
Gly	Glycerin
GM	Gentamicin
G(S)	Radiochemical yield of the scission
G(X)	Radiochemical yield of the crosslinking
HACC	Quaternary ammonium chitosan
HEMA	Hydroxyethyl methacrylate
ITA	Itaconic acid
LAA	L-ascorbic acid
MMA	methyl methacrylate
MOG	Moringa oleifera gum
MMT	Montmorillonite clay
MRSA	Methicillin-resistant Staphylococcus aureus
NEO	Neomycin
PAA	Poly(acrylic acid)
PAAM	Poly(acrylamide)

PEGDA	Poly(ethylene glycol)diacrylate
PEG	Poly(ethylene glycol)
PEO	Poly(ethylene oxide)
PG	Prodigiosin
PGA	Polyglutamic acid
PHEMA	Poly(2-hydroxyethyl methacrylate)
PLGA	Poly(lactic-co-glycodic acid)
PLST	Plasticized starch
PVA	Polyvinyl alcohol
PVP	Polyvinylpyrrolidone
SA	Sodium alginate
SF	Silk fibroin
SSD	Silver sulfadiazine
SD-Na	Sulfadiazine sodium
TA	Tannic acid
TG	Tragacanth gum
TS	tensile strength
ws	water soluble

References

1. Zhao, X.; Wu, H.; Guo, B.; Dong, R.; Qiu, Y.; Ma, P.X. Antibacterial Anti-Oxidant Electroactive Injectable Hydrogel as Self-Healing Wound Dressing with Hemostasis and Adhesiveness for Cutaneous Wound Healing. *Biomaterials* **2017**, *122*, 34–47. [CrossRef] [PubMed]
2. Yuan, N.; Xu, L.; Xu, B.; Zhao, J.; Rong, J. Chitosan Derivative-Based Self-Healable Hydrogels with Enhanced Mechanical Properties by High-Density Dynamic Ionic Interactions. *Carbohydr. Polym.* **2018**, *193*, 259–267. [CrossRef]
3. Song, P.; Wang, H. High-Performance Polymeric Materials through Hydrogen-Bond Cross-Linking. *Adv. Mater.* **2020**, *32*, 1901244. [CrossRef] [PubMed]
4. Figueroa-Pizano, M.D.; Vélaz, I.; Peñas, F.J.; Zavala-Rivera, P.; Rosas-Durazo, A.J.; Maldonado-Arce, A.D.; Martínez-Barbosa, M.E. Effect of Freeze-Thawing Conditions for Preparation of Chitosan-Poly (Vinyl Alcohol) Hydrogels and Drug Release Studies. *Carbohydr. Polym.* **2018**, *195*, 476–485. [CrossRef] [PubMed]
5. Liu, X.; Duan, L.; Gao, G. Rapidly Self-Recoverable and Fatigue-Resistant Hydrogels Toughened by Chemical Crosslinking and Hydrophobic Association. *Eur. Polym. J.* **2017**, *89*, 185–194. [CrossRef]
6. Deng, A.; Yang, Y.; Du, S.; Yang, X.; Pang, S.; Wang, X.; Yang, S. Preparation of a Recombinant Collagen-Peptide (RHC)-Conjugated Chitosan Thermosensitive Hydrogel for Wound Healing. *Mater. Sci. Eng. C* **2021**, *119*, 111555. [CrossRef] [PubMed]
7. Chao, G.; Deng, H.; Huang, Q.; Jia, W.; Huang, W.; Gu, Y.; Tan, H.; Fan, L.; Liu, C.; Huang, A.; et al. Preparation and Characterization of PH Sensitive Semi-Interpenetrating Network Hydrogel Based on Methacrylic Acid, Bovine Serum Albumin (BSA), and PEG. *J. Polym. Res.* **2006**, *13*, 349–355. [CrossRef]
8. Pham, T.-N.; Jiang, Y.-S.; Su, C.-F.; Jan, J.-S. In Situ Formation of Silver Nanoparticles-Contained Gelatin-PEG-Dopamine Hydrogels via Enzymatic Cross-Linking Reaction for Improved Antibacterial Activities. *Int. J. Biol. Macromol.* **2020**, *146*, 1050–1059. [CrossRef] [PubMed]
9. Buwalda, S.J.; Boere, K.W.M.; Dijkstra, P.J.; Feijen, J.; Vermonden, T.; Hennink, W.E. Hydrogels in a Historical Perspective: From Simple Networks to Smart Materials. *J Control. Release* **2014**, *190*, 254–273. [CrossRef]
10. Akibumi, D. Gel Formation of Aqueous Solution of Polyvinyl Alcohol Irradiated by Gamma Rays from Cobalt-60. *J. Phys. Soc. Jpn.* **1958**, *13*, 722–727. [CrossRef]
11. Ikada, Y.; Mita, T.; Horii, F.; Sakurada, I.; Hatada, M. Preparation of Hydrogels by Radiation Technique. *Radiat. Phys. Chem.* **1977**, *9*, 633–645. [CrossRef]
12. Krahl, F.; Arndt, K.-F. Synthesis of Microgels by Radiation Methods. In *Chemical Design of Responsive Microgels*; Pich, A., Richtering, W., Eds.; Springer: Berlin/Heidelberg, Germany, 2011; pp. 95–128. ISBN 978-3-642-16379-1.
13. Sakurada, I.; Ikada, Y. Effects of Gamma Radiation on Polymer in Solution.(IV): Crosslinking and Degradation of Poly (Acrylic Acid) in Aqueous Solution. *Bull. Inst. Chem. Res. Kyoto Univ.* **1963**, *41*, 103–113.
14. Kaetsu, I. Immobilization of Biofunctional Substances. *Radiat. Phys. Chem.* **1981**, *18*, 343–356. [CrossRef]
15. Kaetsu, I. Recent Progress on the Immobilization of Biocomponent by Radiation Polymerization and the Application to Biomedical Uses. *Radiat. Phys. Chem.* **1985**, *25*, 517–528. [CrossRef]
16. Hoffman, A.S. A Review of the Use of Radiation plus Chemical and Biochemical Processing Treatments to Prepare Novel Biomaterials. *Radiat. Phys. Chem.* **1981**, *18*, 323–342. [CrossRef]
17. Rosiak, J.; Burozak, K.; Pękala, W. Polyacrylamide Hydrogels as Sustained Release Drug Delivery Dressing Materials. *Radiat. Phys. Chem.* **1983**, *22*, 907–915. [CrossRef]
18. Rosiak, J.; Rucinska-Rybus, A.; Pekala, W. Method of Manufacturing Hydrogel Dressings. U.S. Patent 4,871,490, 3 October 1989.

19. Benamer, S.; Mahlous, M.; Boukrif, A.; Mansouri, B.; Youcef, S.L. Synthesis and Characterisation of Hydrogels Based on Poly(Vinyl Pyrrolidone). *Nucl. Instrum. Methods Phys. Res. B* **2006**, *248*, 284–290. [CrossRef]
20. Darwis, D.; Hilmy, N.; Hardiningsih, L.; Erlinda, T. Poly(N-Vinylpyrrolidone) Hydrogels: 1.Radiation Polymerization and Crosslinking of N-Vinylpyrrolidone. *Radiat. Phys. Chem.* **1993**, *42*, 907–910. [CrossRef]
21. Kaplan Can, H. Synthesis of Persulfate Containing Poly (N-Vinyl-2-Pyrrolidone) (PVP) Hydrogels in Aqueous Solutions by γ-Induced Radiation. *Radiat. Phys. Chem.* **2005**, *72*, 703–710. [CrossRef]
22. Özyürek, C.; Çaykara, T.; Kantoğlu, Ö.; Güven, O. Radiation Synthesis of Poly(N-Vinyl-2-Pyrrolidone-g-Tartaric Acid) Hydrogels and Their Swelling Behaviors. *Polym. Adv. Technol.* **2002**, *13*, 87–93. [CrossRef]
23. Dafader*, N.C.; Haque, M.E.; Akhtar, F. Synthesis of Hydrogel from Aqueous Solution of Poly(Vinyl Pyrrolidone) with Agar by Gamma-Rays Irradiation. *Polym. Plast. Technol. Eng.* **2005**, *44*, 243–251. [CrossRef]
24. Dergunov, S.A.; Nam, I.K.; Mun, G.A.; Nurkeeva, Z.S.; Shaikhutdinov, E.M. Radiation Synthesis and Characterization of Stimuli-Sensitive Chitosan–Polyvinyl Pyrrolidone Hydrogels. *Radiat. Phys. Chem.* **2005**, *72*, 619–623. [CrossRef]
25. Dergunov, S.A.; Mun, G.A. γ-Irradiated Chitosan-Polyvinyl Pyrrolidone Hydrogels as PH-Sensitive Protein Delivery System. *Radiat. Phys. Chem.* **2009**, *78*, 65–68. [CrossRef]
26. Sohail, K.; Ullah Khan, I.; Shahzad, Y.; Hussain, T.; Ranjha, N. PH-Sensitive Polyvinylpyrrolidone-Acrylic Acid Hydrogels: Impact of Material Parameters on Swelling and Drug Release. *Braz. J. Pharm. Sci.* **2014**, *50*, 173–184. [CrossRef]
27. Abd Alla, S.G.; Nizam El-Din, H.M.; El-Naggar, A.W.M. Structure and Swelling-Release Behaviour of Poly(Vinyl Pyrrolidone) (PVP) and Acrylic Acid (AAc) Copolymer Hydrogels Prepared by Gamma Irradiation. *Eur. Polym. J.* **2007**, *43*, 2987–2998. [CrossRef]
28. Gottlieb, R.; Schmidt, T.; Arndt, K.-F. Synthesis of Temperature-Sensitive Hydrogel Blends by High-Energy Irradiation. *Nucl. Instrum. Methods Phys. Res. B* **2005**, *236*, 371–376. [CrossRef]
29. Awadallah-F, A. Five Years in Vitro Study of (Poly Vinyl Alcohol/Poly Vinyl Pyrrolidone/Poly Acrylic Acid) Hydrogel to Mimic the Knee Joint Meniscus. *Polym. Adv. Technol.* **2014**, *25*, 581–587. [CrossRef]
30. Razzak, M.T.; Zainuddin; Erizal; Dewi, S.P.; Lely, H.; Taty, E.; Sukirno. The Characterization of Dressing Component Materials and Radiation Formation of PVA–PVP Hydrogel. *Radiat. Phys. Chem.* **1999**, *55*, 153–165. [CrossRef]
31. Himly, N.; Darwis, D.; Hardiningsih, L. Poly(n-Vinylpyrrolidone) Hydrogels: 2.Hydrogel Composites as Wound Dressing for Tropical Environment. *Radiat. Phys. Chem.* **1993**, *42*, 911–914. [CrossRef]
32. Lugão, A.B.; Rogero, S.O.; Malmonge, S.M. Rheological Behaviour of Irradiated Wound Dressing Poly(Vinyl Pyrrolidone) Hydrogels. *Radiat. Phys. Chem.* **2002**, *63*, 543–546. [CrossRef]
33. Pekel Bayramgil, N.; Yoshii, F.; Kume, T.; Güven, O. Radiation Crosslinking of Biodegradable Hydroxypropylmethylcellulose. *Carbohyd. Polym.* **2004**, *55*, 139–147. [CrossRef]
34. Rashid, T.U.; Rahman, M.M.; Kabir, S.; Shamsuddin, S.M.; Khan, M.A. A New Approach for the Preparation of Chitosan from γ-Irradiation of Prawn Shell: Effects of Radiation on the Characteristics of Chitosan. *Polym. Int.* **2012**, *61*, 1302–1308. [CrossRef]
35. Wang, M.; Xu, L.; Hu, H.; Zhai, M.; Peng, J.; Nho, Y.; Li, J.; Wei, G. Radiation Synthesis of PVP/CMC Hydrogels as Wound Dressing. *Nucl. Instrum. Methods Phys. Res. B* **2007**, *265*, 385–389. [CrossRef]
36. Park, K.R.; Nho, Y.C. Preparation and Characterization by Radiation of Hydrogels of PVA and PVP Containing Aloe Vera. *J. Appl. Polym. Sci.* **2004**, *91*, 1612–1618. [CrossRef]
37. Nho, Y.C.; Park, K.R. Preparation and Properties of PVA/PVP Hydrogels Containing Chitosan by Radiation. *J. Appl. Polym. Sci.* **2002**, *85*, 1787–1794. [CrossRef]
38. Singh, R.; Singh, D. Radiation Synthesis of PVP/Alginate Hydrogel Containing Nanosilver as Wound Dressing. *J. Mater. Sci. Mater. Med.* **2012**, *23*, 2649–2658. [CrossRef]
39. Song, Y.; Xu, L.; Xu, L.; Deng, L. Radiation Cross-Linked Gelatin/Sodium Alginate/Carboxymethylcellulose Sodium Hydrogel for the Application as Debridement Glue Paste. *Polym. Bull.* **2022**, *79*, 725–742. [CrossRef]
40. Abd El-Rehim, H.A.; Hegazy, E.A.; Khalil, F.H.; Hamed, N.A. Radiation Preparation of Drug Carriers Based Polyacrylic Acid (PAAc) Using Poly(Vinyl Pyrrolidone) (PVP) as a Template Polymer. *Nucl. Instrum. Methods Phys. Res. B* **2007**, *254*, 105–112. [CrossRef]
41. Kadłubowski, S.; Henke, A.; Ulański, P.; Rosiak, J.M. Hydrogels of Polyvinylpyrrolidone (PVP) and Poly(Acrylic Acid) (PAA) Synthesized by Radiation-Induced Crosslinking of Homopolymers. *Radiat. Phys. Chem.* **2010**, *79*, 261–266. [CrossRef]
42. Tavakoli, S.; Klar, A.S. Advanced Hydrogels as Wound Dressings. *Biomolecules* **2020**, *10*, 1169. [CrossRef]
43. Kofinas, P.; Athanassiou, V.; Merrill, E.W. Hydrogels Prepared by Electron Irradiation of Poly(Ethylene Oxide) in Water Solution: Unexpected Dependence of Cross-Link Density and Protein Diffusion Coefficients on Initial PEO Molecular Weight. *Biomaterials* **1996**, *17*, 1547–1550. [CrossRef] [PubMed]
44. Jabbari, E.; Nozari, S. Swelling Behavior of Acrylic Acid Hydrogels Prepared by γ-Radiation Crosslinking of Polyacrylic Acid in Aqueous Solution. *Eur. Polym. J.* **2000**, *36*, 2685–2692. [CrossRef]
45. Peppas, N.A.; Merrill, E.W. Crosslinked Poly(Vinyl Alcohol) Hydrogels as Swollen Elastic Networks. *J. Appl. Polym. Sci.* **1977**, *21*, 1763–1770. [CrossRef]
46. Arndt, K.-F.; Schmidt, T.; Reichelt, R. Thermo-Sensitive Poly(Methyl Vinyl Ether) Micro-Gel Formed by High Energy Radiation. *Polymer* **2001**, *42*, 6785–6791. [CrossRef]

47. Raza, M.A.; Park, S.H. Irradiated Ch/GG/PVP-Based Stimuli-Responsive Hydrogels for Controlled Drug Release. *J. Appl. Polym. Sci.* **2020**, *137*, 49041. [CrossRef]
48. Güven, O.; Şen, M.; Karadağ, E.; Saraydın, D. A Review on the Radiation Synthesis of Copolymeric Hydrogels for Adsorption and Separation Purposes. *Radiat. Phys. Chem.* **1999**, *56*, 381–386. [CrossRef]
49. Krömmelbein, C.; Mütze, M.; Konieczny, R.; Schönherr, N.; Griebel, J.; Gerdes, W.; Mayr, S.G.; Riedel, S. Impact of High-Energy Electron Irradiation on Mechanical, Structural and Chemical Properties of Agarose Hydrogels. *Carbohydr. Polym.* **2021**, *263*, 117970. [CrossRef]
50. Mohamad, N.; Buang, F.; Mat Lazim, A.; Ahmad, N.; Martin, C.; Mohd Amin, M.C.I. Characterization and Biocompatibility Evaluation of Bacterial Cellulose-Based Wound Dressing Hydrogel: Effect of Electron Beam Irradiation Doses and Concentration of Acrylic Acid. *J. Biomed. Mater. Res. B Appl. Biomater.* **2017**, *105*, 2553–2564. [CrossRef]
51. Demeter, M.; Călina, I.; Scărișoreanu, A.; Micutz, M.; Kaya, M.A. Correlations on the Structure and Properties of Collagen Hydrogels Produced by E-Beam Crosslinking. *Materials* **2022**, *15*, 7663. [CrossRef]
52. Călina, I.; Demeter, M.; Scărișoreanu, A.; Sătulu, V.; Mitu, B. One Step E-Beam Radiation Cross-Linking of Quaternary Hydrogels Dressings Based on Chitosan-Poly(Vinyl-Pyrrolidone)-Poly(Ethylene Glycol)-Poly(Acrylic Acid). *Int. J. Mol. Sci.* **2020**, *21*, 9236. [CrossRef]
53. Yoshii, F.; Zhanshan, Y.; Isobe, K.; Shinozaki, K.; Makuuchi, K. Electron Beam Crosslinked PEO and PEO/PVA Hydrogels for Wound Dressing. *Radiat. Phys. Chem.* **1999**, *55*, 133–138. [CrossRef]
54. Yoshii, F.; Makuuchi, K.; Darwis, D.; Iriawan, T.; Razzak, M.T.; Rosiak, J.M. Heat Resistance Poly(Vinyl Alcohol) Hydrogel. *Radiat. Phys. Chem.* **1995**, *46*, 169–174. [CrossRef]
55. Wu, M.; Bao, B.; Yoshii, F.; Makuuchi, K. Irradiation of Crosslinked, Poly(Vinyl Alcohol) Blended Hydrogel for Wound Dressing. *J. Radioanal. Nucl. Chem.* **2001**, *250*, 391–395. [CrossRef]
56. Lopérgolo, L.C.; Lugão, A.B.; Catalaini, L.H. Development of a Poly(N-Vinyl-2-Pyrrolidone)/Poly (Ethylene Glycol) Hydrogel Membrane Reinforced with Methyl Methacrylate-Grafted Polypropylene Fibers for Possible Use as Wound Dressing. *J. Appl. Polym. Sci.* **2002**, *86*, 662–666. [CrossRef]
57. Lugão, A.B.; Machado, L.D.B.; Miranda, L.F.; Alvarez, M.R.; Rosiak, J.M. Study of Wound Dressing Structure and Hydration/Dehydration Properties. *Radiat. Phys. Chem.* **1998**, *52*, 319–322. [CrossRef]
58. Higa, O.Z.; Rogero, S.O.; Machado, L.D.B.; Mathor, M.B.; Lugão, A.B. Biocompatibility Study for PVP Wound Dressing Obtained in Different Conditions. *Radiat. Phys. Chem.* **1999**, *55*, 705–707. [CrossRef]
59. Ajji, Z.; Mirjalili, G.; Alkhatab, A.; Dada, H. Use of Electron Beam for the Production of Hydrogel Dressings. *Radiat. Phys. Chem.* **2008**, *77*, 200–202. [CrossRef]
60. Abd El-Mohdy, H.L.; Hegazy, E.-S.A. Preparation of Polyvinyl Pyrrolidone-Based Hydrogels by Radiation-Induced Crosslinking with Potential Application as Wound Dressing. *J. Macromol. Sci. A* **2008**, *45*, 995–1002. [CrossRef]
61. Nizam El-Din, H.M.; Ibraheim, D.M. Biological Applications of Nanocomposite Hydrogels Prepared by Gamma-Radiation Copolymerization of Acrylic Acid (AAc) onto Plasticized Starch (PLST)/Montmorillonite Clay (MMT)/Chitosan (CS) Blends. *Int. J. Biol. Macromol.* **2021**, *192*, 151–160. [CrossRef]
62. Zhou, Y.; Zhao, Y.; Wang, L.; Xu, L.; Zhai, M.; Wei, S. Radiation Synthesis and Characterization of Nanosilver/Gelatin/Carboxymethyl Chitosan Hydrogel. *Radiat. Phys. Chem.* **2012**, *81*, 553–560. [CrossRef]
63. Leawhiran, N.; Pavasant, P.; Soontornvipart, K.; Supaphol, P. Gamma Irradiation Synthesis and Characterization of AgNP/Gelatin/PVA Hydrogels for Antibacterial Wound Dressings. *J. Appl. Polym. Sci.* **2014**, *131*, 41138. [CrossRef]
64. Micic, M.; Milic, T.V.; Mitric, M.; Jokic, B.; Suljovrujic, E. Radiation Synthesis, Characterisation and Antimicrobial Application of Novel Copolymeric Silver/Poly(2-Hydroxyethyl Methacrylate/Itaconic Acid) Nanocomposite Hydrogels. *Polym. Bull.* **2013**, *70*, 3347–3357. [CrossRef]
65. Khampieng, T.; Brikshavana, P.; Supaphol, P. Silver Nanoparticle-Embedded Poly(Vinyl Pyrrolidone) Hydrogel Dressing: Gamma-Ray Synthesis and Biological Evaluation. *J. Biomater. Sci. Polym. Ed.* **2014**, *25*, 826–842. [CrossRef] [PubMed]
66. Sandri, G.; Bonferoni, M.C.; Rossi, S.; Ferrari, F.; Mori, M.; Cervio, M.; Riva, F.; Liakos, I.; Athanassiou, A.; Saporito, F.; et al. Platelet Lysate Embedded Scaffolds for Skin Regeneration. *Expert Opin. Drug. Deliv.* **2015**, *12*, 525–545. [CrossRef] [PubMed]
67. Choipang, C.; Chuysinuan, P.; Suwantong, O.; Ekabutr, P.; Supaphol, P. Hydrogel Wound Dressings Loaded with PLGA/Ciprofloxacin Hydrochloride Nanoparticles for Use on Pressure Ulcers. *J. Drug. Deliv. Sci. Technol.* **2018**, *47*, 106–114. [CrossRef]
68. Huang, X.; Zhang, Y.; Zhang, X.; Xu, L.; Chen, X.; Wei, S. Influence of Radiation Crosslinked Carboxymethyl-Chitosan/Gelatin Hydrogel on Cutaneous Wound Healing. *Mater. Sci. Eng. C* **2013**, *33*, 4816–4824. [CrossRef]
69. Fan, L.; Yang, H.; Yang, J.; Peng, M.; Hu, J. Preparation and Characterization of Chitosan/Gelatin/PVA Hydrogel for Wound Dressings. *Carbohydr. Polym.* **2016**, *146*, 427–434. [CrossRef]
70. Guo, W.; Yang, M.; Liu, S.; Zhang, X.; Zhang, B.; Chen, Y. Chitosan/Polyvinyl Alcohol/Tannic Acid Multiple Network Composite Hydrogel: Preparation and Characterization. *Iran Polym. J.* **2021**, *30*, 1159–1168. [CrossRef]
71. Ghobashy, M.M.; Elbarbary, A.M.; Hegazy, D.E.; Maziad, N.A. Radiation Synthesis of PH-Sensitive 2-(Dimethylamino)Ethyl Methacrylate/ Polyethylene Oxide/ZnS Nanocomposite Hydrogel Membrane for Wound Dressing Application. *J. Drug. Deliv. Sci. Technol.* **2022**, *73*, 103399. [CrossRef]

72. Del Prado Audelo, M.; Gomez Saldivar, F.; Pérez Díaz, M.; Sanchez Sanchez, R.; González Torres, M.; González Del Carmen, M.; Figueroa González, G.; Sharifi Rad, J.; Reyes Hernández, O.; Cortés, H.; et al. Radiation-Induced Graft Polymerization of Elastin onto Polyvinylpyrrolidone as a Possible Wound Dressing. *Cell Mol. Biol.* **2021**, *67*, 64–72. [CrossRef]
73. Singh, B.; Rajneesh. Gamma Radiation Synthesis and Characterization of Gentamicin Loaded Polysaccharide Gum Based Hydrogel Wound Dressings. *J. Drug. Deliv. Sci. Technol.* **2018**, *47*, 200–208. [CrossRef]
74. Liu, B.; Huang, W.; Yang, G.; An, Y.; Yin, Y.; Wang, N.; Jiang, B. Preparation of Gelatin/Poly (γ-Glutamic Acid) Hydrogels with Stimulated Response by Hot-Pressing Preassembly and Radiation Crosslinking. *Mater. Sci. Eng. C* **2020**, *116*, 111259. [CrossRef]
75. Escutia-Guadarrama, L.; Morales, D.; Pérez-Calixto, D.; Burillo, G. Development of Polyphenol-Functionalized Gelatin-Poly(Vinylpyrrolidone) IPN for Potential Biomedical Applications. *Polymers* **2022**, *14*, 4705. [CrossRef] [PubMed]
76. Fan, L.; Yang, J.; Wu, H.; Hu, Z.; Yi, J.; Tong, J.; Zhu, X. Preparation and Characterization of Quaternary Ammonium Chitosan Hydrogel with Significant Antibacterial Activity. *Int. J. Biol. Macromol.* **2015**, *79*, 830–836. [CrossRef] [PubMed]
77. Moise, V.; Vasilca, S.; Baltac, A.; Pintilie, C.; Virgolici, M.; Cutrubinis, M.; Kamerzan, C.; Dragan, D.; Ene, M.; Albota, F.; et al. Physicochemical Study for Characterization of Lyophilized Collagens Irradiated with Gamma Radiation and for Optimization of Medical Device Manufacturing Process. *Radiat. Phys. Chem.* **2020**, *170*, 108658. [CrossRef]
78. Singh, B.; Kumar, A. Graft and Crosslinked Polymerization of Polysaccharide Gum to Form Hydrogel Wound Dressings for Drug Delivery Applications. *Carbohydr. Res.* **2020**, *489*, 107949. [CrossRef]
79. Park, J.-S.; Kuang, J.; Gwon, H.-J.; Lim, Y.-M.; Jeong, S.-I.; Shin, Y.-M.; Seob Khil, M.; Nho, Y.-C. Synthesis and Characterization of Zinc Chloride Containing Poly(Acrylic Acid) Hydrogel by Gamma Irradiation. *Radiat. Phys. Chem.* **2013**, *88*, 60–64. [CrossRef]
80. Masyruroh, S.; Lasmawati, D.; Suryani, N.; Sumarlin, L.; Nurlidar, F. Release of Ciprofloxacin from Gamma-Irradiated PEGDA-Chitosan Hydrogel. *AIP Conf. Proc.* **2020**, *2296*, 020084. [CrossRef]
81. Tomić, S.L.; Mićić, M.M.; Dobić, S.N.; Filipović, J.M.; Suljovrujić, E.H. Smart Poly(2-Hydroxyethyl Methacrylate/Itaconic Acid) Hydrogels for Biomedical Application. *Radiat. Phys. Chem.* **2010**, *79*, 643–649. [CrossRef]
82. Varshney, L. Role of Natural Polysaccharides in Radiation Formation of PVA–Hydrogel Wound Dressing. *Nucl. Instrum. Methods Phys. Res. B* **2007**, *255*, 343–349. [CrossRef]
83. Alcântara, M.T.S.; Lincopan, N.; Santos, P.M.; Ramirez, P.A.; Brant, A.J.C.; Riella, H.G.; Lugão, A.B. Simultaneous Hydrogel Crosslinking and Silver Nanoparticle Formation by Using Ionizing Radiation to Obtain Antimicrobial Hydrogels. *Radiat. Phys. Chem.* **2020**, *169*, 108777. [CrossRef]
84. Nam, S.Y.; Nho, Y.C.; Hong, S.H.; Chae, G.T.; Jang, H.S.; Suh, T.S.; Ahn, W.S.; Ryu, K.E.; Chun, H.J. Evaluations of Poly(Vinyl Alcohol)/Alginate Hydrogels Cross-Linked by γ-Ray Irradiation Technique. *Macromol. Res.* **2004**, *12*, 219–224. [CrossRef]
85. Abd El-Mohdy, H.L. Radiation Synthesis of Nanosilver/Poly Vinyl Alcohol/Cellulose Acetate/Gelatin Hydrogels for Wound Dressing. *J. Polym. Res.* **2013**, *20*, 177. [CrossRef]
86. Afshari, M.J.; Sheikh, N.; Afarideh, H. PVA/CM-Chitosan/Honey Hydrogels Prepared by Using the Combined Technique of Irradiation Followed by Freeze-Thawing. *Radiat. Phys. Chem.* **2015**, *113*, 28–35. [CrossRef]
87. El Salmawi, K.M. Gamma Radiation-Induced Crosslinked PVA/Chitosan Blends for Wound Dressing. *J. Macromol. Sci. A* **2007**, *44*, 541–545. [CrossRef]
88. Nacer Khodja, A.; Mahlous, M.; Tahtat, D.; Benamer, S.; Larbi Youcef, S.; Chader, H.; Mouhoub, L.; Sedgelmaci, M.; Ammi, N.; Mansouri, M.B.; et al. Evaluation of Healing Activity of PVA/Chitosan Hydrogels on Deep Second Degree Burn: Pharmacological and Toxicological Tests. *Burns* **2013**, *39*, 98–104. [CrossRef]
89. Yang, X.; Yang, K.; Wu, S.; Chen, X.; Yu, F.; Li, J.; Ma, M.; Zhu, Z. Cytotoxicity and Wound Healing Properties of PVA/Ws-Chitosan/Glycerol Hydrogels Made by Irradiation Followed by Freeze–Thawing. *Radiat. Phys. Chem.* **2010**, *79*, 606–611. [CrossRef]
90. Gwon, H.-J.; Lim, Y.-M.; An, S.-J.; Youn, M.-H.; Han, S.-H.; Chang, H.-N.; Nho, Y.-C. Characterization of PVA/Glycerin Hydrogels Made by γ-Irradiation for Advanced Wound Dressings. *Korean J. Chem. Eng.* **2009**, *26*, 1686–1688. [CrossRef]
91. Fadl, F.; Abdel Ghaffar, A.; El-Sawy, N. Radiation Preparation and Antibacterial Activity of Poly(N-Vinyl Pyrrolidone/Chitosan/Itaconic Acid/ZnO) Nanocomposite Hydrogel for Controlled Release of Amoxicillin. *Polym. Polym. Compos.* **2022**, *30*, 1–13.
92. Razzak, M.T.; Darwis, D.; Zainuddin; Sukirno. Irradiation of Polyvinyl Alcohol and Polyvinyl Pyrrolidone Blended Hydrogel for Wound Dressing. *Radiat. Phys. Chem.* **2001**, *62*, 107–113. [CrossRef]
93. Eid, M.; El-Arnaouty, M.; Salah, M.; Soliman, E.-S.; Hegazy, E.-S. Radiation Synthesis and Characterization of Poly(Vinyl Alcohol)/Poly(N-Vinyl-2-Pyrrolidone) Based Hydrogels Containing Silver Nanoparticles. *J. Polym. Res.* **2012**, *19*, 1–10. [CrossRef]
94. Park, K.R.; Nho, Y.C. Preparation and Characterization by Radiation of Poly(Vinyl Alcohol) and Poly(N-Vinylpyrrolidone) Hydrogels Containing Aloe Vera. *J. Appl. Polym. Sci.* **2003**, *90*, 1477–1485. [CrossRef]
95. El-Arnaouty, M.B.; Eid, M.; Salah, M.; Hegazy, E.-S.A. Preparation and Characterization of Poly Vinyl Alcohol/Poly Vinyl Pyrrolidone/Clay Based Nanocomposite by Gamma Irradiation. *J. Macrom. Sci. A* **2012**, *49*, 1041–1051. [CrossRef]
96. Yunoki, S.; Kohta, M.; Ohyabu, Y.; Sekiguchi, M.; Kubo, T.; Iwasaki, T. Electrostatic Immobilization of Cetylpyridinium Chloride to Poly(Vinyl Alcohol) Hydrogels for the Simple Fabrication of Wound Dressings with the Suppressed Release of Antibacterial Agents. *J. Appl. Polym. Sci.* **2014**, *131*, 40456. [CrossRef]
97. Yu, H.; Xu, X.; Chen, X.; Hao, J.; Jing, X. Medicated Wound Dressings Based on Poly(Vinyl Alcohol)/Poly(N-Vinyl Pyrrolidone)/Chitosan Hydrogels. *J. Appl. Polym. Sci.* **2006**, *101*, 2453–2463. [CrossRef]

98. Nho, Y.-C.; Lim, Y.-M.; Gwon, H.-J.; Choi, E.-K. Preparation and Characterization of PVA/PVP/Glycerin/Antibacterial Agent Hydrogels Using γ-Irradiation Followed by Freeze-Thawing. *Korean J. Chem. Eng.* **2009**, *26*, 1675–1678. [CrossRef]
99. Park, K.R.; Nho, Y.C. Synthesis of PVA/PVP Hydrogels Having Two-Layer by Radiation and Their Physical Properties. *Radiat. Phys. Chem.* **2003**, *67*, 361–365. [CrossRef]
100. Yang, X.; Liu, Q.; Chen, X.; Yu, F.; Zhu, Z. Investigation of PVA/Ws-Chitosan Hydrogels Prepared by Combined γ-Irradiation and Freeze-Thawing. *Carbohydr. Polym.* **2008**, *73*, 401–408. [CrossRef]
101. Yang, X.; Yang, K.; Yu, F.; Chen, X.; Wu, S.; Zhu, Z. Preparation of Novel Bilayer Hydrogels by Combination of Irradiation and Freeze–Thawing and Their Physical and Biological Properties. *Polym. Int.* **2009**, *58*, 1291–1298. [CrossRef]
102. Singh, D.; Singh, A.; Singh, R. Polyvinyl Pyrrolidone/Carrageenan Blend Hydrogels with Nanosilver Prepared by Gamma Radiation for Use as an Antimicrobial Wound Dressing. *J. Biomater. Sci. Polym. Ed.* **2015**, *26*, 1269–1285. [CrossRef]
103. Awanthi, D.; Silva, D.A.; Hettiarachchi, B.; Nayanajith, L.D.C.; Prasada, I.M.; Motha, J.T.S. Development of a PVP/Kappa-Carrageenan/PEG Hydrogel Dressing for Wound Healing Applications in Sri Lanka. *J. Natl. Sci. Found.* **2011**, *39*, 25–33. [CrossRef]
104. Şen, M.; Avcı, E.N. Radiation Synthesis of Poly(N-Vinyl-2-Pyrrolidone)–κ-Carrageenan Hydrogels and Their Use in Wound Dressing Applications. I. Preliminary Laboratory Tests. *J. Biomed. Mater. Res. A* **2005**, *74A*, 187–196. [CrossRef]
105. Soler, D.M.; Rodríguez, Y.; Correa, H.; Moreno, A.; Carrizales, L. Pilot Scale-up and Shelf Stability of Hydrogel Wound Dressings Obtained by Gamma Radiation. *Radiat. Phys. Chem.* **2012**, *81*, 1249–1253. [CrossRef]
106. Ajji, Z.; Othman, I.; Rosiak, J.M. Production of Hydrogel Wound Dressings Using Gamma Radiation. *Nucl. Instrum. Methods Phys. Res. B* **2005**, *229*, 375–380. [CrossRef]
107. Temirkhanova, G.E.; Burasheva, G.; Abilov, Z.A.; Irmukhametova, G.; Mun, G.; Beksultanov, Z.I.; Myktybayeva, Z.K.; Shnaukshta, V.S. Creation of Polymer Hydrogel Dressings with Herbal Medicinal Substance "Alkhydin" and Their Properties. *Eurasian Chem. Technol. J.* **2017**, *19*, 57. [CrossRef]
108. Uttayarat, P.; Eamsiri, J.; Tangthong, T.; Suwanmala, P. Radiolytic Synthesis of Colloidal Silver Nanoparticles for Antibacterial Wound Dressings. *Adv. Mater. Sci. Eng.* **2015**, *2015*, 376082. [CrossRef]
109. Elkenawy, N.M.; Karam, H.M.; Aboul-Magd, D.S. Development of Gamma Irradiated SSD-Embedded Hydrogel Dyed with Prodigiosin as a Smart Wound Dressing: Evaluation in a MDR Infected Burn Rat Model. *Int. J. Biol. Macromol.* **2022**, *211*, 170–182. [CrossRef]
110. Singh, B.; Pal, L. Radiation Crosslinking Polymerization of Sterculia Polysaccharide–PVA–PVP for Making Hydrogel Wound Dressings. *Int. J. Biol. Macromol.* **2011**, *48*, 501–510. [CrossRef]
111. Singh, B.; Varshney, L.; Francis, S.; Rajneesh. Synthesis and Characterization of Tragacanth Gum Based Hydrogels by Radiation Method for Use in Wound Dressing Application. *Radiat. Phys. Chem.* **2017**, *135*, 94–105. [CrossRef]
112. Singh, B.; Singh, J.; Rajneesh. Application of Tragacanth Gum and Alginate in Hydrogel Wound Dressing's Formation Using Gamma Radiation. *Carbohyd. Polym. Technol. Appl.* **2021**, *2*, 100058. [CrossRef]
113. Singh, B.; Varshney, L.; Francis, S.; Rajneesh. Designing Tragacanth Gum Based Sterile Hydrogel by Radiation Method for Use in Drug Delivery and Wound Dressing Applications. *Int. J. Biol. Macromol.* **2016**, *88*, 586–602. [CrossRef] [PubMed]
114. Raafat, A.I.; El-Sawy, N.M.; Badawy, N.A.; Mousa, E.A.; Mohamed, A.M. Radiation Fabrication of Xanthan-Based Wound Dressing Hydrogels Embedded ZnO Nanoparticles: In Vitro Evaluation. *Int. J. Biol. Macromol.* **2018**, *118*, 1892–1902. [CrossRef] [PubMed]
115. Gulrez, S.K.H.; Al-Assaf, S.; Phillips, G.O. Hydrogels: Methods of Preparation, Characterisation and Applications. In *Progress in Molecular and Environmental Bioengineering—From Analysis and Modeling to Technology Applications*; Carpi, A., Ed.; IntechOpen: Rijeka, Croatia, 2011.
116. Sen, M.; Çopuroğlu, M. A Comparative Study of Gamma Irradiation of Poly(Ethylene-Co-Vinyl Acetate) and Poly(Ethylene-Co-Vinyl Acetate)/Carbon Black Mixture. *Mater. Chem. Phys.* **2005**, *93*, 154–158. [CrossRef]
117. Katayama, T.; Nakauma, M.; Todoriki, S.; Phillips, G.O.; Tada, M. Radiation-Induced Polymerization of Gum Arabic (Acacia Senegal) in Aqueous Solution. *Food Hydrocoll.* **2006**, *20*, 983–989. [CrossRef]
118. Nagasawa, N.; Yagi, T.; Kume, T.; Yoshii, F. Radiation Crosslinking of Carboxymethyl Starch. *Carbohydr. Polym.* **2004**, *58*, 109–113. [CrossRef]
119. Demeter, M.; Călina, I.; Vancea, C.; Şen, M.; Kaya, M.G.A.; Mănăilă, E.; Dumitru, M.; Meltzer, V. E-Beam Processing of Collagen-Poly(N-Vinyl-2-Pyrrolidone) Double-Network Superabsorbent Hydrogels: Structural and Rheological Investigations. *Macromol. Res.* **2019**, *27*, 255–267. [CrossRef]
120. Mozalewska, W.; Czechowska-Biskup, R.; Olejnik, A.K.; Wach, R.A.; Ulański, P.; Rosiak, J.M. Chitosan-Containing Hydrogel Wound Dressings Prepared by Radiation Technique. *Radiat. Phys. Chem.* **2017**, *134*, 1–7. [CrossRef]
121. Charlesby, A. (Ed.) Chapter 2—Radiation Units. In *Atomic Radiation and Polymers*; International Series of Monographs on Radiation Effects in Materials; Pergamon Press: New York, NY, USA, 1960; Volume 1, pp. 16–24. ISBN 978-1-4831-9776-0.
122. Darwis, D.; Erizal; Abbas, B.; Nurlidar, F.; Putra, D.P. Radiation Processing of Polymers for Medical and Pharmaceutical Applications. *Macromol. Symp.* **2015**, *353*, 15–23. [CrossRef]
123. Olejniczak, J.; Rosiak, J.; Charlesby, A. Gel/Dose Curves for Polymers Undergoing Simultaneous Crosslinking and Scission. *Int. J. Radiat. Appl. Instrum. C* **1991**, *37*, 499–504. [CrossRef]

124. Wach, R.A.; Mitomo, H.; Nagasawa, N.; Yoshii, F. Radiation Crosslinking of Methylcellulose and Hydroxyethylcellulose in Concentrated Aqueous Solutions. *Nucl. Instrum. Methods Phys. Res. B* **2003**, *211*, 533–544. [CrossRef]
125. Rosiak, J.M.; Janik, I.; Kadlubowski, S.; Kozicki, M.; Kujawa, P.; Stasica, P.U.; Ulanski, P. *Radiation Synthesis and Modification of Polymers for Biomedical Applications*; International Atomic Energy Agency (IAEA): Vienna, Austria, 2002.
126. Buxton, G.V.; Greenstock, C.L.; Helman, W.P.; Ross, A.B. Critical Review of Rate Constants for Reactions of Hydrated Electrons, Hydrogen Atoms and Hydroxyl Radicals (OH/O− in Aqueous Solution. *J. Phys. Chem. Ref. Data* **1988**, *17*, 513–886. [CrossRef]
127. Demeter, M.; Virgolici, M.; Vancea, C.; Scarisoreanu, A.; Kaya, M.G.A.; Meltzer, V. Network Structure Studies on γ–Irradiated Collagen–PVP Superabsorbent Hydrogels. *Radiat. Phys. Chem.* **2017**, *131*, 51–59. [CrossRef]
128. Awadallah-F, A. Synergistic Effect of Poly(Acrylamide)-Incorporated Poly(L-Ascorbic Acid) Hydrogels in Controlled Release and Wound Dressings. *Des. Monomers. Polym.* **2014**, *17*, 466–480. [CrossRef]
129. Czechowska-Biskup, R.; Rokita, B.; Ulanski, P.; Rosiak, J.M. Radiation-Induced and Sonochemical Degradation of Chitosan as a Way to Increase Its Fat-Binding Capacity. *Nucl. Instrum. Methods Phys. Res. B* **2005**, *236*, 383–390. [CrossRef]
130. Rosiak, J.M.; Olejniczak, J. Medical Applications of Radiation Formed Hydrogels. *Radiat. Phys. Chem.* **1993**, *42*, 903–906. [CrossRef]
131. Caccavo, D.; Cascone, S.; Lamberti, G.; Barba, A.A. Hydrogels: Experimental Characterization and Mathematical Modelling of Their Mechanical and Diffusive Behaviour. *Chem. Soc. Rev.* **2018**, *47*, 2357–2373. [CrossRef]
132. Piepenbrock, M.-O.M.; Lloyd, G.O.; Clarke, N.; Steed, J.W. Metal- and Anion-Binding Supramolecular Gels. *Chem. Rev.* **2010**, *110*, 1960–2004. [CrossRef]
133. Yu, G.; Yan, X.; Han, C.; Huang, F. Characterization of Supramolecular Gels. *Chem. Soc. Rev.* **2013**, *42*, 6697–6722. [CrossRef]
134. Şen, M.; Hayrabolulu, H. Radiation Synthesis and Characterisation of the Network Structure of Natural/Synthetic Double-Network Superabsorbent Polymers. *Radiat. Phys. Chem.* **2012**, *81*, 1378–1382. [CrossRef]
135. Medina-Cruz, D.; Saleh, B.; Vernet-Crua, A.; Ajo, A.; Roy, A.K.; Webster, T.J. Chapter 22—Drug-Delivery Nanocarriers for Skin Wound-Healing Applications. In *Wound Healing, Tissue Repair, and Regeneration in Diabetes*; Bagchi, D., Das, A., Roy, S., Eds.; Academic Press: Cambridge, MA, USA, 2020; pp. 439–488. ISBN 978-0-12-816413-6.
136. Singh, B.; Kumar, A. Radiation Formation of Functionalized Polysaccharide-Protein Based Skin Mimicking Semi- Inter Penetrating Network for Biomedical Application. *Int. J. Biol. Macromol.* **2016**, *92*, 1136–1150. [CrossRef]
137. Urkimbayeva, P.I.; Abilova, G.; Zh B, K.; Kenessova, Z.; Yessirkepova, A.; Samenova, N.; Bekbayeva, L. Development of a Methodology for the Study of Polymer Wound Coatings for Application Characteristics. *Egypt J. Chem.* **2020**, *64*, 1957–1964. [CrossRef]
138. Barba, B.J.D.; Oyama, T.G.; Taguchi, M. Simple Fabrication of Gelatin–Polyvinyl Alcohol Bilayer Hydrogel with Wound Dressing and Nonadhesive Duality. *Polym. Adv. Technol.* **2021**, *32*, 4406–4414. [CrossRef]
139. Treloar, L.R.G. *The Physics of Rubber Elasticity*; Oxford University Press: New York, NY, USA, 1975.
140. Canal, T.; Peppas, N.A. Correlation between Mesh Size and Equilibrium Degree of Swelling of Polymeric Networks. *J. Biomed. Mater. Res.* **1989**, *23*, 1183–1193. [CrossRef]
141. Rahman, M.S.; Islam, M.M.; Islam, M.S.; Zaman, A.; Ahmed, T.; Biswas, S.; Sharmeen, S.; Rashid, T.U.; Rahman, M.M. Morphological Characterization of Hydrogels. In *Cellulose-Based Superabsorbent Hydrogels*; Mondal, M.I.H., Ed.; Springer International Publishing: Cham, Switzerland, 2019; pp. 819–863. ISBN 978-3-319-77830-3.
142. Apkarian, R.P.; Wright, E.R. Cryo and Cryo-Etch Methods for Quality Preservation of Hydrogels Imaged at High Magnification by Low Temperature SEM. *Micros Microanal.* **2005**, *11*, 1088–1089. [CrossRef]
143. Ngadaonye, J.; Geever, L.; Killion, J.; Higginbotham, C. Development of Novel Chitosan-Poly (N, N-Diethylacrylamide) IPN Films for Potential Wound Dressing and Biomedical Applications. *J. Polym. Res.* **2013**, *20*, 161. [CrossRef]
144. Xu, H.; Adolfsson, K.; Xie, L.; Hassanzadeh, S.; Pettersson, T.; Hakkarainen, M. Zero-Dimensional and Highly Oxygenated Graphene Oxide for Multifunctional Poly(Lactic Acid) Bionanocomposites. *ACS Sustain. Chem. Eng.* **2016**, *4*, 5618–5631. [CrossRef]
145. Lamke, L.-O.; Nilsson, G.E.; Reithner, H.L. The Evaporative Water Loss from Burns and the Water-Vapour Permeability of Grafts and Artificial Membranes Used in the Treatment of Burns. *Burns* **1977**, *3*, 159–165. [CrossRef]
146. Wu, P.; Fisher, A.C.; Foo, P.P.; Queen, D.; Gaylor, J.D.S. In Vitro Assessment of Water Vapour Transmission of Synthetic Wound Dressings. *Biomaterials* **1995**, *16*, 171–175. [CrossRef]
147. Yamdej, R.; Pangza, K.; Srichana, T.; Aramwit, P. Superior Physicochemical and Biological Properties of Poly(Vinyl Alcohol)/Sericin Hydrogels Fabricated by a Non-Toxic Gamma-Irradiation Technique. *J. Bioact Compat. Polym.* **2016**, *32*, 32–44. [CrossRef]
148. Zafalon, A.; Santos, V.; Esposito, F.; Lincopan, N.; Rangari, V.; Lugao, A.; Parra, D. Synthesis of Polymeric Hydrogel Loaded with Antibiotic Drug for Wound Healing Applications. In *The Minerals, Metals & Materials Series*; Springer: Cham, Switzerland, 2018; pp. 165–176.
149. Sun, L.; Du, Y.; Fan, L.; Chen, X.; Yang, J. Preparation, Characterization and Antimicrobial Activity of Quaternized Carboxymethyl Chitosan and Application as Pulp-Cap. *Polymer* **2006**, *47*, 1796–1804. [CrossRef]
150. El-Araby, E.; Raafat, A.; Elsonbaty, S.M. Radiation Synthesis of Psyllium/Frankincense Essential Oil–Based Wound Dressing Hydrogels: Antimicrobial, Antioxidant and Wound Healing Performance. *Arab J. Nucl. Sci. Appl.* **2021**, *54*, 19–36. [CrossRef]

151. Masud, R.A.; Islam, M.S.; Haque, P.; Khan, M.N.I.; Shahruzzaman, M.; Khan, M.; Takafuji, M.; Rahman, M.M. Preparation of Novel Chitosan/Poly (Ethylene Glycol)/ZnO Bionanocomposite for Wound Healing Application: Effect of Gentamicin Loading. *Materialia* **2020**, *12*, 100785. [CrossRef]
152. Nho, Y.C.; Moon, S.-W.; Lee, K.-H.; Park, C.W.; Suh, T.S.; Jung, Y.J.; Ahn, W.S.; Chun, H.J. Evaluations of Poly(Vinyl Alcohol) Hydrogels Cross-Linked under γ-Ray Irradiation. *J. Ind. Eng. Chem.* **2005**, *11*, 159–164.
153. Klosinski, K.; Girek, M.; Wach, R.; Czechowska-Biskup, R.; Arkuszewski, P.; Szymański, P.; Pasieka, Z. Synthesis and Potential Cytotoxicity Evaluation of Carboxymethyl Chitosan Hydrogels. *Prog. Chem. Appl. Chitin. Deriv.* **2017**, *XXII*, 82. [CrossRef]

gels

Article

Babchi Oil-Based Nanoemulsion Hydrogel for the Management of Psoriasis: A Novel Energy Economic Approach Employing Biosurfactants

Aftab Alam [1,*], Mohammed H. Alqarni [1], Ahmed I. Foudah [1], Mohammad Raish [2] and Mohamad Ayman Salkini [1]

[1] Department of Pharmacognosy, College of Pharmacy, Prince Sattam Bin Abdulaziz University, Al Kharj 11942, Saudi Arabia
[2] Department of Pharmaceutics, College of Pharmacy, King Saud University, Riyadh 11451, Saudi Arabia
* Correspondence: a.alam@psau.edu.sa

Abstract: The current research aimed to assess the Babchi oil nanoemulsion-based hydrogel prepared using biosurfactants through a low-energy emulsification process for the topical management of psoriasis. The emulsification capacity and solubilities of many nanoemulsion constituents such as surfactants, co-surfactants, and oil were considered to determine the range of concentration of the constituents. Pseudoternary phase diagrams were created using the method of titration. Nanoemulgel structure, morphology, micromeritics, conductivity, and viscosity were all optimized. The assessment of the Babchi oil nanoemulgel included particle size, polydispersity index (PDI), drug content, pH, spreadability, rheological management, ex vivo drug study, 2,2-diphenyl-1-picrylhydrazyl (DPPH) scavenging ability, in vitro drug release, release kinetics, and dermatokinetics. The selected ratios of the surfactant mixture (Smix) taken were 3:1. The entrapment efficiency estimated was 91.298%. The zeta potential of Babchi oil was observed to be −24.93 mV at 25 °C with water as a dispersant, viscosity as 0.887 cP, and material absorption as 0.01 nm. The size distribution of the particle was 108 nm by the intensity and the conductivity observed was 0.03359 mS/cm. The cumulative amount of Babchi oil penetrated and fluxed by nanoemulgel was considered larger ($p \leq 0.05$) than the conventional formulations. Skin retention was observed to be good with decreased lag time. The formulation followed the Higuchi Korsmeyer for Fickian Peppas model for in vitro drug release studies. The oil was most effective on the epidermal layer of the skin for treatment. It was established that the Babchi oil nanoemulgel formulation had superior permeability capabilities for topical and transdermal administration and is a viable alternative to traditional formulations.

Keywords: Babchi oil; nanoemulsion-based hydrogel; nanoemulgel; low energy emulsification; psoriasis

check for updates

Citation: Alam, A.; Alqarni, M.H.; Foudah, A.I.; Raish, M.; Salkini, M.A. Babchi Oil-Based Nanoemulsion Hydrogel for the Management of Psoriasis: A Novel Energy Economic Approach Employing Biosurfactants. *Gels* **2022**, *8*, 761. https://doi.org/10.3390/gels8120761

Academic Editor: Shige Wang

Received: 29 October 2022
Accepted: 21 November 2022
Published: 23 November 2022

1. Introduction

Psoriasis is an inherited skin condition that presents with persistent non-contagious intense itching. Symptoms of this condition include deformation, inflammation, and scaly and thickened skin [1]. The global statistics of recorded cases of psoriasis are about two to five percent [2–4]. The condition progresses in many ways, and its differentiation is noted by the undergoing inflammation, rash localization, irritation of the local area, intensity of other traits, and its occurrence. This condition is classified into four types: erythroderma, pustular, and guttate, along with a persistent plaque. It manifests during the early forties and is observed equally in both genders [5,6]. Longer periods of therapy are required to treat psoriasis as it is a chronic disorder. Cyclosporines, methotrexate, and retinoids are the commonly used anti-psoriatic medications. Still, several adverse effects occur due to these drugs that include renal and liver dysfunction, loss of hair, stomach discomfort, and inflammation of lips [1,7].

Approved topical therapies for the treatment of psoriasis, such as dithranol, emollients, and coal tar, are usually safe; however, they are minimally effective. Medical advancements are exploring several methods for the treatment of this condition, such as steroidal lotions, oral and injectable medications, etc. Although these medications provide momentary relief, they do not treat the symptoms as a whole because they lack a secure and optimum vehicle that can properly transport the anti-psoriatic properties of the drug for maximum treatment benefit [4,8]. To overcome the limits of traditional therapies, researchers have turned to nanoscience and nanomedicines to improve the effectiveness and lessen the undesirable adverse effects of anti-psoriatic medications. These medicines have become valuable because of their increased bioavailability, low prescription doses, and nanosized delivery of drugs. Similar to these methods, nanoemulsion is also the technique that employs a colloidal approach to administering nanosized particles of active drug moieties to the afflicted skin areas as the drugs have large surface areas.

A nanoemulsion is a transparent, stable kinetic dispersion of two insoluble phases of oil and water in the presence of surfactant particles ranging in size from 5 to 200 nm [9,10]. The usage of nanoemulsion as a vehicle for anti-psoriatic medications is favourable since it does not exhibit flocculation, internal creaming, deposition, or coarsening, which are prevalent in macroemulsions [11]. The ability to permeate is strong in nanoemulsions, along with increased drug-loading potential when used topically [4,12]. The choice of the right surfactants and oils is critical for an optimum nanoemulsion. The potential usage of biosurfactants in substitution of synthesized equivalents is considered better in regards to expense and price, durability, and compatibility with the environment. Biosurfactants are agents that are active on the surface and provide a natural alternative to their chemical equivalents. These are less harmful and also perform well during high temperature and pH conditions. Biosurfactants are significantly durable as bacterial growth is quick, and the source is numerous and producible [13]. Hydrogels are a potential type of material made from natural or artificial polymer that has three-dimensional network structures with large water content [14]. It is the most ideal biomaterial for the creation of surface coatings in the prevention and treatment of multi-drug resistant infections due to their high hydrophilicity, complex three-dimensional network, cell adhesion, and distinctive biocompatibility [15–20].

Plant-derived essential oils are fragrant liquids that have high volatility. Their abundant chemicals, such as hydrocarbons, terpenoids, coumarins, and phenols, are observed and used for their range of therapeutic properties [21]. Babchi oil produced from *P. coryfolia* (*Psoralea coryfolia* L.) is widely recognized among essential oils for its antioxidant, antibacterial, anticancer, immunomodulatory, anti-inflammatory, and antifungal effects [22]. The herb is found to grow annually in warm climates in countries such as China, India, South Africa, and Pakistan. It has become a standard treatment in the Chinese and Ayurveda fields of therapy [23].

Babchi has been used to treat a range of skin ailments, including psoriasis, leukoderma, and leprosy [24]. Isopsoralen, psoralen, and bakuchiol are the main components obtained from Babchi oil and are recognized for their various properties. Although the benefits of Babchi oil are abundant, it has poor physical qualities such as degradation susceptibility, sensitivity to light, hydrophobicity, and extreme viscosity, which has reduced the practical usage of the oil and has been confined to medicines. Recent trends in using Babchi oil include formulations such as emulgel [25] and other forms; however, formulation of the oil into a nanoemulsion-based hydrogel has not been reported. This study aims to investigate Babchi oil nanoemulsion-based hydrogel or nanoemulgel with low energy emulsification method for the management of psoriasis.

2. Results and Discussion

2.1. Component Screening

For development of the system for nanoemulsion for delivering the drug transdermally, certain factors, such as the solubility of the drug and its components, are required for the drug to reach and permeate within the skin. The solubility of Babchi oil was investigated.

2.2. The Behavior of Phase and Nanoemulsion Optimization

Constructing diagrams for pseudoternary phases is required because they help in finding the existing nanoemulsion ranges. Diagrams of the phase represent the translucent regions of nanoemulsion. Visual observation of the other regions on the diagram of the phases is of traditional and turbid type [26]. Separate ratio diagrams for the surfactant to co-surfactant of the pseudoternary phase were constructed to determine the regions of O/W of nanoemulsion and its formulations for optimization as shown in Figure 1.

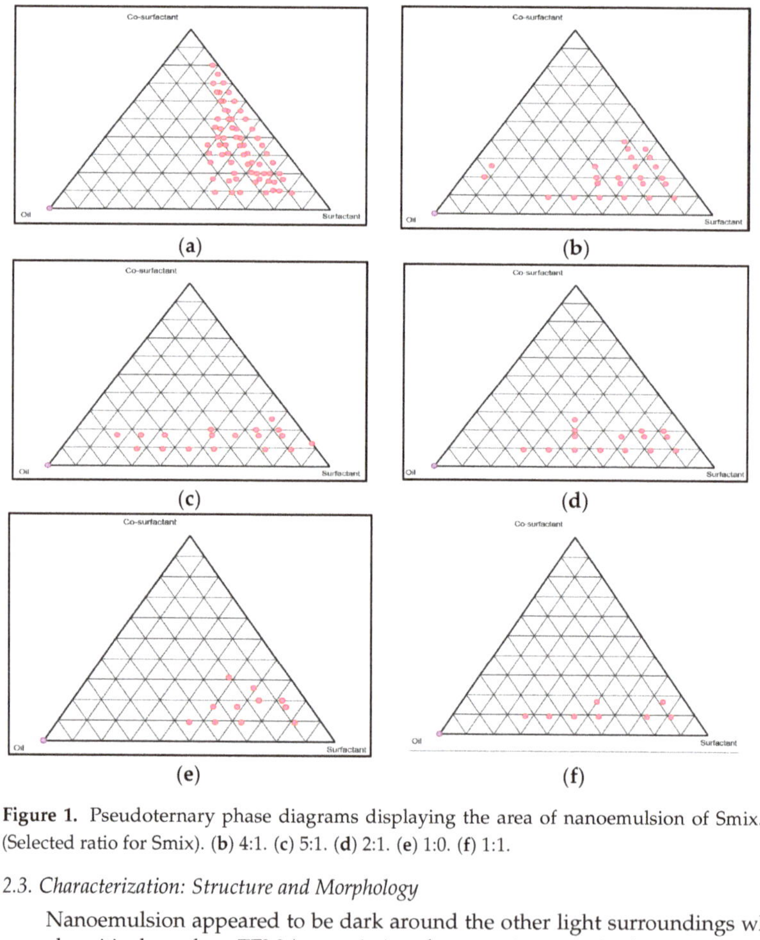

Figure 1. Pseudoternary phase diagrams displaying the area of nanoemulsion of Smix. (**a**) 3:1 (Selected ratio for Smix). (**b**) 4:1. (**c**) 5:1. (**d**) 2:1. (**e**) 1:0. (**f**) 1:1.

2.3. Characterization: Structure and Morphology

Nanoemulsion appeared to be dark around the other light surroundings when observed positively under a TEM (transmission electron microscope). The average size of the droplet was smaller than that of 100 nm from the sample. The droplets were observed to be within the range of ≤200 nm of nanosize, and therefore the emulsion that was prepared was considered to be nanoemulsion. Figure 2 shows the TEM of the spreadability of Babchi oil nanoemulgel.

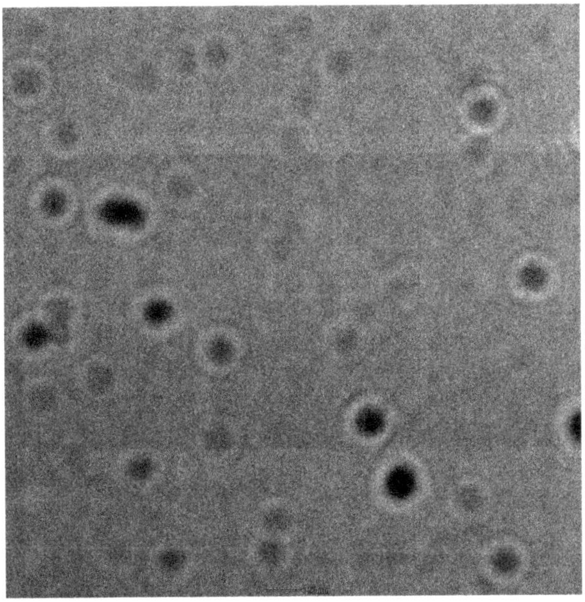

Figure 2. TEM measurement (scale set at 50 nm).

2.4. Micromeritics

To determine the effective and safer dose, it is crucial to characterize the nanoemulsion sizes [27]. It was found that the current formulations fell into the nanosize range. This was also observed through the decreased values of the index of polydispersity. The index for polydispersity is defined as the SD (standard deviation) ratio to the mean size of droplets and denotes the homogeneity of the size inside the formulation.

2.5. Determination of Conductance and Viscosity

The nanoemulsion viscosities were determined to be minimal, with a range of 9.8 $\pm$ 0.42 to 36.1 $\pm$ 0.63 mPaS. This behaviour of nanoemulsions demonstrated that they were unsuitable for topical applications. Therefore, it justified the integration of nanoemulsion within the matrix of gel, resulting in a high viscosity value in the formulated nanoemulgel. The nanoemulsion conductance was observed to be high with a value of 48.2 $\pm$ 0.03 to 161.1 $\pm$ 0.13 μs/cm which resulted in the formulation being of the nanoemulsion O/W category.

2.6. Nanoemulgel Formulation

Even though the formulations were discovered to be in the range suitable for nanosize, their poor viscosities hampered their application and were thus deemed unsuitable for cutaneous usage. To improve the nanoemulsion application, respective viscosities were increased by inserting nanoemulsions within the gel matrix of 940 carbomers, leading to formations of nanoemulgel that were appropriate, homogenous, yet extremely viscous enough to be dermally administered.

2.6.1. Nanoemulgel pH Determination and Concentration of Drug

Values of pH for all nanoemulgels separately were observed at 6.5 $\pm$ 0.15 to 6.9 $\pm$ 0.47 in the neutral range. This allowed the formulated mixture to be used on the skin easily.

2.6.2. Ability to Spread

Due to the utilization of the formulated mixture, the spreading ability of nanoemulgel preparation was assessed. Its texture is pleasant for skin inflammation as it distributes quickly, with maximal drag and slip. The spreading ability of the formulations was determined to be from the 5.6 ± 0.21 to 6.2 ± 0.12 gcm/S ranges. The increased radius indicates a greater ability to spread.

2.6.3. Measurement of Rheological Behaviour

The ability to spread and flow along with the release of oil is all governed by rheological properties. The release of oil and its substances through the formulated mixture is primarily regulated by the components present in it. The index used for measuring the consistency was 1 s^{-1} at the shear rate and it equaled the apparent viscosity. The index of consistency of the formulated mixture was observed to be 0.33. The index of flow was defined as the measurement of the system's divergence from Newtonian behavior (n = 1). The value of "n" lesser than 1 implies pseudoplastic flow or narrowing of shear, whereas a value greater than 1 suggests a narrowing of shear or dilatants. The index of flow indicated the ability to flow the formulated mixture from the container.

In general, a lesser index of flow results in a thick base. The index of flow for nanoemulgel was 0.33, presenting characteristics of pseudoplastic flow. Pseudoplasticity is usually caused by a network of colloidal structures which align themselves in the shear direction. This lowers the viscosity and increases the rate of shear. The flow characteristic of pseudoplastic nature validates the designed system's need to exert some force of expulsion.

2.7. *Applying Central Composite Rotatable Design for Formulation*

Effect of Independent Variables on Particle Size (nm), Zeta Potential, and Time Phases

Three-dementional response surface graphs present the effect of Babchi oil and Smix concentration on particle size (nm), zeta potential (mV), and entrapment efficiency of 91.298% EE and are presented in Figure 3 and Table 1. The zeta potential of Babchi oil was observed to be −24.93 mV at 25 °C with water as a dispersant, viscosity as 0.887 cP, and material absorption as 0.01 nm as observed in Figure 3a,b. The zeta potential distribution conductivity observed was 0.03359 mS/cm. Figure 3c shows the size distribution of 108 nm by intensity. Figure 4a depicts the phase time graph of zeta potential and Figure 4b,c depict the total counts of zeta potential distribution graph and Babchi size distribution and intensity graph, respectively.

$$\text{Particle size (nm) (Y1)} = 194.03 + 63.96A - 10.75B - 5.40AB + 9.80A^2 - 1.80B^2;$$

$$\text{Zeta Potential (mV) (Y2)} = -22.91 - 0.91A + 5.13B - 0.13AB + 2.89A^2 + 0.63B^2;$$

$$\text{Entrapment Efficiency (\%) (Y3)} = 88.78 + 11.59A + 0.46B - 0.15AB - 5.84A^2 - 5.33B^2.$$

2.8. *Studies for Permeability of Skin*

Babchi oil nanoemulgel formulation had higher drug profile transfer throughout the skin of rats with $p < 0.0001$ and flux of 4.21 ± 2.25 μg/cm^2/h compared to conventional formulation with the flux of 3.06 ± 3.12 μg/cm^2/h. The presence of Babchi oil in the nanoemulgel system might contribute to the significant disparity in the percentage of drug distribution. The Babchi oil permeability was increased by enhancing the solubility of the surfactant during the formulation. Due to the compact design of the developed systems, they were capable of deeper penetrating the intrinsic layers of skin and resulting in greater absorption of the drug. In comparison to standard formulations, nanoemulgel was proven to improve penetration rates within deep layers of skin and also reduce lag time. The total amount of oil penetration, skin retention, flux, lag time, LAE, and enhancement ratio was determined.

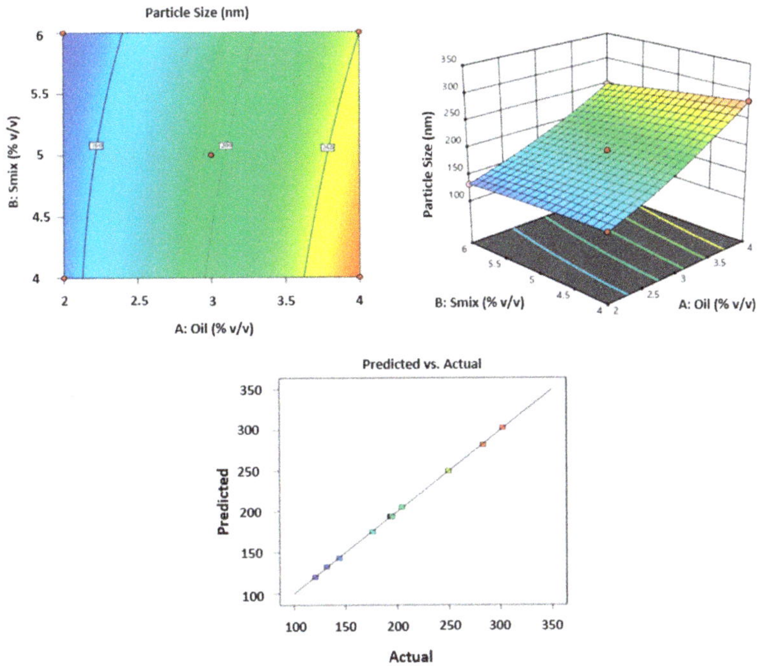

(a)

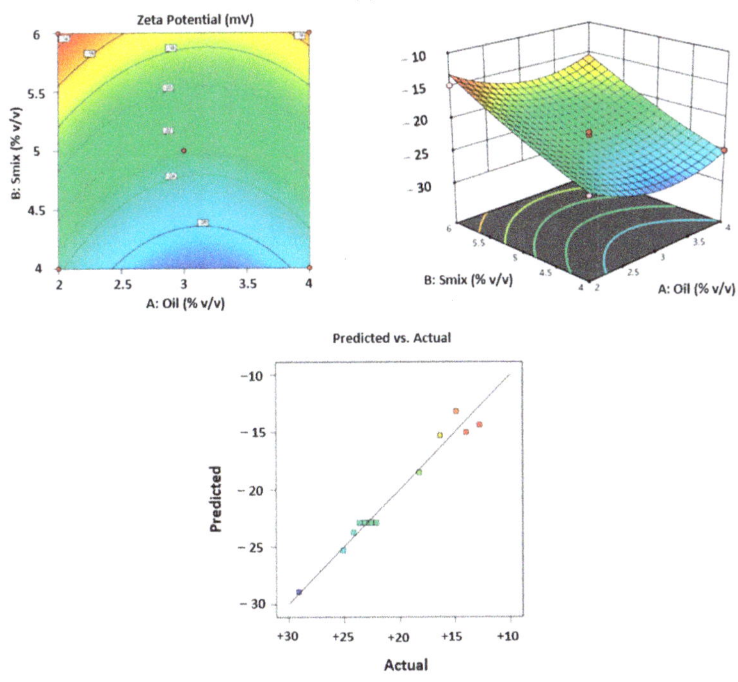

(b)

Figure 3. *Cont.*

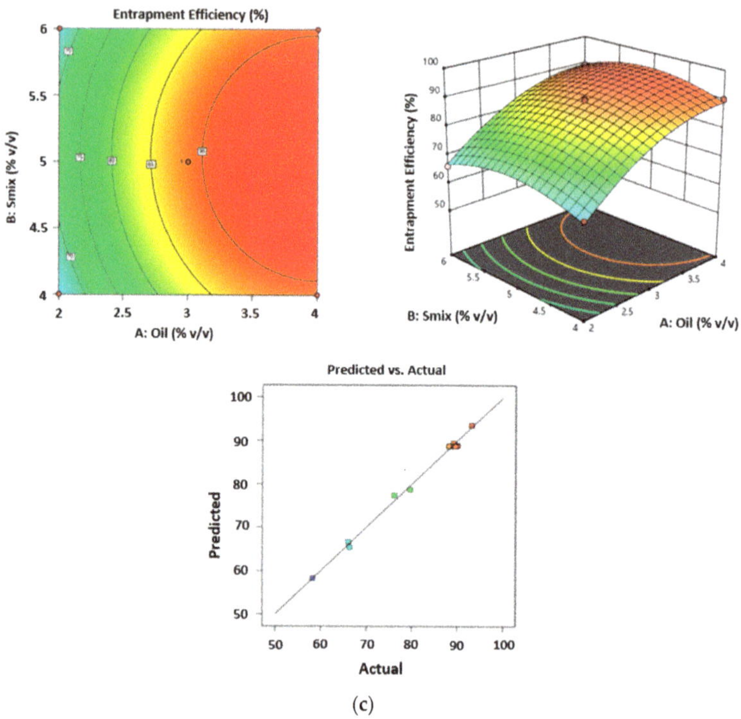

Figure 3. 3D Response surface graphs showing the effect of oil and Smix concentration on (**a**) Particle Size (nm), (**b**) Zeta Potential (mV) and (**c**) Entrapment Efficiency (%EE).

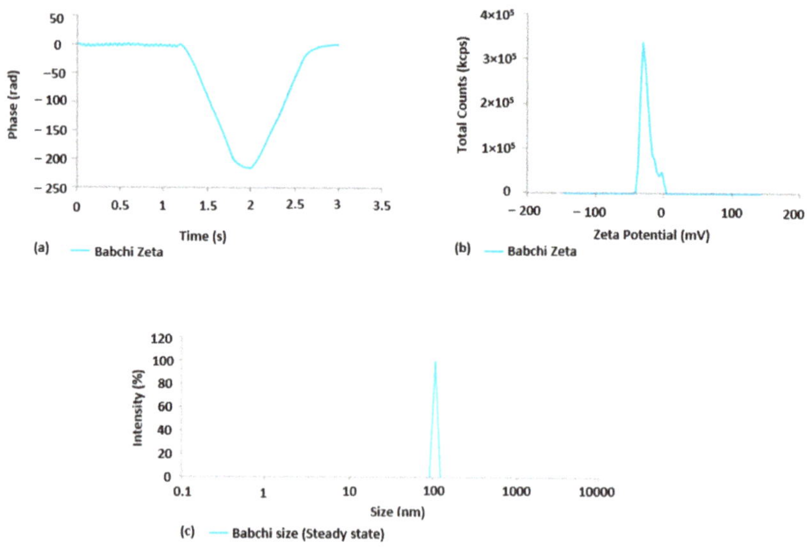

Figure 4. Zeta potential data analysis, (**a**) phase–time graph of Babchi oil zeta potential, (**b**) total count of Babchi oil zeta potential distribution graph, (**c**) Babchi size distribution and intensity graph.

Table 1. Responses obtained using CCRD.

Run	Oil % v/v	S_{mix} % v/v	Particle Size (nm)	Zeta Potential	EE (%)
1	3	5	193.43	−22.57	89.43
2	4	6	249.35	−16.36	89.11
3	3	5	194.34	−23.13	88.09
4	4	4	283.13	−25.08	90.07
5	1.5	5	120.34	−13.98	58.21
6	3	5	193.25	−22.98	88.61
7	3	5	194.12	−23.62	89.75
8	2	6	131.87	−14.87	65.98
9	3	6.4	176.21	−12.76	79.64
10	4.4	5	302.45	−18.25	93.12
11	3	3.5	204.54	−29.08	76.12
12	3	5	195.04	−22.11	88.04
13	2	4	144.04	−24.13	66.31

2.9. *In Vitro Release Studies*

The nanoemulgel formulation was compared with the conventional formulation to observe the percentage of drug release in the epidermal layer. Figure 5 depicts the results obtained during the in vitro release study of the optimized nanoemulsion and its suspension. It was observed that the percentage of release from the optimized nanoemulsion was greater than 81.26 in comparison to the suspension of 38.19. It was observed that Babchi oil nanoemulgel produced a higher percentage of drug release in comparison to the conventional formulation and constantly increased with time. The different models used for the analysis of in vitro release such as the zero-order release model as depicted in Figure 6a produced higher effects and also increased with time. The y value taken was 0.0004x + 0.2893 and the R^2 value taken was 0.7464 for plotting the graph. The first order release model was plotted with a y value of −0.0005x + 1.8667 and an R^2 value of 0.9023 which is depicted in Figure 6b. It was observed that analysis with this model decreased the log percentage of the remaining drug with time. Figure 6c shows the Higuchi model shown with a y value of 0.0219x + 0.0846 and an R^2 value of 0.9185. It was observed that this analysis model increased the fraction of drug release with time. Finally, the Korsmeyer Peppas model was implemented with a y value of 0.5103x + 0.4043 and an R^2 value of 0.9414 and was plotted in Figure 6d. It was observed that Babchi oil nanoemulgel followed the Korsmeyer Peppas model and produced the most significant results of drug release with time. Higuchi model was considered the best with the Korsmeyer Peppas model producing the most potent diffusion. Figure 7 depicts the results obtained for studies of absorption (nM). The percentage of drug release by time graph of the nanoemulsion formulation was plotted and depicted in Figure 8.

2.10. *DPPH Scavenging Activity of Babchi Oil Nanoemulgel*

The lowering property of DPPH produced by the antioxidant was measured by observing a decrease at 517 nm of absorbance and the colorless occurrence of color violet was noted. Babchi oil nanoemulgel's potential for the activity of antioxidants was reduced a little compared to the conventional formulation. The Babchi oil nanoemulgel produced an antioxidant activity of 69.1% while the conventional formulation produced 91.25%. Through these values, the potential for antioxidant activity of Babchi oil nanoemulgel was demonstrated as it did not change through entrapment within the formulation.

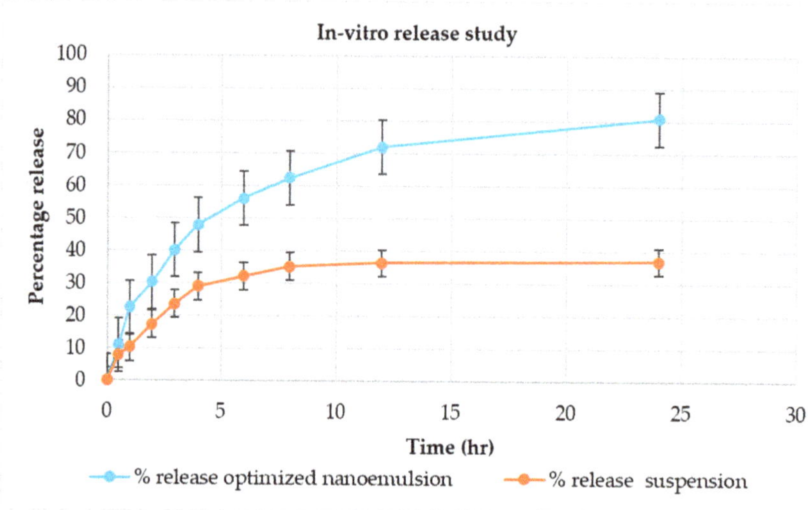

Figure 5. In vitro release study of optimized nanoemulsion and release suspension.

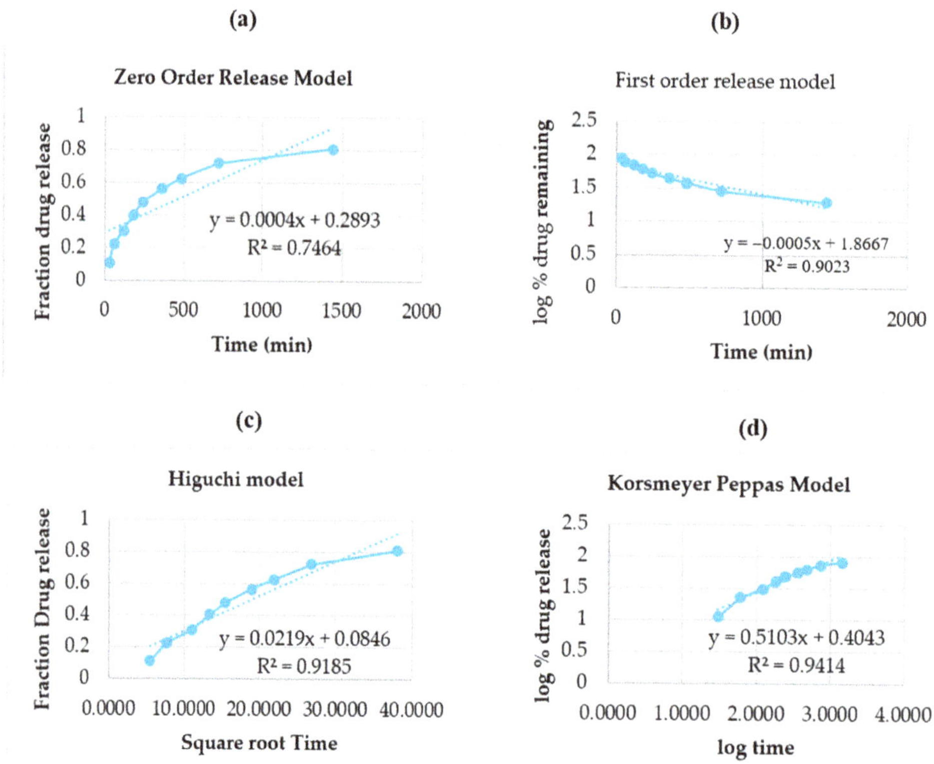

Figure 6. Different models used for in vitro release studies: (**a**) Zero order release model (**b**) First order release model (**c**) Higuchi model (**d**) Korsmeyer Peppas model.

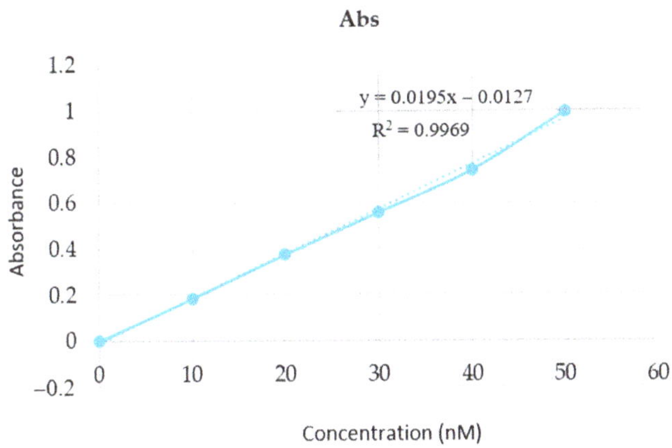

Figure 7. Absorption observed at different concentrations of nanoemulgel.

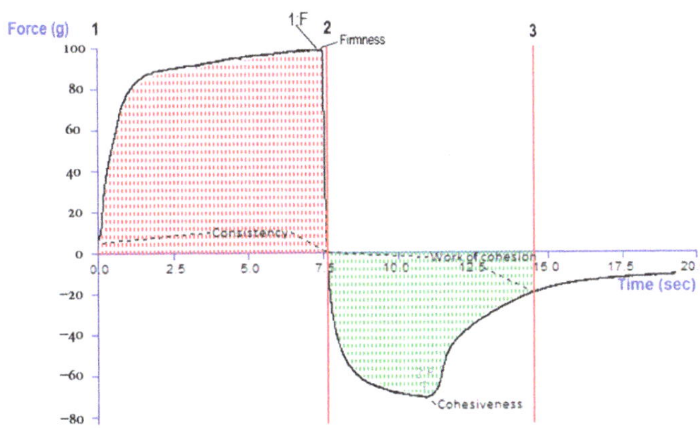

Figure 8. Release kinetics of nanoemulgel observed.

2.11. Dermatokinetic Study

The quantity of Babchi oil present in the epidermis and dermis on rat skin following the application of the nanoemulgel formulation at different intervals of time was analyzed by ANOVA for statistical analysis to compare the epidermal and dermal layers of rat skin. Table 2 depicts the dermatokinetic study chart. It was revealed that the nanoemulgel formulation had a higher value of C skin max with 220.049 ± 5.00 μg/cm^2 while the conventional medicine was lower on the epidermis. It was observed that the conventional formulation produced a higher value of C skin max with 375.918 ± 11.00 μg/cm^2 in comparison to the nanoemulgel formulation on the dermis. The area under the curve (AUC) of the formulation was highest in the epidermal layer of the nanoemulgel formulation with a value of 872.4898 μg/cm^2 while the conventional formulation presented with a value of 114.569 μg/cm^2. The dermis layer presented a higher value of conventional formulation with the value of 375.918 μg/cm^2 while the nanoemulgel produced only 164.875 μg/cm^2. The Ke value nanoemulgel formulation on epidermal and dermal layers was 0.1329 μg/cm^2 and 0.1393 μg/cm^2, respectively, while the conventional formulation produced 0.1377 μg/cm^2 and 0.1391 μg/cm^2, respectively. This suggested better penetration of nanoemulgel formulation and its elasticity along with the activator when the nanoemulgel is applied topically. Figure 9 depicts the images obtained through CLSM within the perpendicular cross-section of optimal skin surface of rats while Figure 10a,b depicts the effects of Babchi oil nanoemulgel on epidermis and dermis concentration, respectively.

2.12. Studies for Stability

When the nanoemulgel formulations were centrifuged, no drug precipitation or phase separation was detected, indicating that the produced nanoemulgel was physically stable. Although these were exposed to thaw or freeze cycles, they did not exhibit symptoms of cracks or breaks. The findings of stability investigations demonstrated that nanoemulgels remain transparent, even 3 months later at different exposed temperatures such as 25 °C ± 2 °C, 37 °C ± 0.1 °C, along with 4 ± 0.2 °C. Even during stability investigation, each of the formulated mixtures was observed to be stable in terms of pH (6.8), transparency, the content of oil, and separation of phases. The increased viscosity of nanoemulgel may potentially limit Brownian mobility.

Table 2. Dermatokinetic study chart.

Skin Part	Type	T_{max}	C_{max}	AUC	Ke
Epidermis	Formulation	2	220.049	872.4898	0.13295
	Conventional	2	114.569	472.9198	0.137736
Dermis	Formulation	2	164.875	644.4886	0.139326
	Conventional	2	375.918	375.9179	0.139155

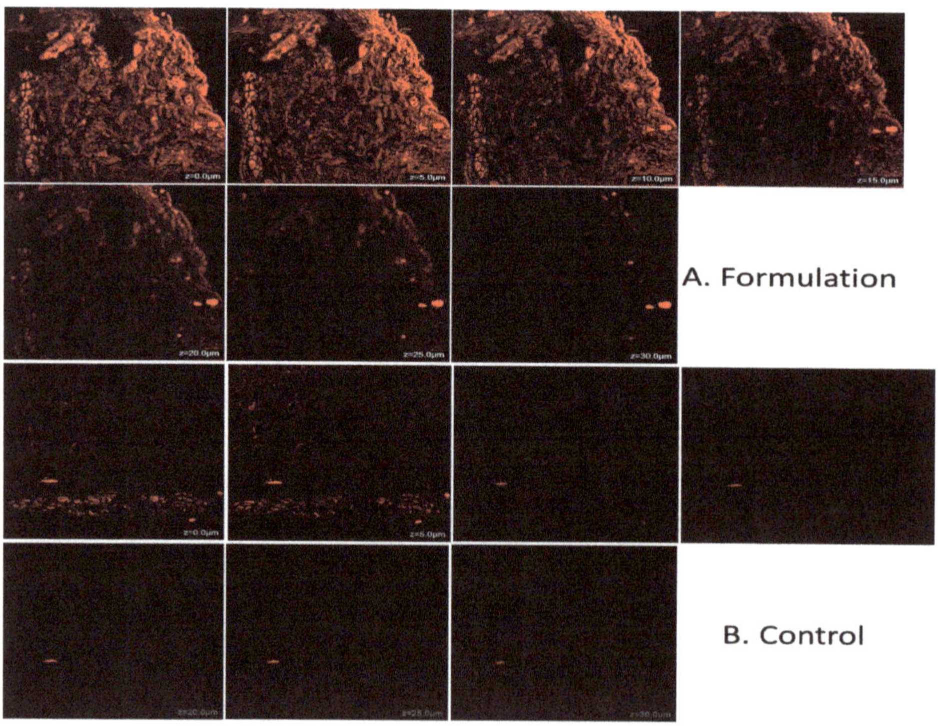

Figure 9. Images of CLSM within the perpendicular cross-section of optimal skin surface of rats.

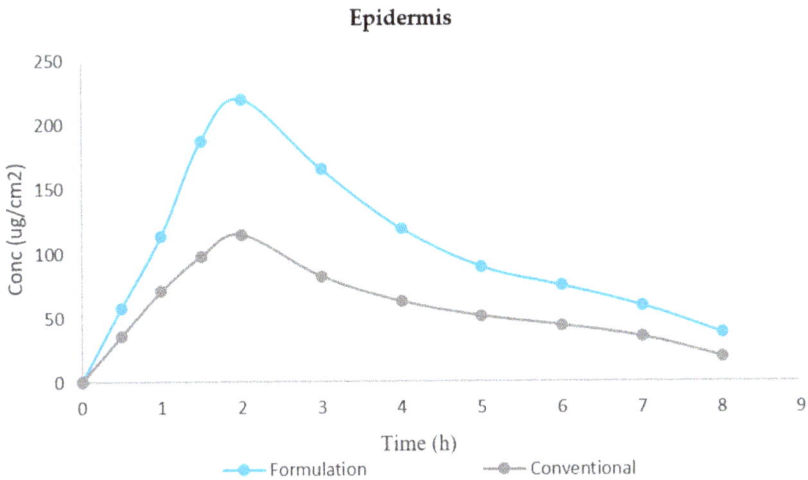

(a)

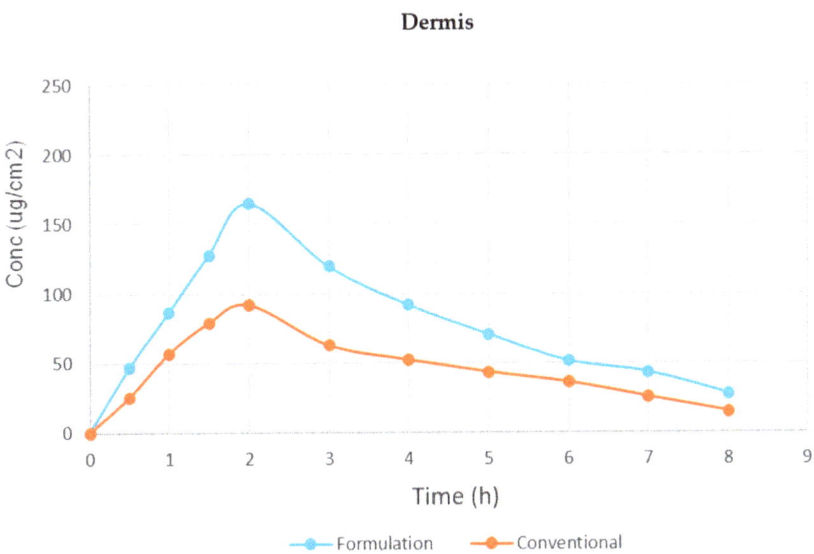

(b)

Figure 10. (**a**) Effect of Babchi oil nanoemulgel on epidermis concentration; (**b**) Effect of Babchi oil nanoemulgel on dermis concentration.

Nanoemulgels have great capacities for solubilization and for boosting permeability; due to this, nanoemulgel was used as a vehicle for distributing drug transdermally [28]. Aside from these facts, a topical approach for administration provides the opportunity to avoid the issues related to persistent oral dosing. The innovative techniques of nanoemulgel composed of combinations of aqueous and gel-based, surfactant and co-surfactants gels along with oil create multiple components for the drug loading system. Therefore, the nanoemulsions are created by determining the range of concentrations of each component. The size of the droplet, the conductance of the nanoemulsion, viscosity, and structural morphology were developed and optimized. Images observed through the TEM indicated a round shape. The formulations observed were in the range of 10–100 nm nano size. PDI value showed the droplets to be of consistent size throughout the formulations. O/W structure was observed which validated the nanoemulsion for its increased values of conductivity. Values obtained for viscosity were increased which made it suitable for skin application. The nanoemulsions that were induced within the base of the gel resulted in the formulation of the nanoemulgel. These nanoemulgel formulations had minimal loss of Babchi oil molecules. The pH of the formulation was similar to the skin pH, and therefore the formulation presented to be of minimal irritation in nature. The drag and slip phenomena with increased diameters were observed when formulations spreadability was checked. Higher viscosity content was observed along with a pseudoplastic manner when produced by the rheogram which ensured the flow would not occur within the system that was developed. In addition, collapsible tube and container filling would have required an ejection yield value [29].

The nanoemulgels formulated underwent studies for permeability in which Babchi oil nanoemulgel proved to be of higher significance as the permeation observed was higher and the components of the formulation correlated with each other. When the co-surfactant was reduced and the content of oil was increased, the permeability of the formulation was observed to be greater. Babchi oil and its permeability to penetrate the lipid bilayers of the skin occur due to the synergistic effect as it disrupts the order and improvises the thermodynamic effect of the formulation due to lesser concentration of surfactant in the skin [30]. Through this study, we also observed the co-surfactant effect when used as an enhancer for permeation when compared with the formulations that were optimized and those that were prepared without the usage of co-surfactants as the amount of drug permeation was low cumulatively during the flux. It was also observed that retention of formulation on the skin was improved along with the lag time. In comparison with the parameters for permeability, the optimized Babchi oil nanoemulgel had increased permeation of drug amount along with the flux and ratio for enhancement and high retention by the skin than the other plain and marketed formulations. The formulation's AUC was higher in the epidermis, indicating improved formulation absorption and flow ability.

In this study, the nanoemulgel formulated presented with good potential for permeation without any induction of chemically synthesized enhancers. This reduced the cause of skin irritations and hence, the novelty of the study lies here because the constituents such as surfactant, co-surfactant, and oil that are present in the nanoemulgel act as enhancers for permeation. Studies for stability were performed at the refrigerator and normal temperatures which indicated the Babchi oil formulation was stable and there were no observed changes in pH or the content of the components in the formulation. These factors during the study provided an improved illustration of drug stability and the effect of Babchi oil. Therefore, this study provides sufficient data to provide evidence for the use of Babchi oil nanoemulgel in the treatment use of psoriasis.

3. Conclusions

Babchi oil nanoemulgel has appropriate viscosity and was recently created for transdermal administration. The interaction between skin and nanoemulgels influences the rhamnolipid with incredibly precise penetration and enhances its potential, contributing significantly to drug absorption through the skin. In comparison to conventional pharmaceuticals, this Babchi oil-modified nanoemulgel demonstrated a significant improvement in drug penetration, showing that these nanoemulsion systems are possible carriers or vehicles for surrogate Babchi oil delivery.

4. Materials and Methods

4.1. Materials

Babchi oil and rhamnolipid along with Propylene glycol was obtained from Sigma Merck. Other materials such as ethanol and required chemicals were also purchased from Sigma Merck. The reagents used in the study were analytically graded.

4.2. Experimental Methods

4.2.1. Solubility of Babchi Oil

The solubility of Babchi oil in surfactants (rhamnolipid) and co-surfactants (propylene glycol) was observed by dissolving the oil within these in excessive amounts. To achieve equilibrium, the sample was constantly swirled for ten minutes in a vortex mixer and preserved at room temperature in an isothermal shaker for seventy-two hours. The samples that were equilibrated were then centrifuged at 3000 rpm for fifteen minutes. The supernatant acquired underwent filtration from a 0.45 μm filter membrane and was diluted using the mobile phase. The content of the drug was determined by using a UV-VIS spectrophotometer of Shimadzu-1700, Japan set at 260 nm.

4.2.2. Screening of Nanoemulsion Components

Based on solubility experiments, Babchi oil has the best solubilization ability. Surfactant and co-surfactant screening were performed using the percentage of transmittance. The capacity of surfactants (rhamnolipid) to emulsify was tested by the addition of 300 mg into the Babchi oil. To accomplish homogeneity, the sample was carefully heated at forty to forty-five degrees Celsius for thirty seconds. In order to formulate a fine emulsion, a measurement of 50 mg of the sample was taken and then blended with 50 mL of double distilled water. The overall turbidity of the final sample was assessed visually. After two hours, transmittance tests through the UV-VIS spectrophotometer of Shimadzu-1700, Japan were performed on the emulsions at 700 nm and a blank of water with double distillation as a narrow spectrum focused more on particles with shorter wavelengths. Out of a variety of co-surfactants that were screened, propylene glycol was used for the nanoemulsion formulation as it is considered safe. Propylene glycol (mg), rhamnolipid (mg), and Babchi oil (mg) formulations were created and analyzed in a similar method as indicated in the screening technique of surfactants.

4.2.3. Construction of Phase Diagrams

Titration methods were used to construct diagrams for pseudoternary phases. To create the surfactant, co-surfactant, and oil phase we used rhamnolipid, propylene glycol, and Babchi oil. Ratios of surfactant and co-surfactant weights that were used in the study were 2:1, 1:1, and 1:0 and underwent optimization to determine the exact ratio at which maximum area for nanoemulsion existence could be achieved.

These formulations underwent titration in presence of water through micro syringing drops until the separation of phase or turbidity began. Within the other batch, similar ratios of surfactant, co-surfactant, and water were made and parameters were visualized. The formulations were briskly agitated for long enough periods for homogenization to occur in both situations. The endpoints of the formulations were monitored visually through a darkened screen by lighting the formulations using white light. To ensure repeatability, the tests were carried out thrice. The percent of mass components of Babchi oil and the combination of surfactant and co-surfactant along with the water was determined and represented on coordinates in the triangular form to build the diagrams for the pseudoternary phase from the endpoint.

4.2.4. Nanoemulsion Formulation

Babchi oil was added to the samples of surfactant and co-surfactant with different ratios that were observed from diagrams of the pseudoternary phase. Drop by drop, an adequate quantity of water was poured into the sample. The Babchi oil nanoemulsion was thus created by constantly stirring the formulation at the temperature of the room. The nanoemulsions that were prepared were kept at room temperature for future research.

4.2.5. Nanoemulsion Optimization

Optimization of nanoemulsion was performed through 3 levels and 3-factor software of Design expert, namely Box–Behnken version 12 design of Stat-Ease, USA. The impacts of several processing factors such as concentration, sonication, zeta potential, size of particles, and efficiency of entrapment along with the in vitro release of nanoemulgel were comprehensively examined. The different levels used were observed with axial-α, low, medium, high, axial + α of oil and Smix with dependent variables to be particle size, zeta potential (ZP), and entrapment efficiency. Table 3 depicts the variables used during the CCRD to optimize the formulation. The design included a formulation run with several combinations at three points of center to evaluate the impact of the variables. Several models are provided by polynomial equations along with response plots for the surface to assess the influence of different variables and factors such as quadratic and linear factors. Of all the models, both individual and combined influence on dependent variables was best demonstrated by the quadratic model.

Table 3. Variables used in CCRD to optimize the formulation.

Factors	Levels Used				
Independent Variable	Axial −α	Low (−1)	Medium (0)	High (+1)	Axial + α
A—Oil (% v/v)	1.58	2	3	4	4.41
B—S_{mix} (% v/v)	3.58	4	5	6	6.41
Dependent Variable			Constraints used		
R1—Particle size (nm)			Minimum		
R2—ZP			Maximum		
R3—Entrapment efficiency (%)			Maximum		

Nanoemulsion Structure and Morphology

TEM (Transmission electron microscope) of Hitachi H7500, Japan was used to evaluate the shape and small structures of loaded nanoemulsion drugs. Water was used to dissolve the nanoemulsion compositions in a 1:10 ratio. Thereafter, a dissolved nanoemulsion droplet was implemented on the holey grid of film, dyed, and dried with a 1% aqueous solution of phosphotungstic acid.

Nanoemulsion Micromeritics

Dynamic light dispersion with zeta-sizer HSA 3000 was used to analyze the size of globules of the nanoemulsion and index of polydispersity (PDI) of Malvern Instruments Ltd., UK. Before determining the PDI and size of globules, all the formulations were sonicated.

Nanoemulsion Conductivity and Viscosity

At twenty-five degrees, the electrical conductance of the nanoemulsion was measured through an EC Testr 11+, USA conductance meter. This test was carried out thrice for reproducibility.

The nanoemulsion prepared was observed for viscosity using the viscometer of Brookfield DV-II+ Pro without dilution. The formulation was placed within the beaker for five minutes before being measured at 0.5, 1, 2.5, and 5 rpm with a spindle. The accompanying dial was read on and recorded from the viscometer.

Nanoemulsion-Based Hydrogel (Nanoemulgel) Formulation

Nanoemulgel was formed when each of the formulations was observed to be nanosized and so integrated inside the matrix of gel. The matrix of the gel base was chosen to be carbomer 940. The phase for oil was created by combining Babchi oil, rhamnolipid, and propylene glycol. Swelling of carbomer 940 was performed in a small amount of water for twenty-four hours to generate solutions of high viscosity. During this duration, the oily phase was progressively added into the viscous carbomer 940 formulations during the magnetic stir. After adjusting the pH from 6 to 9 using triethanolamine, nanoemulgel was produced.

4.2.6. Nanoemulsion Characterization

Particle Size and Polydispersity Index (PDI)

Rhamnolipid as surfactant and propylene glycol as cosurfactant was mixed with distilled water to produce an O/W type of nanoemulsion with Babchi oil. The nanoemulsion was prepared and the combination was sonicated for twenty minutes in an ice bath with the usage of a sonicator of 20 kHz with a peak power of 750 Watts. To detect the diameter of the mean droplet, the Litesizer-500 particle size analyzer was used with PDI.

Drug Content Determination

By dissolving 100 mg of the created nanoemulgel in 10 mL of distilled water, the quantity of medication included in the nanoemulgel was evaluated. This combination was tested using a UV spectrophotometer at 260 nm against distilled water as a blank control. It was then quantified by spreadability.

The drug content was determined by the following formula:

$$\text{Concentration} = \text{Absorbance}/(\text{E1 cm1\%}) \times \text{Dilution factor} \times 10.$$

pH Determination

As the formulation would be administered to the skin, pH monitoring was required to guarantee that it was non-irritating. A digital pH meter was used to analyze the formulation pH and was measured at room temperature.

Spreadability

The potential spreading capacity of the nanoemulgels was measured 48 h after preparation. The spreadability was measured by evaluating the nanoemulgel diameter after spreading it across two glass plates for one minute. The mass of nanoemulgel was positioned upon the glass slide with one a pre-marked circle diameter of one centimeter, and a second glass slide had been placed. The diameter was observed to be grown as a result of the masses introduced, causing the gels to spread. The formulation may be used to calculate spreadability:

$$\text{S} = (\text{m} \cdot \text{l})/\text{t}.$$

The initial S is the ability to spread, m is the higher slide weight, l is the upper slide length, and t is the required time.

Measurements of Viscosity and Rheological Management

The produced preparations were evaluated for viscosity through spindle number 4 of Brookfield DV-II+ Pro viscometer at various angular speeds at 31.0 ± 0.1 °C. The rheological behavior of the formulation of nanoemulgel was assessed using plate and cone configurations of a 40 mm cone with a 2.5-degree cone angle. Rheology experiments were carried out at 25 °C with shear rates ranging from 53.21–496.5 s^{-1}.

The power law equation was used to determine the consistency and flow indexes:

$$\text{Ţ} = Krn.$$

The symbol Ţ refers to shear stress, r refers to shear rate, K refers to the index of consistency, and n is the index of flow.

Log analysis on both sides is expressed as follows:

$$\text{Log } \tau = \log K + n \log r.$$

Therefore, the slope of the graph of logs of shear stress versus logs of shear rate was selected as the flow index, while the inverse function of the Y-intercept provided the index for consistency.

4.2.7. Ex Vivo Drug Studies on Permeation

Ex vivo studies for permeation were conducted using the Franz cell for diffusion, which is a proven approach for predicting the delivery of drugs through the skin. The skin of Wistar rats was removed for this research. The sacrificed rats' hair from their dorsal part was detached using a surgical blade of 24 numbering in the region from tail to head. The shaved skin was split, and unwanted fat along with the connective tissues was excluded with a scalpel. The skin was further removed and cleaned with saline and inspected for its integrity to be used. The skin of the rats was placed on an assembly cell with a 10 cm^2 high area for diffusion, with the stratum corneum facing the donor segment and the dermal side facing the receiver segment.

4.2.8. Determination of Nanoemulgel Drug Content

Through the lower back direction using 24 surgical blades, the animal was sacrificed through the dorsal side. A knife was used to remove superfluous fat and connective tissue from the shaved area of the animal skin. The removed skin was cleaned with normal saline before being inspected for integrity and then utilized.

4.2.9. DPPH Scavenging Activity of Nanoemulgel

The DPPH (2,2 diphenyl-1-picryl hydrazyl) technique developed by Williams et al. was used to assess the overall radical scavenging activity of Babchi oil emulsions before and following encapsulation [31]. The donating capacity of electrons of the anti-oxidants causes the violet hue of the solution to become colorless at normal temperature. Upon dissolution of the sample of 0.5 within methanol of 3 mL, the resultant solution was processed using the methanolic solution of 0.3 mL for DPPH. During the progression of the reaction, this formulation was held inside a darkened room for an hour. The changes in the color indicated the properties of antioxidants in the formulation due to the donating capacity of hydrogen. The sample of 0.3 mL and 3.3 mL of methanol was included within the blank while 0.3 mL of DPPH reagent along with 3.5 mL of methanol was included in the control. At 517 nm, the formulations were examined in a spectrophotometer.

4.2.10. In Vitro Release and Permeation Studies

The process of dialysis bag was used to examine the in vitro drug release from conventional and Babchi oil nanoemulgels. The dialysis bags that were activated beforehand were carefully secured after adding 1 mL of the conventional and similarly 1 mL Babchi oil into the gel. The bag of dialysis was then immersed in a 200 mL medium of phosphate buffer along with methanol for dissolving at pH 6.8. This was maintained at room temperature and stirred at 400 rpm. At preset intervals, HPLC (High-Performance Liquid Chromatography) was used to evaluate all the formulations. Different models such as the zero-order release model, first-order release model, Higuchi model, and Korsmeyer Peppas model were used for analysis. The total quantities of Babchi oil released from the membrane were plotted as diffusion areas per time.

4.2.11. Release of Babchi Oil Nanoemulgel

The process involved entailed replacing a dialysis bag with the rat skin section. Any extra hairs were removed from the rat skin and were washed with a solution of Tyrode. These skin segments were immersed in 1 mL of Babchi oil nanoemulgel and conventional formulation. The study was conducted with the placement of applied rat skin in 100 mL of solution of Tyrode at room temperature with constant agitation on Hanson Research SR8 plus of California, USA at the speed of 100 rpm. This was observed at various time points, and a sample of 2 mL was removed and a quantity similar to the solution of Tyrode was included as a preservative. The Babchi oil concentration was observed at 218 nm through a spectrophotometer and the procedures were repeated thrice for reproducibility.

4.2.12. Dermatokinetic Studies

Application of Babchi oil nanoemulgel on rat skin was observed with Franz Diffusion Cell (FDC) according to the literature search on in vitro skin permeation studies. Using the tool, we analyzed the content and concentration of Babchi oil and its formulated mixture at different periods of 0, 0.5, 1, 1.5, 2, 3, 4, 5, 6, 7, and 8 hours with the entire skin obtained from FDC. Excess nanoformulation was removed from the skin and rinsed using saline of pH 7.4. It was then dipped in the mildly warm water of temperature 60 °C for three minutes. We then separated the dermis and epidermis layers of skin using forceps. These layers were cut into small parts and kept in methanol (5 mL) for a day to obtain the Babchi oil content. The solution of methanol left after removing the layers underwent membrane filtration and the Babchi oil content was measured through HPLC. Separate concentrations per cm^2 of Babchi oil from dermis and epidermis layers were observed with time and parameters of T skin max, C skin max, AUC 0–8 h, and Ke were analyzed.

4.2.13. Stability Studies

Assessment of stability of the created formulation was performed by keeping it for three months at different temperatures of 30 ± 2 °C and 40 ± 2 °C with a humidity level of 60 ± 5%. According to the Iqubal et al. method, the samples were observed at different intervals of 0, 1, 2, and within 3 months to observe their appearance, separation of phases, size of globules, EE and PDI. These measurements were done thrice to determine the repeatability [32].

4.2.14. Statistical Analysis

The measurements obtained for the experiment were in triplicate values and were presented in values of mean ± SD (standard deviation). The analysis of statistical difference was performed on the flux at the stable stage and the permeation of ex vivo on intervals that were pre-determined from the formulations. Unpaired *t*-test was utilized with a p-value < 0.05 as the significance level.

Author Contributions: Conceptualization, A.A. and M.R.; methodology, M.R., M.A.S. and A.A.; software, M.R.; validation, M.H.A., A.I.F. and M.R.; formal analysis, A.A. and M.A.S.; investigation, A.A. and M.R.; resources, M.H.A., A.I.F. and A.A.; data curation, M.R. and M.A.S.; writing—original draft preparation, A.A.; writing—review and editing, M.H.A., A.I.F. and M.R.; visualization, M.H.A. and A.I.F.; supervision, A.A. and M.A.S.; project administration, A.A. and M.R.; funding acquisition, A.A. and A.I.F. All authors have read and agreed to the published version of the manuscript.

Funding: This research was funded by the Deputyship for Research and Innovation, Ministry of Education in Saudi Arabia through the project number-IF-PSAU-2022/03/22592 and the APC was funded by IF-PSAU.

Institutional Review Board Statement: Not applicable.

Informed Consent Statement: Not applicable.

Data Availability Statement: Not applicable.

Conflicts of Interest: There is no conflict of interest.

References

1. Rendon, A.; Schäkel, K. Psoriasis Pathogenesis and Treatment. *Int. J. Mol. Sci.* **2019**, *20*, 1475. [CrossRef] [PubMed]
2. Affandi, A.M.; Khan, I.; Saaya, N.N. Epidemiology and Clinical Features of Adult Patients with Psoriasis in Malaysia: 10-Year Review from the Malaysian Psoriasis Registry (2007–2016). *Dermatol. Res. Pract.* **2018**, *2018*, 4371471. [CrossRef]

3. Krueger, J.G.; Bowcock, A. Psoriasis Pathophysiology: Current Concepts of Pathogenesis. *Ann. Rheum. Dis.* **2005**, *64*, ii30–ii36. [CrossRef] [PubMed]
4. Salim, N.; Ahmad, N.; Musa, S.H.; Hashim, R.; Tadros, T.F.; Basri, M. Nanoemulsion as a Topical Delivery System of Antipsoriatic Drugs. *RSC Adv.* **2016**, *6*, 6234–6250. [CrossRef]
5. Sarkar, S. A Treatise on Topical Corticosteroid in Dermatology. *Indian J. Dermatol.* **2018**, *63*, 530. [CrossRef]
6. Campalani, E.; Allen, M.H.; Fairhurst, D.; Young, H.S.; Mendonca, C.O.; Burden, A.D.; Griffiths, C.E.M.; Crook, M.A.; Barker, J.N.W.N.; Smith, C.H. Apolipoprotein E Gene Polymorphisms Are Associated with Psoriasis but Do Not Determine Disease Response to Acitretin. *Br. J. Dermatol.* **2006**, *154*, 345–352. [CrossRef]
7. Warren, R.B.; Griffiths, C.E.M. Systemic Therapies for Psoriasis: Methotrexate, Retinoids, and Cyclosporine. *Clin. Dermatol.* **2008**, *26*, 438–447. [CrossRef] [PubMed]
8. Pradhan, M.; Singh, D.; Singh, M.R. Novel Colloidal Carriers for Psoriasis: Current Issues, Mechanistic Insight and Novel Delivery Approaches. *J. Control. Release* **2013**, *170*, 380–395. [CrossRef]
9. Chen, D.; Wan, Z.; Zhou, Y.; Liang, X.; Ding, M.; Zhong, J.; Ji, Z.; Xiang, W. Controllable Synthesis of Lanthanide Upconversion Nanomaterials through Impurity Doping. *Fabr. Self-Assem. Nanobiomater. Appl. Nanobiomater.* **2016**, *1*, 211–241. [CrossRef]
10. Solans, C.; Izquierdo, P.; Nolla, J.; Azemar, N.; Garcia-Celma, M.J. Nano-Emulsions. *Curr. Opin. Colloid Interface Sci.* **2005**, *10*, 102–110. [CrossRef]
11. Amsterdam, R. *Basic Physical Sciences for the Formulation of Cosmetic Products*; Elsevier: Amsterdam, The Netherlands, 2017; pp. 39–76.
12. Mason, T.G.; Wilking, J.N.; Meleson, K.; Chang, C.B.; Graves, S.M. Nanoemulsions: Formation, Structure, and Physical Properties. *J. Phys. Condens. Matter* **2006**, *18*, R635. [CrossRef]
13. Md, F. Biosurfactant: Production and Application. *J. Pet. Environ. Biotechnol.* **2012**, *3*, 124. [CrossRef]
14. Alam, A.; Mustafa, G.; Agrawal, G.P.; Hashmi, S.; Khan, R.A.; Aba Alkhayl, F.F.; Ullah, Z.; Ali, M.S.; Elkirdasy, A.F.; Khan, S. A Microemulsion-Based Gel of Isotretinoin and Erythromycin Estolate for the Management of Acne. *J. Drug Deliv. Sci. Technol.* **2022**, *71*, 103277. [CrossRef]
15. Gutekunst, S.B.; Siemsen, K.; Huth, S.; Möhring, A.; Hesseler, B.; Timmermann, M.; Paulowicz, I.; Mishra, Y.K.; Siebert, L.; Adelung, R.; et al. 3D Hydrogels Containing Interconnected Microchannels of Subcellular Size for Capturing Human Pathogenic Acanthamoeba Castellanii. ACS Biomater. *Sci. Eng.* **2019**, *5*, 1784–1792. [CrossRef]
16. Zhang, Y.; Li, Y.; Thakur, V.K.; Gao, Z.; Gu, J.; Kessler, M.R. High-Performance Thermosets with Tailored Properties Derived from Methacrylated Eugenol and Epoxy-Based Vinyl Ester. *Polym. Int.* **2018**, *67*, 544–549. [CrossRef]
17. Pandele, A.M.; Neacsu, P.; Cimpean, A.; Staras, A.I.; Miculescu, F.; Iordache, A.; Voicu, S.I.; Thakur, V.K.; Toader, O.D. Cellulose Acetate Membranes Functionalized with Resveratrol by Covalent Immobilization for Improved Osseointegration. *Appl. Surf. Sci.* **2018**, *438*, 2–13. [CrossRef]
18. Thakur, S.; Govender, P.P.; Mamo, M.A.; Tamulevicius, S.; Mishra, Y.K.; Thakur, V.K. Progress in Lignin Hydrogels and Nanocomposites for Water Purification: Future Perspectives. *Vacuum* **2017**, *146*, 342–355. [CrossRef]
19. Alam, A.; Jawaid, T.; Alsanad, S.M.; Kamal, M.; Rawat, P.; Singh, V.; Alam, P.; Alam, P. Solubility Enhancement, Formulation Development, and Antibacterial Activity of Xanthan-Gum-Stabilized Colloidal Gold Nanogel of Hesperidin against Proteus Vulgaris. *Gels* **2022**, *8*, 655. [CrossRef] [PubMed]
20. Alqarni, M.H.; Foudah, A.I.; Alam, A.; Salkini, M.A.; Muharram, M.M.; Labrou, N.E.; Kumar, P. Development of Gum-Acacia-Stabilized Silver Nanoparticles Gel of Rutin against Candida Albicans. *Gels* **2022**, *8*, 472. [CrossRef]
21. Wadhwa, G.; Kumar, S.; Mittal, V.; Rao, R. Encapsulation of Babchi Essential Oil into Microsponges: Physicochemical Properties, Cytotoxic Evaluation and Anti-Microbial Activity. *J. Food Drug Anal.* **2019**, *27*, 60–70. [CrossRef] [PubMed]
22. Chopra, B.; Dhingra, A.K.; Dhar, K.L. *Psoralea Corylifolia* L. (Buguchi)—Folklore to Modern Evidence: Review. *Fitoterapia* **2013**, *90*, 44–56. [CrossRef] [PubMed]
23. Alam, F.; Khan, G.N.; Asad, M.H.H. Bin *Psoralea Corylifolia* L: Ethnobotanical, Biological, and Chemical Aspects: A Review. *Phyther. Res.* **2018**, *32*, 597–615. [CrossRef]
24. Sah, P.; Agarwal, D.; Garg, S.P. Isolation and Identification of Furocoumarins from the Seeds of *Psoralea Corylifolia* Linn. *Indian J. Pharm. Sci.* **2006**, *68*, 768. [CrossRef]
25. Marwaha, T.K. Formulation design and evaluation of Herbal anti psoriatic emulgel. *J. Pharm. Sci. Innov.* **2013**, *2*, 30–42. [CrossRef]
26. Chen, H.; Chang, X.; Du, D.; Li, J.; Xu, H.; Yang, X. Microemulsion-Based Hydrogel Formulation of Ibuprofen for Topical Delivery. *Int. J. Pharm.* **2006**, *315*, 52–58. [CrossRef] [PubMed]
27. Aboofazeli, R.; Barlow, D.J.; Lawrence, M.J. Particle Size Analysis of Concentrated Phospholipid Microemulsions: I. Total Intensity Light Scattering. *AAPS PharmSci* **2000**, *2*, 27–39. [CrossRef]
28. Singh, B.; Singh, R.; Bandyopadhyay, S.; Kapil, R.; Garg, B. Optimized Nanoemulsifying Systems with Enhanced Bioavailability of Carvedilol. *Colloids Surf. B Biointerfaces* **2013**, *101*, 465–474. [CrossRef] [PubMed]
29. Lee, C.H.; Moturi, V.; Lee, Y. Thixotropic Property in Pharmaceutical Formulations. *J. Control. Release* **2009**, *136*, 88–98. [CrossRef]
30. Ngawhirunpat, T.; Worachun, N.; Opanasopit, P.; Rojanarata, T.; Panomsuk, S. Cremophor RH40-PEG 400 Microemulsions as Transdermal Drug Delivery Carrier for Ketoprofen. *Pharm. Dev. Technol.* **2013**, *18*, 798–803. [CrossRef]

31. Brand-Williams, W.; Cuvelier, M.E.; Berset, C. Use of a Free Radical Method to Evaluate Antioxidant Activity. *LWT-Food Sci. Technol.* **1995**, *28*, 25–30. [CrossRef]
32. Iqubal, M.K.; Iqubal, A.; Imtiyaz, K.; Rizvi, M.M.A.; Gupta, M.M.; Ali, J.; Baboota, S. Combinatorial Lipid-Nanosystem for Dermal Delivery of 5-Fluorouracil and Resveratrol against Skin Cancer: Delineation of Improved Dermatokinetics and Epidermal Drug Deposition Enhancement Analysis. *Eur. J. Pharm. Biopharm.* **2021**, *163*, 223–239. [CrossRef] [PubMed]

MDPI
St. Alban-Anlage 66
4052 Basel
Switzerland
www.mdpi.com

Gels Editorial Office
E-mail: gels@mdpi.com
www.mdpi.com/journal/gels

www.ingramcontent.com/pod-product-compliance
Lightning Source LLC
Chambersburg PA
CBHW040814250926
51850CB00007BA/1166

9783036590578